Hans Troger/Alois Steindl

Nonlinear Stability and Bifurcation Theory

An Introduction for Engineers and Applied Scientists

Springer-Verlag Wien New York

Univ.-Prof. Dr. Hans Troger
Univ.-Ass. Dr. Alois Steindl
Technical University Vienna
Austria

Printed on acid-free paper

With 141 Figures

Cover-design: T. Erben, Wien

ISBN-13: 978-3-211-82292-0   e-ISBN-13: 978-3-7091-9168-2
DOI: 10.1007/978-3-7091-9168-2

Preface

Every student in engineering or in other fields of the applied sciences who has passed through his curriculum knows that the treatment of nonlinear problems has been either avoided completely or is confined to special courses where a great number of different ad-hoc methods are presented. The wide-spread believe that no straightforward solution procedures for nonlinear problems are available prevails even today in engineering circles. Though in some courses it is indicated that in principle nonlinear problems are solveable by numerical methods the treatment of nonlinear problems, more or less, is considered to be an art or an intellectual game. A good example for this statement was the search for Ljapunov functions for nonlinear stability problems in the seventies.

However things have changed. At the beginning of the seventies, starting with the work of V. I. Arnold, R. Thom and many others, new ideas which, however, have their origin in the work of H. Poincaré and A. A. Andronov, in the treatment of nonlinear problems appeared. These ideas gave birth to the term *Bifurcation Theory*. Bifurcation theory allows to solve a great class of nonlinear problems under variation of parameters in a straightforward manner.

The bifurcation approach to nonlinear problems allows to solve one of the key problems in system theory, namely, the problem of loss of stability of a given stable state of a dynamical or statical system under changing parameter values. Here not only the parameter value for which the loss of stability occurs is of interest but also the system's behavior after loss of stability which is called the postbifurcation behavior.

Though *Elementary Catastrophe Theory* introduced by R. Thom in the late sixties turned out later on not to be as generally applicable as it was assumed to be by many of its enthusiastic protagonists, who applied it to many problems in various fields in the sciences, it had still a major impact on the general approach how to treat nonlinear problems. This is because Elementary Catastrophe Theory showed that for a certain special class of systems (statical systems governed by potentials) where only a restricted number of essential parameters was involved in the stability problem (given by the codimension of the problem) only very few qual-

v

itatively different types of loss of stability of the considered state can occur (given by the classification theorem).

The work of V. I. Arnold, especially, gave insight on how *Elementary Catastrophe Theory* fitted into the more general framework of *Bifurcation Theory*. Important extensions both in the theory and applications were given by M. Golubitsky and his coworkers.

The first text books appeared in the late seventies and in the early eighties. Among them were books by M. Golubitsky, D. Schaeffer, T. Poston, I. Stewart, J. Guckenheimer, P. Holmes, S. N. Chow, J. K. Hale, G. Iooss, D. D. Joseph, B. D. Hassard, N. D. Kazarinov, R. Gilmore, H. Wan, P. T. Saunders, R. Seydel and S. Wiggins to name a few.

However, almost all of these books were written for readers who are assumed to possess a certain mathematical background which not always can be assumed for engineers, engineering students and applied scientists. In addition the treatment in some of these books focuses on questions which often are not too relevant to somebody coming from engineering or the applied sciences and who simply wants to solve a stability problem.

Hence, the aim of our book is to explain to a reader, who is assumed to possess only the minimum of mathematical background acquired by undergraduate courses, how to solve in a straightforward manner nonlinear stability problems. For this purpose we treat several problems from mechanics in varying degrees of details. We believe that these problems from mechanics should be understandable also for readers with little or even no knowledge in mechanics. As the subject has been growing at a great pace and in order to keep the size of the book within certain limits, in some places we do not go into great detail because almost everything we present here is neatly explained in more detail in articles or books. Therefore, we think we can afford to present theorems without proofs and refer for the proofs to the literature. In addition, often we only indicate the use of several of the concepts, and again, we give the interested reader the references where in books or articles a more detailed treatment can be found.

The main part of the book is devided into six chapters where, as we believe, the most important and also most elementary concepts of nonlinear stability and bifurcation theory are collected. In several appendices we explain both some mathematical concepts and also some state equations of slender elastic structures which are used in the text. We have included these derivations because we believe that the theoretical treatment of a nonlinear problem first of all must take care of the knowledge of the range of validity of the governing equations.

It is our pleasant duty to thank many colleagues for looking into earlier versions of the manuscript among them K. Kirchgässner, S. Shaw, H. J. Weinitschke and J. Falzarano who made many valuable comments. We also thank those students at the Technische Universität Wien who read parts of the manuscript and helped to eliminate errors. Further we thank all those persons who worked at the Technische Universität Wien in this field over the last decade and who have contributed to our knowledge. Among them R. Schnabl, R. Weiss, W. Herfort, R. Dirl, P. Kasperkovitz, F. Vogl, K. Zeman, R. Scheidl, A. Machinek and A. Stribersky.

Special thanks go to Mrs. K. Mittermayr who produced the LaTeX-file for the camera ready version of the manuscript.

Due to many other obligations and, in the beginning, some technical problems with the drawing of the figures by means of the program AUTO-CAD which, however, allowed us to avoid completely qualitative sketching of figures the finishing of the manuscript was considerably delayed. In this respect we would like to express our gratitude to Springer-Verlag Wien for the great patience and good cooperation.

Vienna, April 1991   *A. Steindl and H. Troger*

Contents

Appendix   287

Chapter 1

Introduction

The collapse of the Tacoma Narrows suspension bridge in the state of Washington in 1940 is one of most spectacular failures of a large engineering structure due to an (aeroelastic) instability ([150]) in a steady wind flow. This example shows that for the practical use of many technical systems stability properties can be a decisive design criterium. Some other examples, where stability properties are important, are slender structures consisting of rods, plates and shells under pressure loading or under loading by flowing fluids, like the liquid flow in a pipe. For vehicles moving at high speed, like truck-trailer combinations or railway trains ([67], [162]), severe operating restrictions can result from stability limits. An important class of stability problems concerns the field of *hydrodynamic stability*. Here the stability of certain flow states of viscous fluids is studied. Two important examples are the convective motion in a horizontal layer heated from below (Bénard problem) and the flow in the gap between two coaxial cylinders rotating at different angular velocities (Couette problem) ([68], [64]). Here not only the stability of certain states is of interest but also the qualitative change in the flow pattern under variation of external parameters. Such parameters would be in these two flow examples the temperature difference for the Bénard flow and the angular velocities for the Couette flow. As a final example consider the following question: how should the shape of the hull of a ship be designed in order to guarantee static stability (safety against capsizing) over a wide range of loading conditions ([111], ch. 10)?

Over the past decades, engineers have approached, with great success, many of their stability problems using a linearized stability analysis. In addition, nowadays, if a linear stability analysis does not seem to be sufficient the availability of high speed computers allows to supplement the linear investigation by numerical simulation. Such a numerical simulation allows to check whether a linear analysis provides practically useful results. However, contrary to the widespread believe that a linearized stability analysis together with numerical simulation are a general method of treating stability problems, we claim that this is not the case. There exists a large number of problems where a linearized analysis does not give much information about the behavior of the nonlinear system at all and, hence, a numerical simulation would be very costly without yielding much insight into the qualitative behavior. Alternatively, we want to convince the reader in this Introduction that a wide range of nonlinear problems can be analyzed and solved in a straightforward manner making use only of a moderate mathematical background.

Without going into too much mathematical detail let us ask some questions which will be answered in this section:

(1) What do we mean by a nonlinear stability investigation?

(2) Do problems exist where it is absolutely necessary to perform a nonlinear analysis?

(3) Under which restrictions is it possible to carry out a nonlinear stability analysis in a systematic way?

(4) What is the advantage of a nonlinear analysis over a linear one?

By considering some examples, we shall discuss several relevant points related to a nonlinear stability theory. These examples are the axially compressed rods depicted in Fig. 1.1 ([140]) and the steady state motion of the railway bogy depicted in Fig. 1.2. It is essential for the manner

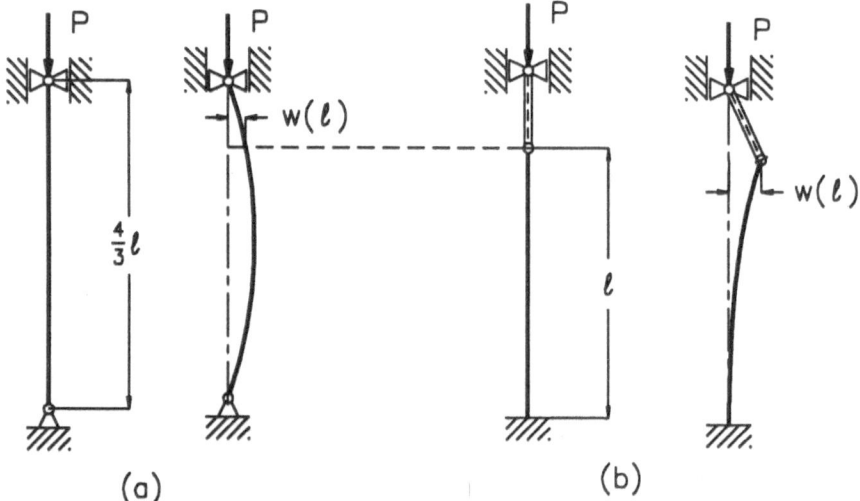

Figure 1.1. Buckling of rods: (a) simply supported ends, (b) clamped at lower end and rigid part at upper end

in which these problems will be studied in the coming chapters that we consider not only a single system, given by a fixed set of parameter values

consisting of design and operating quantities, but a family of systems which, in general, depend on some parameters that can be varied. Among these parameters will always be one or several parameters which will be called *main parameters*. They belong to the perfect system. The remaining components of the parameter vector are called imperfection parameters. Among the main parameters we select one component as *distinguished*. For the rods in Fig. 1.1, this distinguished parameter is the load P, whereas for the motion of the railway bogy of Fig. 1.2 the distinguished parameter is the speed V (Section 3.1.1).

It is important to point out that a system itself cannot be investigated for its stability, but only a certain state of it. This can be an *equilibrium position*, an *oscillatory motion* or a more complicated motion of the system. In the two examples considered, the respective states of interest are the straight static configurations of the two rods and the motion of the bogy with constant speed and no transverse or rotatory deviation, which is also an equilibrium position for the equations of motion.

Furthermore, it is clear that for small values of the load P and the speed V these two equilibrium solutions will be stable. Roughly speaking, we mean by stability (a precise definition will be given in Section 2.3) that small perturbations of the state of the system are only allowed to lead to small bounded motions of the system about the considered state. It is not too difficult to imagine that by increasing the parameter values P and V sooner or later a critical value of the parameters will be reached for which the corresponding equilibrium position becomes unstable: the rod will buckle and for the bogy an unpleasant but well-known oscillatory motion, called hunting motion, will set in.

Mathematical description of physical systems

For a particular problem, the first step in performing a stability analysis of a certain state of a system is to obtain the governing equations which sufficiently accurately describe the behavior of the system. Here, of course, no general advice can be given and the investigators in their respective field of expertise have to decide which mathematical model is most appropriate. Already for the two simple examples considered above the governing equations are qualitatively quite different.

For the buckling problem, it is known from experiments and also from general experience that after loss of stability of the straight configuration the rod diverges into another statical state. Hence, a mathematical description will be sufficient which only takes into account the different possible equilibrium positions of the problem, but ignores the generally fast and strongly damped transitions from one state to the other. Thus, a mathematical model which neglects the dynamics is sufficient. For the

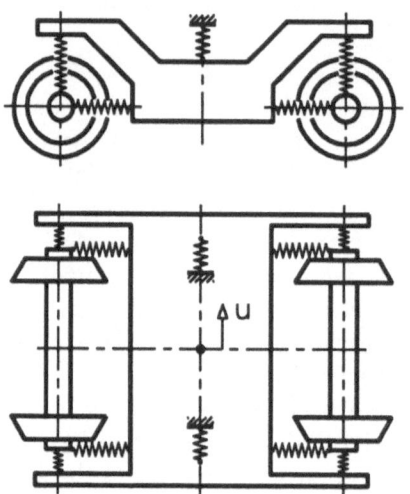

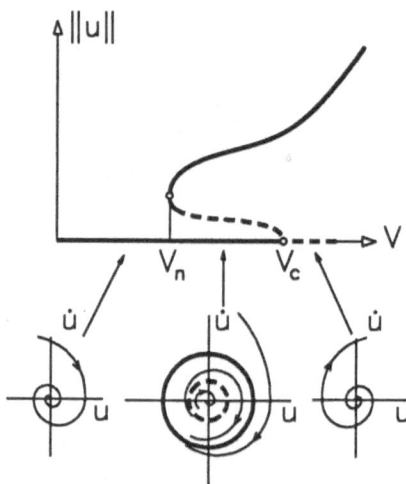

Figure 1.2. Simple mechanical model of a railway bogy

Figure 1.3. Subcritical Hopf bifurcation at V_c for the railway bogy. The practical stability limit is V_n rather than V_c

at loss of stability an oscillatory motion sets in, which requires modelling a dynamical process. Fortunately, for the railway bogy a model can be used which is composed of a set of rigid bodies coupled by hinges, massless springs and dampers. The motion of such a *lumped mass system* can be described by a finite number of variables, leading to a set of *ordinary differential equations*. However, for continuous structures more general variables must be used, which depend both on time and position. In this case *partial differential equations* will be obtained (see the example of the fluid conveying tube in Section 5.3 and Appendix L). The statical buckling problem of a rod, although modelled by a continuous system is still described by an ordinary differential equation. This is because the single independent variable is the coordinate s measured along the axis of the rod (Chapter 2).

The equations of motion of the bogy, therefore, can be given in the form

$$\dot{x} = \mathbf{F}(x, \overline{\lambda}), \qquad (1.1)$$

where $x, \mathbf{F} \in \mathbb{R}^n$ are n-vectors in Euclidean space and $\overline{\lambda} \in \mathbb{R}^\ell$ is an ℓ-dimensional parameter vector. In the parameter vector $\overline{\lambda}$ we distinguish two qualitatively different types of parameters. The first set are those parameters which we will call the *main* or *bifurcation parameters* and designate by λ and the second set are the *imperfection parameters* to be designated by e. Among the main parameters λ which correspond to the perfect system (to be defined in Section 3.1) further a *distinguished component* is selected. For this example, the distinguished component of λ is proportional to the speed V of the bogy.

On the other hand, the equation governing the buckling problem (derived in Section 2.2) is given by a boundary value problem

$$G(w, p) = 0 , \tag{1.2}$$

where $w(x)$ is a *field variable* describing the displacement of the beam, p is proportional to the axial load and G designates a differential operator. With the variables selected in Section 2.2, (1.2) can be given in the form

$$G(\varphi, p) = 0 : \begin{cases} \dfrac{d^2\varphi}{ds^2} + p\sin\varphi &= 0, \\[2mm] \dfrac{d\varphi}{ds}(0) = \dfrac{d\varphi}{ds}(L) &= 0 . \end{cases} \tag{1.3}$$

If, for example, the rod of Fig. 1.1a is considered, then $L = 4\ell/3$. Concerning (1.2) and (1.3) we have been a little bit sloppy in our notation because in (1.2) the boundary conditions are included in the operator whereas in (1.3) they are not.

Besides time-continuous systems like (1.1) it will often turn out to be very useful to be able to describe the behavior of a dynamical system by a time-discrete system such as

$$\boldsymbol{x}_{t+1} = \mathbf{P}(\boldsymbol{x}_t, \overline{\lambda}) \tag{1.4}$$

where $\boldsymbol{x}, \mathbf{P} \in \mathbb{R}^n$, $\overline{\lambda} \in \mathbb{R}^\ell$ and the discrete time step is normalized to one. Such systems arise quite naturally either from periodic solutions of (1.1) or from problems with periodic excitations if they are cut in phase space by means of a *Poincaré mapping* (Section 2.1).

Linear and nonlinear stability

After having decided on how to describe the behavior of the system, the state to be investigated for its stability must be determined. Often it is a simple equilibrium $\boldsymbol{x}_0$ (in particular $\boldsymbol{x}_0 = \boldsymbol{0}$ for (1.1) and $w_0 = 0$ for (1.2)) or a periodic solution expressed by the fixed point $\boldsymbol{x}_{t0} = \boldsymbol{0}$ of (1.4). In these cases either

$$\mathbf{F}(\boldsymbol{x}_0, \lambda) = 0, \quad G(w_0, p) = 0 \quad \text{or} \quad \boldsymbol{x}_{t0} = \mathbf{P}(\boldsymbol{x}_{t0}, \lambda) \tag{1.5}$$

holds. By λ we mean the vector of the main parameters (see p. 47 and page 252). We call the system in which only the main parameters are included the *perfect system*.

Next, a linearized stability analysis is performed. For this purpose, the governing equations are expanded in a power series with respect to the considered state. This process explicitly yields the linear part. For (1.1) one obtains for the state x_0 with $x = x_0 + \xi$

$$\dot{\xi} = \mathbf{A}(\lambda)\xi + \mathbf{f}(\xi, \lambda) \tag{1.6}$$

where

$$\mathbf{A}(\lambda) := \left. \frac{\partial F_i}{\partial x_j} \right|_{x_0} \tag{1.7}$$

is the Jacobian and $\mathbf{f}$ contains higher order terms of at least second order. An expansion of the form (1.6) requires the right-hand side of (1.1) to be differentiable. If, for example, in the description of the system non-differentiable functions appear – this, for example, would be the case if clearances were present in an oscillatory system – the theory to be presented in this book will not be applicable (but see [132] for an alternative treatment). From (1.6) the linearized stability problem is obtained by setting $\mathbf{f}(\xi, \lambda) = \mathbf{0}$.

For equation (1.2), the expansion is performed by taking the *Fréchet* derivative (Section 2.2), resulting in

$$G_w(w_0, p)v = N(v, p) , \tag{1.8}$$

where the linear part is on the left-hand side and the nonlinear terms are on the right hand side. From (1.8), the linear eigenvalue problem

$$G_w(w_0, p)v = 0 \tag{1.9}$$

follows.

Similarly, for (1.4), a point mapping

$$\xi_{t+1} = \mathbf{A}(\lambda)\xi_t + \mathbf{f}(\xi_t, \lambda) \tag{1.10}$$

is obtained, with $\mathbf{A}$ defined analogously to (1.7).

According to a theorem by *Ljapunov* ([83] p. 48), the state x_0 of (1.1) is asymptotically stable, if all eigenvalues of $\mathbf{A}(\lambda)$ according to (1.6) have a negative real part. This is easy to understand, because for small deviations of the variables from the considered state the linear terms in the equations will dominate the behavior (Section 6.1). Since any solution of a linear system can be given by elementary functions ([6]), the negative real part of the eigenvalues results in *exponentially decaying* motions. However, it cannot be obtained from a linearized analysis, whether the stable state has a practically sufficiently large *domain of attraction*. This question can only be answered by a nonlinear theory. Engineers usually estimate the domain of attraction by simulation. That is, the full nonlinear system (1.1) must be solved numerically for a great number of different initial conditions, from which the border of the domain of attraction of a stable state can be constructed.

If at least one of the eigenvalues of $\mathbf{A}(\lambda)$ has a positive real part, the considered state is unstable.

The linearization does not determine stability, if some of the eigenvalues have a zero real part and all the others have a negative real part. This case is called the *critical case* in the sense of Ljapunov. If one studies the stability of a solution of a system for fixed values of the parameters then at a first glance, one could think that this case is of no practical significance, because it can be made to disappear by a small shift in the values of the parameters. However, if, as it is done here, the stability problem is treated in a parametrized family by the methods of bifurcation theory, the critical case in the sense of Ljapunov will occur quite naturally. This can be explained by means of the example of the railway bogy. That is, we study the stability behavior of the equilibrium position for a range of constant speeds V. For low speed V, it is natural to assume from an engineering point of view that the straight motion $x_0 \equiv \mathbf{0}$ of the bogy is asymptotically stable. Hence, all the eigenvalues of (1.7) have a negative real part. Now the speed of the bogy is increased *quasistatically*. That is, for increasing, but always constant values of V the eigenvalues of $\mathbf{A}$ are calculated. Again, it is natural to assume from an intuitive point of view that a critical value V_c of V will be reached. This occurs when for the first time one or several eigenvalues with zero real part (called *critical eigenvalues*) appear, whereas the remaining eigenvalues still have negative real parts. In traditional nonlinear stability theory, as mentioned before, such a case has been called a critical case in the sense of Ljapunov. In this case nonlinear behavior dominates the stability of the solution. In modern nonlinear stability theory, such a case is called *loss of stability* of a given state or *bifurcation* case.

A similar situation appears also for static buckling problems for which a loss of stability occurs, even if the approach is slightly different. We explain this for the example of the beam buckling problem. From (1.3), it is obvious that the state $\varphi_0(s) \equiv 0$, the stability of which should be investigated, satisfies (1.3) for any values p_0. The basic question to be asked is: can through any such a solution pair (φ_0, p_0) a locally unique solution $\varphi(p)$ be constructed? This question can be answered by the *implicit function theorem* (Section 3.2.1) which is closely related to the linearized problem (1.9). If the linear operator G_φ is invertible at (φ_0, p_0) then there exists locally a smooth curve $\varphi(p)$ through (φ_0, p_0) and bifurcation cannot occur. Hence, for the stability problem the case is interesting where G_φ is not invertible for a certain value $p = p_c$ and (1.9) has a nontrivial solution $v \neq 0$. Again, the linear eigenvalue problem determines the loss of stability of the given state under variation of the parameter.

Also, for the *point mapping* (1.4) the linearization (1.10) about a fixed point x_{t0} determines the parameter value for which loss of stability occurs. Here, however, the critical eigenvalues are those with absolute value equal to one (Section 3.1.2).

It is important to note already at this point that under variation of
one distinguished parameter (for practical examples the choice of the distinguished parameter is quite natural) in a generic way only a very limited
number of qualitatively different types of loss of stability can occur. This
is a very important fact because only very few mathematically different
cases, therefore, must be studied. To make this statement clearer the
notion of *genericity* and *qualitatively different* types of loss of stability
must be explained. The loss of stability in a one parameter family of systems is generic if it does not change qualitatively, if, for example, other
main parameters in the system are slightly varied. Such a change of parameters in the system is sometimes called a small shift in the family of
systems to bring a special system into a general position. Generically,
in a one parameter dynamical system family like (1.6), only two cases of
loss of stability can occur, namely either the case of a *zero root* or that
of a *purely imaginary pair* of eigenvalues. If in an investigation of a one
parameter family a more complicated structure of the *critical eigenvalues*
is found, then this can be a hint that the system is either not in general
position (see Section 4.1 for several examples, especially Fig. 4.9, where
for $d = d_c$ the system is not in general position if it is considered to
be a one-parameter (V) family) and a small change of other system parameters can make this higher degeneracy disappear or that the system
obeys certain symmetry properties, for which the occurrence of *multiple
eigenvalues* is also generic (Chapter 5).

So far, we have seen that by linearization the parameter values, for
which a loss of stability of the considered state occurs, can be calculated.
Engineers sometimes claim that keeping the parameter values far enough
away from such a critical parameter value would be a meaningful strategy. By means of the two examples (Fig. 1.1 and Fig. 1.2) we will now
show that problems exist for which such a strategy does not provide a
practically meaningful way of analysis of a stability problem.

To make this point clear, the behavior of the rods depicted in Fig. 1.1
under a quasistatically increasing load P is represented in a $P, w(\ell)$ diagram, where $w(\ell)$ designates the deflection of the rods at $x = \ell$. The
result is shown in Fig. 1.4, from which the fundamental difference in the
behavior of the two rods is clear. In case (a) the rod still has some load
carrying capacity, if the load is increased beyond the critical value P_c
which can be calculated from (1.9). The solution bifurcating off P_c is
called *supercritical*. However, in case (b), no adjacent stable equilibria
exist, if P is increased beyond P_c. Now the bifurcation is called *subcritical*. If, instead of the one parameter or perfect system considered
so far, the imperfect system is investigated, the difference between these
two cases becomes even more significant. For the example under consideration, such imperfections can result from either geometric reasons (the

unloaded rod is not perfectly straight) or from imperfections in the application of the load (Fig. 1.6). Such imperfections are always present in technical systems. It can be seen from Fig. 1.4 that for rod (b) of Fig. 1.1 there exists already a catastrophic loss of stability at the value $P = P_T$, which can be significantly smaller than P_c.

Hence, there exists a wide difference in the conclusions which can be made from a nonlinear stability analysis compared to a linear one. Rod (a) of Fig. 1.1, even in its imperfect behavior, can be well described by a linear analysis, whereas this is not true for rod (b). The rod (b), therefore, is said to possess a *critical postbuckling behavior* or to be *imperfection sensitive*.

This leads to the important conclusion that whether a linear analysis is sufficient to describe the stability behavior of a structure or not, can only be determined from a nonlinear analysis!

Although the rod of Fig. 1.1b may appear a little bit academic, a qualitatively similar behavior is found for shell structures under pressure loadings. The analysis of the postbuckling behavior and the imperfection sensitivity (Section 4.3.4) of shell structures creates a major challenge to mathematicians and engineers.

A similar situation is found for the motion of the railway bogy of Fig. 1.2, as well. Representing, for example, the side motion u of the

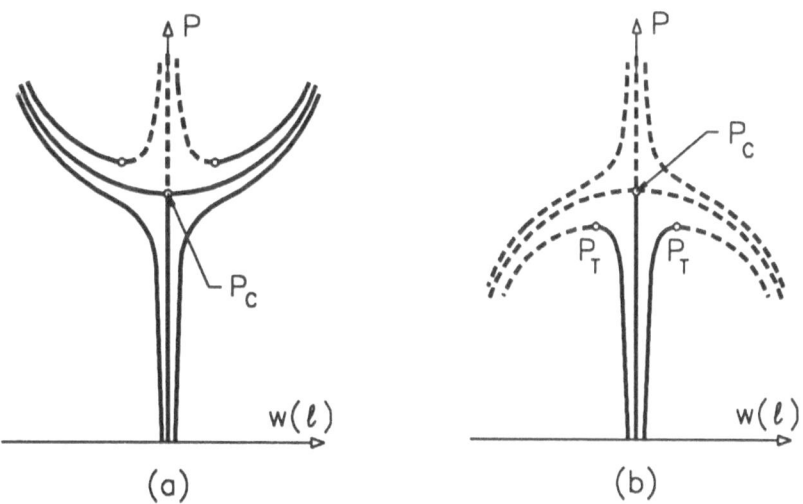

Figure 1.4. Qualitative representation of the nonlinear behavior of the rods of Fig. 1.1: (a) supercritical (stable), (b) subcritical (unstable) postbuckling behavior

center of gravity of the bogy as a function of speed V, Fig. 1.3 is obtained. There exists a basic difference, compared to the behavior of the rods.

Whereas for the rods, after loss of stability of the original equilibrium position, a new equilibrium position is attained, for the bogy after loss of stability of the straight motion an oscillatory motion sets in (that is, a *limit cycle* is created), whose amplitude in its dependence on the parameter V is shown in Fig. 1.3. It is obvious that the value V_c obtained from the linearized analysis, has only a restricted practical significance, because for values of V, for which the inequality $V_n < V < V_c$ holds, already small perturbations of the linearly stable trivial state will be located outside of the *domain of attraction*. In fact the perturbations have only to be larger than the small amplitude of the unstable limit cycle, which is given by the broken line in Fig. 1.3. Practically, not V_c but V_n is the critical speed of the system. The practical stability limit V_n, again, can only be calculated using a nonlinear analysis.

After having achieved some familiarity with nonlinear problems, we will now give a short survey on how to perform a nonlinear analysis in a straightforward manner by means of the methods of bifurcation theory. Such an analysis will lead to the results just presented for the two examples.

Reduction to a low-dimensional problem

One of the main advantages in the application of the methods of bifurcation theory to nonlinear stability problems is that a strong reduction of the dimension of the original system to a low-dimensional bifurcation system is generally possible in the neighborhood of the bifurcation point. The existence and possibility of such a reduction process follows from experimental evidence and has also been intuitively well-known to engineers for a long time. If, for example, an equilibrium position of a continuous (infinite dimensional) system turns unstable, often only few (mostly only one) modes (here the notation mode is also used for an eigenvector) generally determine the spatial structure of the resulting motion.

A crucial question in the treatment of stability problems is: how does one identify these so-called *active or critical modes* describing the post-bifurcation behavior? There are no general rules to answer this question in all circumstances ([1]). However, since this is a very important question, let us make some comments. If the parameter variation is restricted to a local region about the bifurcation point, and if in addition a system with small *aspect ratio* is given (see p. 131) a mathematically correct reduction can be performed. A system with small aspect ratio possesses the property that *critical eigenvalues* (that is, eigenvalues for which a loss of stability occurs) occur only with low (finite) multiplicity. In this case, the active modes are the eigenfunctions corresponding to the critical eigenvalues. For the rod buckling problem the active modes are the deflections shown in Fig. 1.1 and for the bogy it is the eigenspace associated with the critical eigenvalues. The corresponding *bifurcation equations* are algebraic equations or ordinary differential equations in variables which measure the amplitudes of the active modes. Hence, roughly speaking, the spatial structure is fixed and the time evolution is determined by a low-dimensional system of bifurcation equations. For systems with large aspect ratio, a straightforward reduction is not always possible and therefore *"selection principles"* must be used to find out the relevant modes from all possible cases ([20], [113]).

The *dimension* of the bifurcation system is equal to the *multiplicity* of the critical eigenvalue in the case of the buckling problem or to the *number* of eigenvalues with zero real part for the moving bogy, respectively. It is important to point out that this reduction works only locally and it must always be checked whether a sufficiently large domain of parameter variation is achieved for practical applications (see the discussion of the shell buckling problems in Section 4.3.4). To perform this reduction process, the *Ljapunov-Schmidt-reduction* in the statical case (1.3) and the *center manifold theory* in the dynamical cases (1.1) and (1.4) can be used, as is explained in detail in Chapter 3.

The problem becomes much more complicated if the parameter variation is not confined to a local neighborhood of the bifurcation point. Then a correct identification of the active modes becomes very difficult ([1]). There are still two ways of approaching such a problem. One is by means of the *Rayleigh-Ritz-Galerkin* method, which has been used by engineers over many decades with great success. However, this method works well only in the local parameter domain and even here it can still give wrong results. In a parameter domain which exeeds the local region one can only hope to use enough ansatz-functions in order to describe the spatial behavior of the motion of the system correctly. A well-known example where this approach has been applied, is the reduction of the convection (Bénard) problem in a fluid layer heated from below, described by a set of five nonlinear partial differential equations, to the famous Lorenz equations ([127], [50]). However, it is also well-known ([1]) that the parameter which has the greatest influence on the qualitative behavior of the solution of this discrete convection system, is the number of ansatz-functions. Indeed, not a very satisfactory result.

A second class of problems where some guidelines for a reduction can be given is related to completely integrable, partial differential equations. The soliton revolution has uncovered a long list of such problems ([37]). If such a problem is perturbed by some parameters, the solutions of the unperturbed (integrable) problem can be used as appropriate ansatz-functions with good success ([1]).

Ljapunov-Schmidt method

To give a rough idea how the reduction process works in the low aspect ratio case, the statical buckling problem will be considered first. In this case, the usual approach used by engineers, is the Rayleigh-Ritz-Galerkin-method. At a first glance, this method is not very different from the Ljapunov-Schmidt method. However, as already mentioned, problems exist for which the Rayleigh-Ritz-Galerkin-method does not yield the correct result (see p. 239).

In order to obtain the correct bifurcation equation, an ansatz for the solution of (1.3) is made in the form

$$\tilde{\varphi}(s) = qv(s) + \psi(q,s) \ . \tag{1.11}$$

From the solution of the linear eigenvalue problem (1.9) it is known that only simple eigenvalues with the corresponding eigenfunctions $v(s) = \sin n\pi s/L$ (Section 4.3.1) are possible. Hence, only a one-dimensional algebraic bifurcation equation in one variable q will be obtained. The function $v(s)$ in (1.11) is the critical or active mode. The equation for the amplitude q can be determined by the Galerkin method. However, it has been overlooked by most engineers (one of the first exceptions was W. T. Koiter in his thesis ([75])) that the second part in (1.11), which includes all the *noncritical* or *passive* modes, may also contribute to the determination of the equation for q. In order to explain this special point properly, a detailed analysis is necessary, which is given in Section 3.2. Finally, one obtains a scalar equation for the amplitude q from the projection on v

$$\int_0^L G(\tilde{\varphi},p)vds = 0 \ . \tag{1.12}$$

Expanding the sine-function in (1.3) into its power series leads to a polynomial of the form (Section 4.3.2)

$$q^3 + aq + h.o.t. = 0 \ . \tag{1.13}$$

This is the same result as would have been obtained using the Ljapunov-Schmidt method. For the considered example – as is explained in detail in Chapters 3 and 4 – this follows from the fact that the contribution of $\psi(q,s)$ in (1.11) to (1.13) is at least of fourth or of higher order.

Determinacy

The abbreviation *h.o.t.* stands for "higher order terms" (see also p. 29). This means that for a local bifurcation and stability analysis terms of higher than third order in (1.13) do not affect the local stability behavior. The problem of up to which order terms in the bifurcation equation must be retained is a question of *determinacy* which, roughly speaking, can be formulated in the following way: at what point is it safe to truncate a Taylor series? ([141]). For a given polynomial bifurcation system with terms of k-th order one can ask, whether a perturbation with terms of order $k+1$ and higher will change the behavior qualitatively. If such a perturbation does not result in a qualitative change in the behavior, the considered bifurcation system will be called *k-determinate*. For a polynomial in one variable like (1.13), this question is easy to answer because the first nonvanishing nonlinear term decides about determinacy. However, for bifurcation systems in several variables the question of determinacy is a difficult problem (Section 6.3).

Center manifold theory

For the two descriptions of a dynamical system (1.1) and (1.4) and for continuous dynamical systems (Section 3.2.2) the appropriate method of performing the reduction process to a lower dimensional bifurcation system is *center manifold theory* (Section 3.1).

The center manifold is an *invariant manifold* in the phase space which, in those cases we are studying, is attractive to all nearby trajectories. Then the practically interesting local stability behavior is completely governed by the flow on the center manifold. The dimension of the center manifold, and hence also of the bifurcation system, is equal to the number of eigenvalues of (1.7) with zero real part. All remaining eigenvalues have a negative real part, revealing the flow's property to be locally attractive to the center manifold.

After a linear transformation of coordinates $\xi \to y$ in (1.6), which transforms the matrix $\mathbf{A}(\lambda)$ (1.7) into its *Jordan form* (Appendix B), (1.6) can be written in the form

$$\begin{aligned} \dot{y}_c &= \mathbf{J}_c y_c + \mathbf{g}_c(y_c, y_s) \\ \dot{y}_s &= \mathbf{J}_s y_s + \mathbf{g}_s(y_c, y_s), \end{aligned} \tag{1.14}$$

where all eigenvalues of $\mathbf{J}_c$ have a zero real part and all eigenvalues of $\mathbf{J}_s$ have a negative real part. For a small perturbation about the bifurcation point, it would be tempting to assume that the coordinates y_s decay to zero because the linear flow of $(1.14)_2$ is exponentially decaying and only the coordinates y_c dominate the behavior of the system. However, this is in general incorrect. Due to nonlinear couplings, the influence of y_s in the equations for y_c must not be ignored. Hence, the correct way to treat (1.14) is to compute the center manifold $y_s = \mathbf{h}(y_c)$, by expressing the dependence of y_s on y_c from $(1.14)_2$ and then to eliminate y_s from $(1.14)_1$ to obtain the bifurcation equations

$$\dot{y}_c = \mathbf{J}_c y_c + \mathbf{g}_c(y_c, \mathbf{h}(y_c)). \tag{1.15}$$

It is important to note that (1.13) or (1.15) in the neighborhood of the bifurcation point have qualitatively the same solutions as the original equations (1.3) or (1.1), respectively. It should be mentioned that the center manifold reduction works for point mappings and continuous systems as well, as is shown in Sections 3.1.2, 3.2.2, 4.2.4 and 5.3.

Returning to the example of the railway bogy, a complex conjugate eigenvalue crosses the imaginary axis at the critical speed V_c (see Fig. 4.5). Hence a purely imaginary pair of eigenvalues $\pm i\omega$ is obtained. The resulting bifurcation system is two-dimensional (the index c is omitted) and has the form

$$
\begin{aligned}
\dot{y}_1 &= -\omega y_2 + g_1(y_1, y_2) \\
\dot{y}_2 &= \omega y_1 + g_2(y_1, y_2).
\end{aligned}
\tag{1.16}
$$

After having applied the reduction process, the original stability problems of the two rods of Fig. 1.1 and the railway bogy of Fig. 1.2 have been strongly reduced in their complexity. Instead of the infinite-dimensional differential system (1.3), we obtain only a one-dimensional algebraic system (1.13), and instead of the n-dimensional system of ordinary differential equations (1.1) only the two-dimensional system (1.16) must be further investigated. Therefore, we will restrict ourselves in this book to the local case with small aspect ratio, which, as we hope to have convinced the reader by the given examples, is still of practical importance. Only if necessary shall we comment on more complicated cases (Section 4.3.5).

Unfolding

After having performed the reduction process under consideration of the question of determinacy (see p. 12), the original system is replaced, in general, by a lower dimensional bifurcation system. There are two quantities which allow this bifurcation system to be classified.

First, of course, its dimension which, as we recall, is either equal to the multiplicity of the critical eigenvalue of (1.9), or equal to the number of eigenvalues with real part zero for (1.7) or equal to the number of eigenvalues with absolute value one for the linearization of (1.4) about x_{t0}.

Second, one is not only interested in the solutions of (1.13) or (1.16), but one wishes to obtain all solutions of the family of equations to which these equations belong. This is the *unfolding problem* which roughly speaking is: what are the essential perturbations of a singular bifurcation system? The quantity characterizing this property is called *codimension* (Section 6.2) and depends both on the dimension of the bifurcation equations and on the nonlinear terms appearing in them. The nonlinear terms are determined by the (degenerate) linear part and the *normal form transformation* (see p. 17).

To make this more comprehensible, let us consider (1.13). It is explained in Section 6.2.1 that one additional parameter must be introduced in (1.13) to obtain all qualitatively different solutions for an algebraic equation of third order. This is easy to understand from the solution graphs. In Fig. 1.5a, the solution of (1.13) is given, which is the solution of the so-called *perfect system*. That is the system in which only the main parameters are present. This system is *structurally unstable*. That means that small perturbations of the system will result in a qualitative change of the solutions given in Fig. 1.5a. For example, the addition of a small term $b \neq 0$ to (1.13) yields

$$q^3 + aq + b = 0. \tag{1.17}$$

From (1.17), the graphs giving the solution of the *imperfect system* in

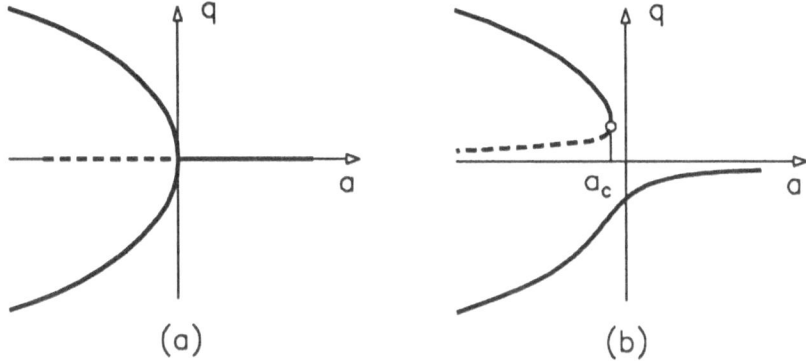

Figure 1.5. Solutions of (1.17): (a) $b = 0$ (pitchfork), (b) $b > 0$

Fig. 1.5b is obtained. These graphs do not change qualitatively under an arbitrary small perturbation of (1.17). Thus, (1.17) is a structurally stable system and it is called a *universal unfolding* of the degenerate system (1.13). By a universal unfolding, a parametrized family is implied. Such a family includes all qualitatively different cases of the considered class with the minimum number of parameters. This minimum number of parameters is called the *codimension* of the problem. Since the *universal unfolding* (1.17) requires two parameters the buckling problem is of codimension two.

Classification

The main impact of the concept of codimension is that for low values of codimension only very few qualitatively different cases of loss of stability can occur. For example, in the case of statical systems up to codimension *four*, only *seven* different cases exist which form the famous list of the *elementary catastrophes* of catastrophe theory ([165], [111], [121], [42], [141]). Therefore, a *classification* of the cases with small codimension can be given. However, it should be brought to attention at this point that from an engineering point of view a different type of unfolding may sometimes be more useful. In this classification, in the unfolding a distinction is made between distinguished parameters and imperfection parameters ([45]). This approach leads to a practically more useful but also larger list of classification of qualitatively different cases and is explained in Section 6.5.3. The theoretical difference between these different types

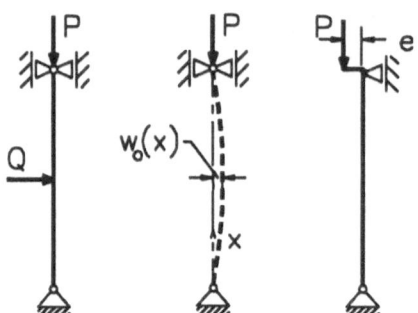

Figure 1.6. Three qualitatively different imperfections taken care of by the parameter b in (1.17).

of unfoldings is founded in different equivalence relations and is shortly explained in Appendix M.

To explain the practical relevance of the universal unfolding given by (1.17), we return to the rod buckling problem, of Fig. 1.1a. In Fig. 1.6 three different types of imperfections are shown. They are caused by completely different effects. The first is due to a transversal loading, the second follows from a geometrically imperfect structure and the third results from an excentric application of the load P. As is shown in Sections 4.3.2 or 6.5.3 the parameter a in (1.13) is related to the load P for the perfect system. Since (1.17) is the universal unfolding of (1.13), all three physically different imperfections are represented by the parameter b as a simple analysis shows. This is a very important result, because the design engineer knows from the outset that he cannot have missed any other possible imperfection, if he bases his design on a universally unfolded bifurcation equation.

A similar situation applies for dynamical systems. For dynamical systems, a complete *classification* up to codimension *two* is given in the literature ([7], [50], Section 6.5.1). There are *two* qualitatively different cases possible if one restricts to codimension *one*, that is, for a one parameter family, and *five* cases for *two* parameter families (codimension *two*). Finally, for point mappings (1.4) *three* different cases of codimension *one*, are possible ([7]). The details, as far as they are relevant in this context, are given in Chapter 6.

With these classified cases, a large number of important practical stability problems can be solved. However, it must be pointed out that certain problems can have higher degeneracies. These higher degenerate cases are not included in the above mentioned classifications. However, other less powerful concepts such as the concept of restricted generic bifurcation ([28], [54], [51]) may sometimes be applicable, as is shown in Section 6.5.4.

Normal form

An important step in the treatment of the stability problem could not be explained by considering (1.13), because it is trivial for this case. This is the transformation to *normal form*. In connection with (1.16) the power and usefulness of the normal form theorem ([7], [50]) can be quite nicely demonstrated. From the symmetry of the railway bogy problem with respect to a deviation to the left or to the right, it is not too difficult to understand that in g_1 and g_2 of (1.16) only terms of *odd* order appear. Therefore, the lowest order terms are of third degree. Hence, for g_1 and g_2 the following expressions (see also (6.14))

$$g_1(y_1, y_2) = a_{130}y_1^3 + a_{121}y_1^2 y_2 + a_{112}y_1 y_2^2 + a_{103}y_2^3 + O(|y_1|^5 + |y_2|^5)$$

$$(1.18)$$

$$g_2(y_1, y_2) = a_{230}y_1^3 + a_{221}y_1^2 y_2 + a_{212}y_1 y_2^2 + a_{203}y_2^3 + O(|y_1|^5 + |y_2|^5)$$

are obtained. The application of the normal form theorem makes use of a nonlinear transformation of coordinates $y \to z$

$$y = z + \mathbf{h}(z). \qquad (1.19)$$

Using (1.19), one tries to eliminate all higher order terms of (1.16) leaving the linear part in (1.16) unchanged. If the real parts of all eigenvalues of the linearized system are different from zero, the nonlinear flow is locally equivalent to its linear one and all nonlinear terms can be eliminated. However in the case of bifurcation typically eigenvalues with a zero real part occur and it is only possible to eliminate some but not all nonlinear terms. Hence, the nonlinear terms can only be transformed into a form as simple as possible still preserving all qualitative properties of the flow. The details of this transformation are given in Section 6.1. For (1.16), a reduction of the eight nonlinear terms of cubic order appearing in (1.18) to four terms

$$\dot{z}_1 = -\omega z_2 + (A_3 z_1 - B_2 z_2)(z_1^2 + z_2^2) + h.o.t.$$
$$\dot{z}_2 = \omega z_1 + (B_2 z_1 + A_3 z_2)(z_1^2 + z_2^2) + h.o.t.$$

$$(1.20)$$

can be achieved. The two constants A_3, B_2 can be expressed by the original coefficients a_{kij} appearing in (1.18). Introducing polar coordinates $z_1 = r \cos \psi$, $z_2 = r \sin \psi$ into (1.20) yields

$$\dot{r} = A_3 r^3 + h.o.t.$$
$$\dot{\psi} = \omega + B_2 r^2 + h.o.t.$$

$$(1.21)$$

The first equation in (1.21) decouples from the second. This decoupling allows to determine easily the stability of the critical point. If $A_3 > 0$, it is unstable and if $A_3 < 0$, it is stable, which follows from the integration of the separable equation $(1.21)_1$ (see p. 51).

Bifurcation diagrams

If we look for all solutions of (1.21), the unfolded one parameter family is given by

$$\dot{r} = ar + A_3 r^3 \tag{1.22}$$

since $(1.21)_1$ has codimension one (Section 6.2.2). From (1.22), the solutions shown in Fig. 1.7 follow immediately. The detailed calculations are given in Section 6.6.

We remark that the direction of the flow in the plates of Fig. 1.3 and Fig. 1.7 depends on whether we depict it in the phase plane (Fig. 1.3) or in a plane with coordinates (z_1, z_2) obtained from the transformation to (6.39).

Furthermore $(1.21)_1$ is one of the two codimension one cases for dynamical systems mentioned before. Low degenerate but practically important cases, which are equivalent to the classified cases listed in Chapter 6, are treated in this book. For these classified cases, so-called *bifur-*

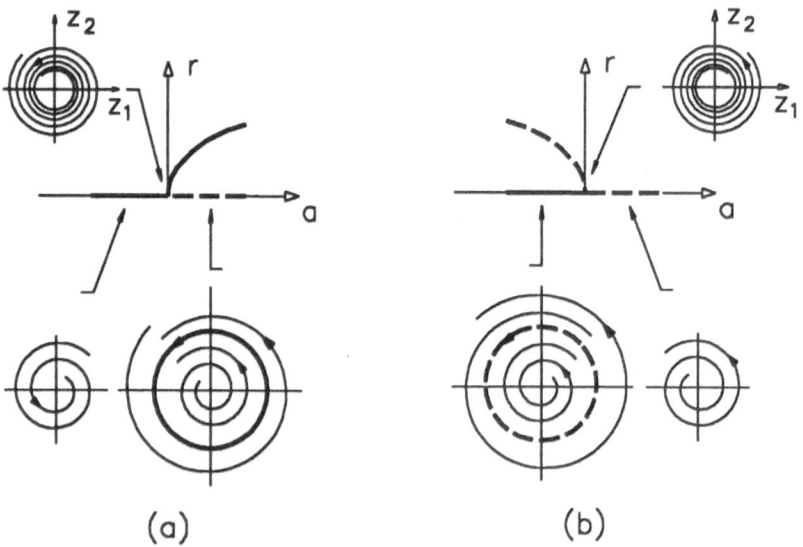

(a)   (b)

Figure 1.7. Hopf bifurcation according to (1.22): (a) supercritical ($A_3 < 0$), (b) subcritical ($A_3 > 0$)

cation diagrams can be calculated. These diagrams give a stratification of the space of parameters into domains of qualitatively similar system behavior. A specific system is given by its parameter values. If it is located in such a domain, it is called *structurally stable, (coarse, robust)* whereas a system located on the boundary between two such domains is not structurally stable. These systems are called structurally unstable or *bifurcation systems*. The space of the bifurcation diagram for (1.22) is the one-dimensional parameter axis a and (1.22) for $a = 0$ is the bifurcation system corresponding to the bifurcation point $a = 0$ on the parameter line. For (1.17), the two-dimensional parameter plane (a, b) is stratified by the semicubical parabola calculated in Section 6.2 and shown in Fig. 6.7.

Correct modelling

Here, another practically important point when dealing with stability problems can be briefly discussed. The structural stability problem deals with perturbations (small changes) of a system, and is closely related to the problem of how to formulate a *correct model* of a system. The main question in this respect is: how many parameters and specifically which parameters have to be included in the model of a system in order to include all qualitatively different types of behavior for systems of a given class? This problem can be quite nicely illustrated by studying again (1.17), which as a universal family contains all possible imperfections for the buckling problem of Fig. 1.1a. As was indicated above, all imperfect cases sketched in Fig. 1.6 are included in the solutions obtained from (1.17).

This question is also discussed in Appendix II, where the form of the *stability boundary* in parameter space is investigated. From the appearence of a nongeneric form of the stability boundary the conclusion can be drawn that the corresponding model has not been correctly formulated.

Higher degenerate systems

Another important feature should be mentioned in this introductory chapter. Sometimes it is worthwhile to study more degenerate bifurcation cases than are expected to occur in practical engineering applications. To explain this idea in more detail, let us consider the example of the railway bogy. From Fig. 1.3 we see that the amplitude curve of the limit cycle oscillations bifurcates subcritically at V_c. As a function of speed V, it has a limit point for $V = V_n$ where an exchange of stability from the unstable to the stable branch for increasing values of $V > V_n$ takes place. In order to calculate the location of the limit point V_n, an oscillatory motion given by the unstable limit cycle must be considered as the basic state for a stability analysis. That is, a nonautonomous (periodic) system must be studied, which is far more complicated to analyse because the calculation of the basic oscillatory state requires the solution of the full system. However, following an idea of *M. Golubitsky*, one may treat such a problem as an autonomous stability problem with its equilibrium position as the basic state by considering it as a multiparameter bifurcation problem. This means that a second parameter besides a, which is proportional to $V - V_c$, must be introduced into the analysis. This second parameter has to be chosen from the physical problem in such a way that it allows to make the coefficient A_3 in $(1.21)_1$ zero. Hence, a more degenerate situation is obtained and the bifurcation equations (1.16) must be calculated at least up to terms of fifth order. The degenerate system now has codimension *two*, which means that a two parameter unfolding is necessary. From this two parameter unfolding and for a small variation of the parameters, a curve qualitatively equivalent to Fig. 1.3 follows. Instead of two local problems about $V = V_c$ and $V = V_n$, respectively (Sections 4.1.1, 4.1.5 and [162]), we obtain one "global" problem that includes both cases.

Optimized systems

Finally, we return to the question of the occurrence of *multiple critical eigenvalues*. It was mentioned above that symmetric systems provide one example. Another class of examples is provided by optimized systems, where multiple eigenvalues are forced by the design of a structure or, for example, by a loading distribution expressed by a special value of a main parameter (see, for example, d_c in Fig. 4.9).

Let us give a simple example from structural mechanics ([150]). If a stiffened panel is loaded under the thrust P, as shown in Fig. 1.8a, two different types of instability can occur. One type of instability occurs, if the size of the stiffeners is large compared to the thickness of the plate. For this case the stiffeners remain straight and local plate buckling will occur, as is shown in Fig. 1.8b. The other type of instability occurs if the size of the stiffeners is small compared to the plate thickness. Then the panel will buckle in the Euler mode (Fig. 1.8c). Both types of instability can be described by a bifurcation equation of the form (1.17) describing the amplitude of the corresponding critical mode. However, a panel that under critical loading buckles in one or the other mode just described is not a satisfactory engineering design, because, if buckling is the design criterium, its weight can yet be reduced without decreasing its performance. For example, in the case of instability by local plate buckling the stiffeners can be reduced in weight until the critical buckling loads both for the local plate mode and the Euler mode are coincident. Clearly, a double critical eigenvalue is now obtained at loss of stability. Hence, in optimized systems more degenerate stability problems typically occur. This has two consequences. One is that these systems are much more difficult to analyse and the other is that their behavior is much more complicated, due to the coupling of several modes. This latter property

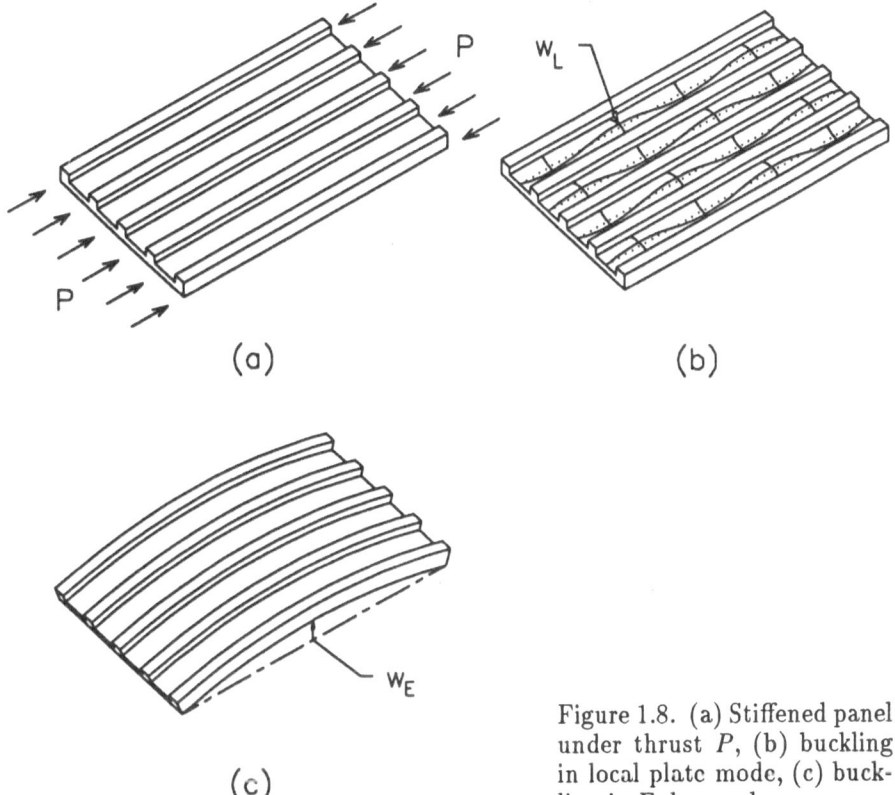

Figure 1.8. (a) Stiffened panel under thrust P, (b) buckling in local plate mode, (c) buckling in Euler mode

means that a failure of an optimized system mostly is desastrous. This is due to the fact that several modes of instability are involved, and hence, a combination of several instability mechanisms is obtained. This is also shown in an increased imperfection sensitivity of such systems.

Summary

At the end of the Introduction, let us summarize the essential steps that must be performed in a nonlinear stability analysis using the methods to be presented in this book:

1. For the perfect system (only with the main parameters), an *inverse linear eigenvalue problem* must be solved to obtain the critical parameter value at which loss of stability of the considered state occurs.

2. A reduction of the original infinite or n-dimensional *perfect* system to a low-dimensional bifurcation system at the critical parameter

value is achieved by means of the *Ljapunov-Schmidt method* or by *center manifold theory*.

3. The order of the nonlinear terms required in the bifurcation equations is precisely specified by the concept of *determinacy*. This is of great practical importance because it allows a Taylor series to be truncated without influencing the result.

4. A strong simplification (elimination of many) of the nonlinear terms in the bifurcation equations is performed by means of *normal form theory*. Normal form theory essentially tries to linearize the equations about an equilibrium position, but may fail to do so when *resonant* terms occur.

5. For bifurcation equations with low *codimension* there exists only a limited number of different cases which are listed in the proper *classifications*. Hence, by a smooth change of coordinates a given problem of low codimension can be transformed into one of the classified cases.

6. Not only the solutions of special degenerate bifurcation equations of the *perfect* system are sought, but all solutions of the bifurcation equations of the class they belong to. This goal is achieved by the concept of *universal unfolding*. For example the concept of universal unfolding allows to treat the problem of *imperfection sensitivity* of structures.

7. The problem of *robustness (coarseness)* or *structural stability* is considered when calculating *bifurcation diagrams*, which are stratifications of the parameter space into domains of qualitatively similar (topologically equivalent) behavior of the system.

Chapter 2

Representation of systems and definition of stability

To keep the contents of the book as general as possible we shall classify systems according to their mathematical description. The proper mathematical description of a system depends on the problem one is interested in studying and on the results one wants to obtain. Hence, dealing with the same problem, completely different descriptions might be adequate if the investigations are aiming in different directions concerning the results. To clarify this point, let us consider two examples.

In truck dynamics various mechanical models are in use. Some lead to systems of ordinary and some to hybrid systems of ordinary and partial differential equations. If, for example, the stability of motion of a heavy truck is to be investigated, engineering experience shows that a mechanical model of coupled rigid bodies, and consequently, a mathematical model of ordinary differential equations will yield accurate results. If, however, for the motion of the same truck, for design reasons the deformation of the chassis of the truck is of interest, a continuous mechanical model of the truck must be used. In general, this model will require a mathematical description consisting of a combination of ordinary and partial differential equations.

Another characteristic example is shell buckling. It is well-known that under static loading fast catastrophic changes in the state of the shell can occur. Sometimes this is called *snap-through* buckling. The main question for an engineer modelling such a phenomenon is whether it is sufficient to know at which parameter values such a fast transition occurs and into which new state the system moves, or whether it is essential to describe also the dynamics of the transition itself. In the former case, a statical description neglecting the dynamics will be sufficient, whereas in the latter case a description of a dynamical process must be given.

Therefore, in the following, two different classes of systems, namely dynamical and statical systems, will be considered.

2.1 Dynamical systems

2.1.1 Time continuous system

Flows

It will turn out to be useful to study not only dynamical systems with a continuous time variable, but also such systems with discrete time variables. As a representative of the first case consider, for example, a vector field $\mathbf{F}$ in $\mathbb{R}^n$, which furnishes the system of ordinary differential equations $(\mathbf{x} \in \mathbb{R}^n)$

$$\dot{\mathbf{x}} = \mathbf{F}(\mathbf{x}). \tag{2.1}$$

This vector field, in general, (for exceptions see [6] ch. 1, sec. 3.5) generates the *flow* $\varphi_t : U \to \mathbb{R}^n$, where $U \subset \mathbb{R}^n$, that is $\varphi_t(\mathbf{x}_0) = \mathbf{x}(\mathbf{x}_0, t)$. The flow φ_t is a smooth function for all values of $\mathbf{x}$ in U and all times t in $[a, b]$ (Fig. 2.1a). In other words the flow gives the evolution in time of all points in the domain of definition. In order to obtain a mathematical expression for the flow of a mechanical problem, where almost always the differential equations, for example (2.1) are given, a solution of these equations would be necessary.

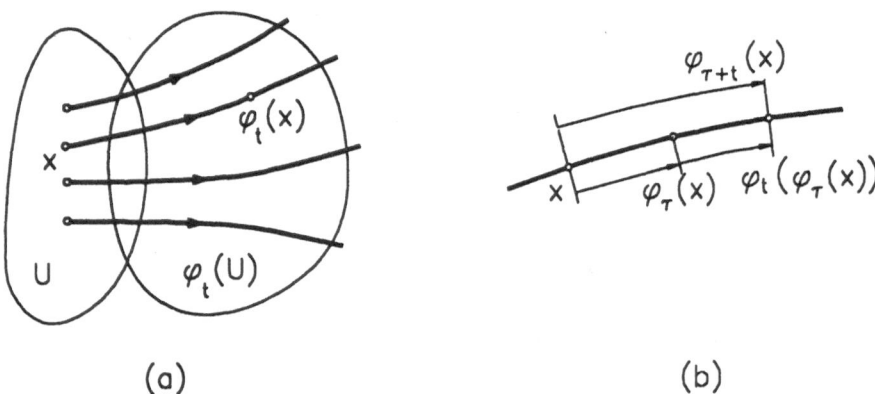

$$\text{(a)} \qquad\qquad\qquad\qquad \text{(b)}$$

Figure 2.1. (a) Flow $\varphi_t(x)$, (b) semi-group property of the flow

In principle, for the dynamical system the description by the flow can be considered to be completely equivalent to the description by the differential equations. Introduction of the flow, however, has the advantage that a unified description of different types of systems can be given.

In the time continuous case one has for $\varphi_t = \varphi(\mathbf{x}, t)$

$$\left. \frac{d}{dt}\varphi(\mathbf{x}, t) \right|_{t=\tau} = \mathbf{F}(\varphi(\mathbf{x}, \tau)). \tag{2.2}$$

The flow φ_t of a dynamical system satisfies the following *semi-group* properties (Fig. 2.1b)

$$\varphi_0(\boldsymbol{x}) = \boldsymbol{x}, \qquad \varphi_{\tau+t}(\boldsymbol{x}) = \varphi_t(\varphi_\tau(\boldsymbol{x})). \tag{2.3}$$

Using the flow concept, it is easy to characterize steady states (equilibrium positions) and periodic solutions. For a steady state $\overline{\boldsymbol{x}}$

$$\varphi_t(\overline{\boldsymbol{x}}) = \overline{\boldsymbol{x}} \qquad \text{for all } t$$

is satisfied and for a periodic solution γ

$$\varphi_{t+T}(\boldsymbol{x}) = \varphi_t(\boldsymbol{x}), \qquad \text{for all } t \text{ and all } \boldsymbol{x} \in \gamma$$

where the period $T > 0$ is the smallest value for which this relation holds.

We consider two examples:

Example 1: The flow is given by

$$\begin{aligned}
\varphi_t(\boldsymbol{x}) &= e^{t\mathbf{A}}\boldsymbol{x} \\
\varphi_t : &\quad \mathbb{R}^n \to \mathbb{R}^n \\
e^{t\mathbf{A}} &= \sum_{k=0}^{\infty} \frac{(\mathbf{A}t)^k}{k!}.
\end{aligned}$$

The group properties (2.3) are easily checked. From (2.2) it follows

$$\frac{d}{dt}\varphi_t(\boldsymbol{x})\Big|_{t=0} = \mathbf{A}e^{t\mathbf{A}}\boldsymbol{x}\Big|_{t=0} = \mathbf{A}\boldsymbol{x}.$$

Thus, this flow is equivalent to solutions of the linear ordinary differential equation

$$\dot{\boldsymbol{x}} = \mathbf{A}\boldsymbol{x}.$$

Example 2: The flow is given by

$$\varphi(\boldsymbol{u}_0, t) = \sum_n \begin{pmatrix} \cos \beta_n t & \frac{1}{\beta_n}\sin \beta_n t \\ -\beta_n \sin \beta_n t & \cos \beta_n t \end{pmatrix} \begin{pmatrix} c_n \sin n\pi x \\ d_n \sin n\pi x \end{pmatrix}$$

$$\varphi(\boldsymbol{u}_0, 0) = \boldsymbol{u}_0 = \sum_n \begin{pmatrix} c_n \sin n\pi x \\ d_n \sin n\pi x \end{pmatrix}. \tag{2.4}$$

The use of elementary trigonometric relations shows that the relations (2.3) are satisfied by (2.4). The flow (2.4) corresponds to the solution of a boundary value problem given by the linear partial differential equation

$$\frac{\partial^4 u}{\partial x^4} + p\frac{\partial^2 u}{\partial x^2} + \frac{\partial^2 u}{\partial t^2} = 0 \qquad t > 0, \quad 0 < x < 1, \qquad (2.5)$$

which describes the oscillations of an axially loaded elastic rod for $p < \pi^2$ with the boundary conditions

$$u(x,t) = \frac{\partial^2 u}{\partial x^2}(x,t) = 0 \qquad \text{for} \quad x = 0,1; \quad t > 0.$$

The initial conditions are

$$u(x,0) \;=\; \sum_n c_n \sin n\pi x,$$

$$\frac{\partial u}{\partial t}(x,0) \;=\; \sum_n d_n \sin n\pi x \qquad 0 \le x \le 1 .$$

The solution of (2.5) with these boundary and initial conditions is given by

$$u(x,t) = \sum_n \left(c_n \sin n\pi x \cos \beta_n t + \frac{d_n}{\beta_n} \sin n\pi x \sin \beta_n t \right)$$

where

$$\beta_n^2 = n^2\pi^2(n^2\pi^2 - p).$$

2.1.2   Time discrete system

If the time variable takes discrete values instead of continuous ones, a discrete flow or, as it is also called, a *point mapping* $\mathbf{P} : \mathbb{R}^n \to \mathbb{R}^n$ is obtained. If $\mathbf{x}_t$ denotes the state of the system at time t and the discrete time interval is normalized to 1, then the state at time $t + 1$ is $\mathbf{x}_{t+1}$ and is given by

$$\mathbf{x}_{t+1} = \mathbf{P}(\mathbf{x}_t). \qquad (2.6)$$

Point mappings occur quite naturally for impulsively loaded systems. A simple example is the plane rigid pendulum with impulsive loading illustrated in Fig. 2.2 ([61]). The equation of motion is

$$I\frac{d^2\psi}{dt^2} + b\frac{d\psi}{dt} + c\psi + \left[P_0 L \sum_{m=-\infty}^{\infty} \delta(t - m\tau) \right] \sin \psi = 0. \qquad (2.7)$$

From (2.7), it can be seen that the nonlinearity gives a contribution only if $t = m\tau$, that is, if an impulse is exerted on the pendulum. Between

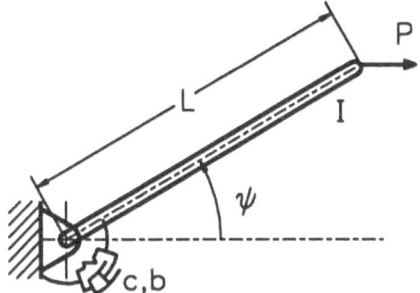

Figure 2.2. Rigid pendulum with impulsive loading $P(t)$, which in a natural way is best described by a point mapping

any two impulses equation (2.7) is linear and the solution can be easily determined. The corresponding point mapping consists of two parts, namely, the solution of the linear differential equation and the impulse, which results in a discontinuous change of the velocity of the pendulum. The corresponding expressions are given in [61]. Applications where this type of mappings are of great practical interest are given in [108].

Poincaré mapping

By far the most important class of discrete dynamical systems is given by Poincaré mappings. These are very convenient in the investigation of systems possessing *periodic* or *recurrent* solutions. A recurrent solution is an orbit in phase space which after some time returns arbitrarily close to its initial position. This property is especially important for chaotic motions and strange attractors, which are introduced on p. 39. For a periodic flow in n-dimensional phase space, the Poincaré mapping (2.6) can be calculated by introducing a $(n-1)$-dimensional surface, which the flow must intersect transversally. Fig. 2.3 shows an example in three-dimensional space in the neighborhood of a periodic solution, which is

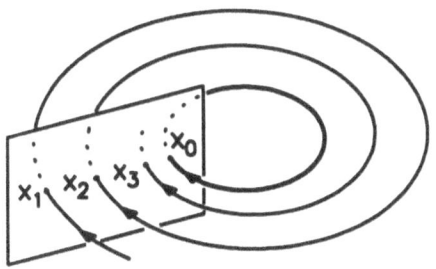

Figure 2.3. Poincaré mapping $x_{t+1} = P(x_t)$ generated by a transversal surface of section in the neighborhood of a periodic solution given by the fixed point x_0 on the surface

represented by the fixed point x_0 on the two-dimensional *Poincaré surface*. It is obvious that the three-dimensional continuous flow is replaced by a two-dimensional point mapping on the surface of intersection. Such a reduction can simplify a stability analysis considerably. For example, if a periodic motion in $\mathbb{R}^3$ is investigated, it is possible to examine its stability simply by studying the stability of the fixed point of the point mapping on the two-dimensional surface. In general, the explicit calculation of the Poincaré mapping is not possible. We give an extensive treatment how to calculate an approximation in Section 4.2.

Example: We consider now the Van der Pol equation

$$\ddot{x} - \varepsilon(1 - x^2)\dot{x} + x = 0 , \tag{2.8}$$

which is an important mathematical model in the theory of self-excited oscillations ([93], [52]). With $x_1 = x$, $x_2 = \dot{x}$, we obtain from (2.8) the system of first order differential equations

$$\begin{aligned} \dot{x}_1 &= x_2 \\ \dot{x}_2 &= \varepsilon(1 - x_1^2)x_2 - x_1. \end{aligned} \tag{2.9}$$

The introduction of polar coordinates r, ψ with $x_1 = r\cos\psi$, $x_2 = r\sin\psi$ in (2.9) yields (see Section 6.1)

$$\begin{aligned} \dot{r} &= \varepsilon(1 - r^2\cos^2\psi)r\sin^2\psi \\ \dot{\psi} &= -1 + \varepsilon(1 - r^2\cos^2\psi)\cos\psi\sin\psi. \end{aligned} \tag{2.10}$$

The exact solution of (2.10) is not known. However, we can calculate an approximation for the variation of r for one cycle provided ε is small. For $\varepsilon = 0$ the solution of (2.10) is $r(t) = r(0) = r_0$ and $\psi_0(t) = \psi(0) - t$, where by shifting the time origin we may choose $\psi(0) = 0$. For $\varepsilon \neq 0$, but small, me make use of the fundamental *theorem* in the theory of ordinary differential equations of the continuous dependence of solutions on parameters ([6]) p. 58). This allows us to write the solution in the form

$$\begin{aligned} r(t, \varepsilon) &= r_0 + \varepsilon r_1(t) + O(\varepsilon^2) \\ \psi(t, \varepsilon) &= -t + \varepsilon\psi_1(t) + O(\varepsilon^2) . \end{aligned} \tag{2.11}$$

Inserting (2.11) into (2.10) yields in the order ε

$$\begin{aligned} \dot{r}_1 &= (1 - r_0^2\cos^2 t)r_0\sin^2 t \\ \dot{\psi}_1 &= -(1 - r_0^2\cos^2 t)\cos t\sin t . \end{aligned} \tag{2.12}$$

Integration of (2.12) over one period yields

$$r_1(2\pi) - r_1(0) = \int_0^{2\pi} r_0(1 - r_0^2 \cos^2 t) \sin^2 t\, dt$$

$$= 2\pi \left(\frac{r_0}{2} - \frac{r_0^3}{8} \right) \tag{2.13}$$

$$\psi_1(2\pi) - \psi_1(0) = -\int_0^{2\pi} (1 - r_0^2 \cos^2 t) \cos t \sin t\, dt = 0 .$$

Since $r_1(0) = 0$, the radius $r(2\pi)$ follows from (2.11) to

$$r(2\pi) = r_0 + \varepsilon r_1 + O(\varepsilon^2) = r_0 + \varepsilon\pi \left(r_0 - \frac{r_0^3}{4} \right) + O(\varepsilon^2) \tag{2.14}$$

where $O(\varepsilon^2)$ designates *h.o.t.* of at least second order in ε.

To express the abbreviation *h.o.t.* in a precise way the order symbols $O(\varepsilon)$ and $o(\varepsilon)$ are defined as follows.

Definition 2.1 (Order symbols)

$$f(\varepsilon) = O(\varepsilon^k) \iff \textit{there exists a constant } K > 0 \textit{ such that}$$
$$f(\varepsilon) \le K\varepsilon^k \textit{ for all small } \varepsilon .$$

$$f(\varepsilon) = o(\varepsilon^k) \iff \lim_{\varepsilon \to 0} \frac{f(\varepsilon)}{\varepsilon^k} = 0 .$$

Neglecting terms $O(\varepsilon^2)$ and $o(\varepsilon)$ may not be the same. In the first case it means that terms starting with ε^2 and of higher order are neglected, whereas in the second case, for example, already terms $\varepsilon^{3/2}$ and higher order terms may be neglected.

Setting in (2.14) $r(2\pi) = r_{t+1}$ and $r(0) = r_0 = r_t$ yields, to first order in ε, the discrete dynamical system

$$r_{t+1} = r_t + \varepsilon\pi \left(r_t - \frac{1}{4} r_t^3 \right) . \tag{2.15}$$

The fixed points of this one-dimensional point mapping follow from setting $r_{t+1} = r_t$ in (2.15). This gives

$$r_t - \frac{1}{4} r_t^3 = 0. \tag{2.16}$$

From (2.16), we find the two fixed points $r_t = 0$ and $r_t = 2$. It is not difficult to check the stability of these fixed points by a linearized stability analysis. Recalling the comment from the Introduction (on p. 7), the fixed points will become unstable if an eigenvalue with absolute value equal to one appears. For $\varepsilon < 0$ it follows that $r_t = 0$ is stable and for $\varepsilon > 0$ it is unstable. Linearization of (2.15) about $r_t = 2$ suggests that this fixed point is stable for $\varepsilon > 0$ and unstable for $\varepsilon < 0$. The conclusion for (2.8) is that the system has an equilibrium position (which follows already from (2.8)) and a limit cycle. When $\varepsilon > 0$ the equilibrium position is unstable and the limit cycle is stable, but for $\varepsilon < 0$ the equilibrium position is stable and the limit cycle is unstable.

These arguments can be made more comprehensible if we use a graphical representation of (2.15), as it is given in Fig. 2.4a for $\varepsilon > 0$ and Fig. 2.4b for $\varepsilon < 0$. Fig. 2.4 has to be interpreted in the following way.

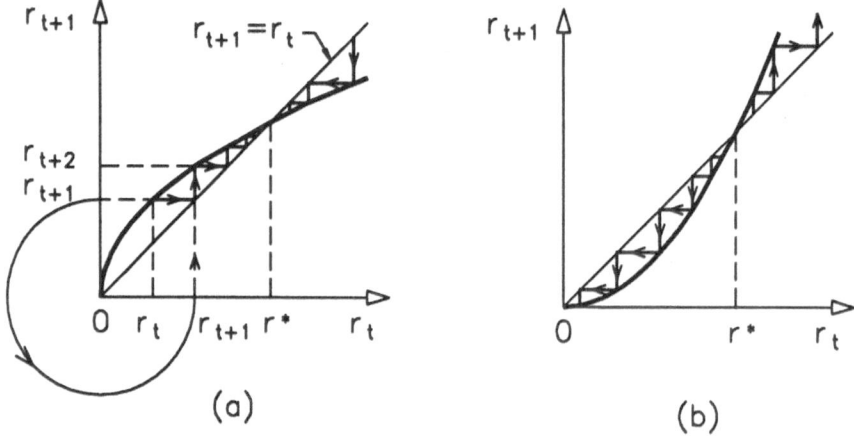

(a)   (b)

Figure 2.4. One-dimensional point mappings (2.15) with (a) unstable origin and stable nontrivial fixed point r^* for $\varepsilon > 0$ and (b) stable origin and unstable fixed point r^* for $\varepsilon < 0$

Starting with an initial value r_t, the corresponding value r_{t+1} is given by the intersection of the vertical line with the graph. In the next iteration this value r_{t+1} on the ordinate axis must be used at the abscissa to obtain r_{t+2} and so on. However, the value of r_{t+2} can also be determined if the horizontal line from r_{t+1} to the identity line $r_{t+1} = r_t$ is drawn and then the vertical to the graph. Hence, in Fig. 2.4 the iteration can be performed quite easily. Furthermore, the stability of the fixed points is easy to check. It is decided by the absolute value of the slope of the tangent to the fixed point. If it is greater (less) than one the fixed point is unstable (stable), respectively. This type of representation is very useful for one-dimensional point mappings and is frequently used for investigations of chaotic systems.

2.2 Statical systems

Potentials

Often the mathematical representation of a stability problem as a dynamical system is unnecessarily complicated and can be replaced by simpler mathematical models, namely by functions or functionals. Forming the gradient or variational derivative of these functions or functionals, respectively, one obtains the corresponding equations to calculate the equilibrium positions, which govern the long term behavior of the system. This type of description of a system is possible if the following points are satisfied: (a) the parameters are varied quasistatically, (b) the forces acting on the system can be derived from a potential and (c) only the equilibrium positions are of interest and not the motion describing the transition from one state to the other one. For many statical buckling problems these points are almost always satisfied.

As an example, we consider a simple mechanical system which displays some essential properties. It is the plane double pendulum (Fig. 2.5) consisting of two rigid rods with two different types of loading. In case (a)

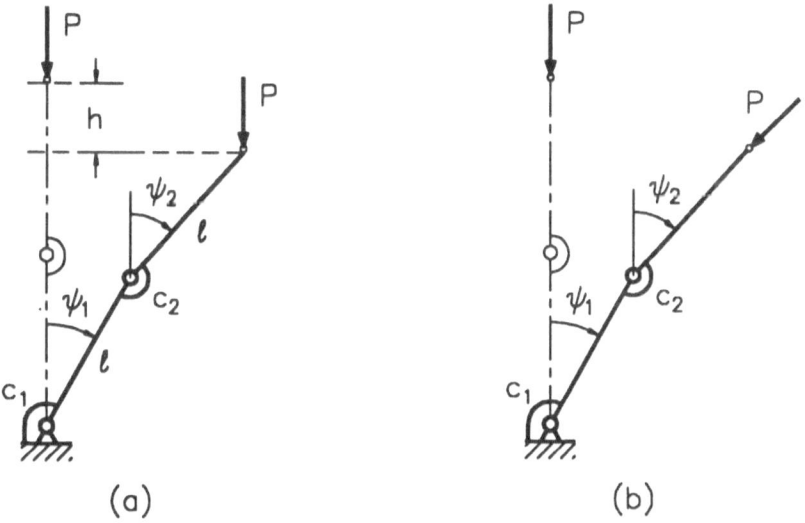

(a)   (b)

Figure 2.5. Double pendulum with (a) conservative (dead) load, (b) nonconservative (follower force) load

the load P always keeps its direction (dead loading), whereas in case (b) the loading is such that P always has the direction of the second bar (follower force loading). It is easy to check that the work of the loading during the deformation in case (a) is given by $A = Ph$. It is independent of the deformation path and only depends upon the vertical displacement. If additionally, linear torsion springs (stiffnesses c_1, c_2) are introduced at

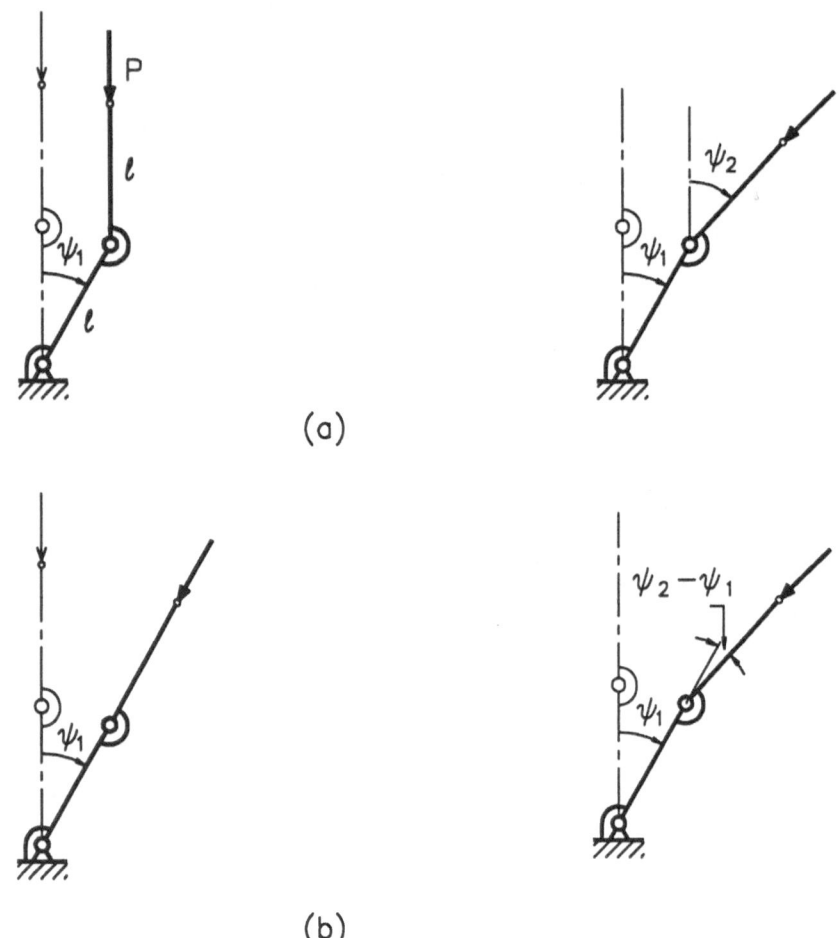

Figure 2.6. Explanation of the path dependence of the work done by a follower force: (a) $A = P\ell(1 - \cos\psi_1)$, (b) $A = 0$

the hinges, one obtains for the potential energy $V = U + W$

$$V(\psi_1, \psi_2, P) = \frac{1}{2}c_1\psi_1^2 + \frac{1}{2}c_2(\psi_2 - \psi_1)^2 - P\ell[(1 - \cos\psi_2) + (1 - \cos\psi_1)],$$
(2.17)

where U and W are the potentials due to internal and external forces, respectively $(dW = -dA)$. In case (b) of Fig. 2.5, no potential function for the load exists, because the work done by a deformation of the system is path dependent. In Fig. 2.6a the deformation from the position $\psi_1 = \psi_2 = 0$ to $\psi_1 \neq 0$ and $\psi_2 \neq 0$ is done in two steps. First only rod 1 is rotated about the angle ψ_1 and ψ_2 is kept to zero. This results in the work $A_1 = P\ell(1 - \cos\psi_1)$. Then ψ_1 is kept fixed and rod 2 is rotated about the angle ψ_2. However, the rotation through the angle ψ_2 does not contribute to the work because the load is displaced perpendicularly to its direction. Hence, the total work done is A_1. In case of the deformation of Fig. 2.6b no work is done at all by P since P is always displaced perpendicularly to its current direction. A system such as in Fig. 2.5b is called nonconservative, and even if the loading is statical, its behavior must be described by its equations of motion.

In Chapter 4 it is shown that after loss of stability of the equilibrium position of such a system, the system may not evolve to an adjacent equilibrium position, but may instead develope an oscillatory motion. Mathematicians call this phenomenon *Hopf bifurcation*, engineers refer to it as *flutter instability*. Though it has not been possible so far to design a simple laboratory experiment to produce exactly a follower force as it is given in Fig. 2.5b ([79]), there is no doubt that such types of forces occur in nature. Examples include pipes conveying fluid and wind excitation processes. Also a rocket mounted at the end of the bar would produce a follower force type of loading.

Let us now return to systems which allow a description by means of a potential function. For a system with n degrees of freedom q_i a potential function

$$V(q_1, \ldots, q_n, \lambda_1, \ldots, \lambda_\ell) , \qquad (2.18)$$

is obtained, which may depend on ℓ parameters λ_j. (2.18) defines a mapping: $\mathbb{R}^{n+\ell} \to \mathbb{R}^1$. The corresponding equilibrium positions can be calculated from the set of n equations

$$\frac{\partial V}{\partial q_i} = 0 \qquad i = 1, \ldots, n. \qquad (2.19)$$

In (2.17) $q_1 = \psi_1$, $q_2 = \psi_2$ and $\lambda_1 = P$.

If, however, the state equation of a continuous system must be derived, for example, for a slender elastic rod (Fig. 2.7), then the situation is a little bit more complicated. Under the assumption of an inextensible axis of the rod ([19]) the potential

$$V = \frac{1}{2}EJ \int_0^L \left(\frac{d\psi}{ds}\right)^2 ds - P\left[L - \int_0^L \cos\psi ds\right] \qquad (2.20)$$

is obtained. Here, the first term represents the strain energy U due to bending, and the second term the potential W of the external loading. EJ is the bending stiffness (see Appendix I). The potential of the external force P is given by $W = -Pu(L)$. Referring to Fig. 2.7 the displacement $u(s)$ is given by $u(s) = s - x$. Noting that $dx = \cos\psi ds$ we obtain

$$u(s) = s - \int_0^s \cos\psi ds$$

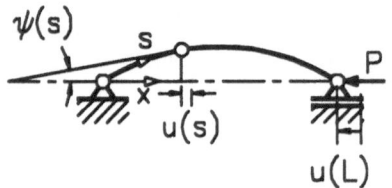

Figure 2.7. Finite deformation of an axially inextensible simply supported rod

and finally (2.20). Expression (2.20) is a *functional* and the relation corresponding to (2.19) is its first variation set to zero:

$$\delta V = 0 . \tag{2.21}$$

Fréchet derivative

To calculate (2.21), it is advantageous to introduce the *Fréchet derivative* f' of a functional f. Generally by a functional $f[u]$, the function of a function is meant.

We do not want to enter into the subtle distinctions between *Gateaux* and *Fréchet* derivatives ([19]) and simply write

$$\delta f = f' \delta u, \tag{2.22}$$

where

$$f'(u)u_1 = \lim_{t \to 0} \frac{f[u + tu_1] - f[u]}{t}.$$

The derivative f' is calculated according to

$$f'u_1 = \frac{\partial}{\partial \varepsilon} f[u + \varepsilon u_1] \Big|_{\varepsilon=0} , \tag{2.23}$$

where u_1 is an arbitrary but fixed function and f' is a linear operator. In a similar way, higher derivatives can also be defined. For the second derivative,

$$f''u_1 u_2 = \frac{\partial^2}{\partial \varepsilon_1 \partial \varepsilon_2} f[u + \varepsilon_1 u_1 + \varepsilon_2 u_2] \Big|_{\varepsilon_1, \varepsilon_2=0} = \frac{\partial}{\partial \varepsilon} (f'[u + \varepsilon u_2] u_1) \Big|_{\varepsilon=0}$$

is obtained. We give now some examples:

(a) $f(u) = u^2$ $f'(u)h = 2uh$ $f''(u)hk = 2hk$ $f'''(u) = 0$

(b) $f(u) = u^3$ $f'(u)h = 3u^2h$ $f''(u)hk = 6uhk$

(c) $f(u) = e^u$ $f'(u)h = e^uh$ $f''(u)hk = e^uhk$

(d) $f(u) = uu'' + u'^2$ $f'(u)h = u''h + uh'' + 2u'h'$

$f''(u)hk = hk'' + h''k + 2h'k'$

(e) $f(u) = (1 + u'^2)^{1/2}$ $f'(u)h = u'h'(1 + u'^2)^{-1/2}$

$f''(u)hk = h'k'(1 + u'^2)^{-3/2}.$

Returning to (2.20) the following expressions

$$\delta V = V'(\psi)\delta\psi = EJ \int_0^L \frac{d\psi}{ds}\left(\frac{d}{ds}\delta\psi\right) ds$$

$$-P \int_0^L \sin\psi\delta\psi ds \qquad (2.24)$$

$$\delta^2 V = V''(\psi)\delta\psi\delta\varrho = EJ \int_0^L \left(\frac{d}{ds}\delta\varrho\right)\left(\frac{d}{ds}\delta\psi\right) ds$$

$$-P \int_0^L \cos\psi\delta\psi\delta\varrho ds \qquad (2.25)$$

$$\delta^3 V = V'''(\psi)\delta\psi\delta\varrho\delta\varphi = P \int_0^L \sin\psi\delta\psi\delta\varrho\delta\varphi ds \qquad (2.26)$$

$$\delta^4 V = V^{IV}(\psi)\delta\psi\delta\varrho\delta\varphi d\chi = P \int_0^L \cos\psi\delta\psi\delta\varrho\delta\varphi\delta\chi ds \qquad (2.27)$$

are obtained. From (2.24) and (2.21) follows the equilibrium condition

$$EJ \int_0^L \frac{d\psi}{ds}\left(\frac{d}{ds}\delta\psi\right) ds - P \int_0^L \sin\psi\delta\psi ds = 0. \qquad (2.28)$$

Partial integration of the first integral in (2.28) yields

$$EJ \left[\frac{d\psi}{ds}\delta\psi\Big|_0^L - \int_0^L \frac{d^2\psi}{ds^2}\delta\psi ds\right] - P \int_0^L \sin\psi\delta\psi ds = 0. \qquad (2.29)$$

To satify (2.29) it is necessary that the boundary term vanishes for all admissible functions $\delta\psi(s)$. This can be achieved either if $\delta\psi = 0$, which is called a kinematic boundary condition and corresponds to clamped ends, or if $\delta\psi \neq 0$, which is the case for the rod in Fig. 2.7, if the natural boundary condition ([38] p. 93)

$$\frac{d\psi}{ds} = 0 \quad \text{for} \quad s = 0 \quad \text{and} \quad s = L \tag{2.30}$$

is satisfied. The physical meaning of (2.30) is that the bending moment vanishes at the ends of the rod. Thus, (2.29) reduces to

$$\int_0^L \left[EJ\frac{d^2\psi}{ds^2} + P\sin\psi \right] \delta\psi ds = 0. \tag{2.31}$$

The application of the *fundamental lemma* of the calculus of variations requires the term in square brackets in (2.31) to vanish. Otherwise we always could find a continuous function $\delta\psi(s)$, such that the integral would not be equal to zero. Therefore, we obtain

$$EJ\frac{d^2\psi}{ds^2} + P\sin\psi = 0. \tag{2.32}$$

Equation (2.32) is the equation analogous to (2.19) and still has to be supplemented with the boundary conditions (2.30).

Equation (2.32) could also be derived, without making use of a variational formulation. In general, such an equation can be written in the form

$$G(u, \lambda) = 0, \tag{2.33}$$

where u is an element of a Banach space H and the parameter $\lambda \in \Lambda \subset \mathbb{R}^1$. G is a mapping (operator): $H \times \Lambda \rightarrow K$, where K is a Banach space, too. The Banach spaces are function spaces, where H contains only functions that satisfy the boundary conditions. For the example (2.32) $H = \{u \in C^2 | u'(0) = u'(L) = 0\}$ contains twice differentiable functions and $K = \{u \in C^0\}$ continuous ones (Appendix A).

2.3   Definitions of stability

As already mentioned in the Introduction, we must distinguish between two basically different concepts. On the one hand, one can investigate how a specific solution of a fixed system behaves under small perturbations of its initial conditions. On the other hand, one can ask for the influence of a small change of the system itself on the stability of its solutions. The first case is the stability problem in the sense of Ljapunov whereas the second is the *coarseness (robustness)* problem or the problem of *structural stability*.

2.3.1 Stability in the sense of Ljapunov

It is essential to point out that, for a given system, the stability of a specific solution is studied. In order to do this, it must be possible to define solutions in the neighborhood of the given solution. This requires the ability to measure distances between two solutions, and thus, requires the introduction of a metric. Specific solutions to be studied for their stability include steady states and non-steady states.

To be able to measure the distance between two solutions, a *metric* ϱ on a function space must be defined. The definitions of the concepts of metric and norm are given in Appendix A. We consider (2.5) as example for the definition of various different norms. Setting $v = \partial u/\partial t$, the vector $\boldsymbol{w} = (u, v)^T$ is introduced. For stability problems of rods the following norms (distances to the undeformed system)

$$\varrho(\boldsymbol{w}, 0) = \left[\int_0^L \left(v^2 + \left(\frac{\partial^2 u}{\partial s^2} \right)^2 + \left(\frac{\partial u}{\partial s} \right)^2 + u^2 \right) ds \right]^{1/2} \quad (2.34)$$

$$\varrho(\boldsymbol{w}, 0) = \sup_x [u] + \left[\int_0^L \left(v^2 + \left(\frac{\partial^2 u}{\partial s^2} \right)^2 \right) ds \right]^{1/2} \quad (2.35)$$

are physically meaningful. Which norm one selects depends on the specific properties of the problem. For example, in a problem with strongly localized deformations the sup-norm will be superior to the integral-norm.

Now we are able to give the *definition of stability* in the sense of Ljapunov ([74]):

Definition 2.2 (Ljapunov stability) *The solution $\varphi_t(u_0, t_0)$ is stable if and only if, for any initial time t_0 and any prescribed positive number $\varepsilon > 0$, there exists a number $\delta > 0$ such that for all times $t > t_0$*

$$\varrho_{t_0}(\varphi_{t_0}(u_0, t_0), \varphi_{t_0}(v_0, t_0)) = \varrho_{t_0}(u_0, v_0) < \delta$$

implies that

$$\varrho(\varphi_t(u_0, t_0), \varphi_t(v_0, t_0)) < \varepsilon.$$

A geometrical interpretation of this definition is given in Fig. 2.8. The two norms ϱ_{t_0} and ϱ need not necessarily be identical.

Definition 2.3 (Asymptotic stability) *If $\varphi_t(u_0, t_0)$ is stable according to Definition 2.2 and in addition, $\varrho(\varphi_t(u_0, t_0), \varphi_t(v_0, t_0)) \to 0$ for $t \to \infty$, then the solution $\varphi_t(u_0, t_0)$ is called asymptotically stable.*

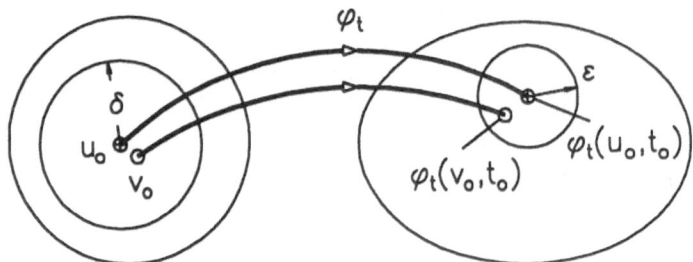

Figure 2.8. Geometric interpretation of the definition of stability in the sense of Ljapunov

It is important to point out that the definition of stability in the sense of Ljapunov can be too restrictive for certain problems, and therefore, it must sometimes be replaced by other definitions of stability. A simple example where this can be demonstrated is the large-amplitude undamped oscillation of a plane pendulum described by the angle $\psi(t)$. If we compare two solutions with the initial conditions $\psi_1(0) = \psi_0$ and $\psi_2(0) = \psi_0 + \delta$ and $\dot{\psi}_1(0) = \dot{\psi}_2(0) = 0$, it is obvious that no matter how small $\delta \neq 0$ is chosen, after a sufficiently long time period the difference between the two motions will exceed any prescribed value of ε, because the period of oscillation depends on the amplitude. This becomes quite clear from the phase plane picture of Fig. 2.9. We see in this figure that

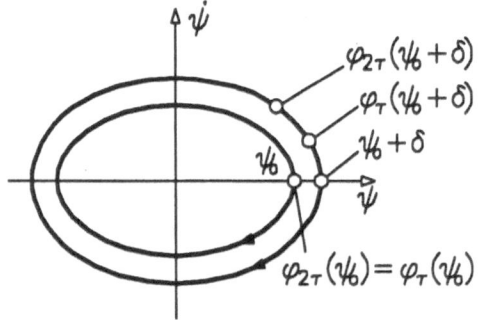

Figure 2.9. Geometric argument to show that the oscillation of a pendulum with constant but large amplitude is unstable in the sense of Ljapunov

after one revolution of the phase point on the inner curve, the phase point on the outer curve has not completely returned to its starting point. With each successive swing of the two pendula, the distance grows. However, if we do not measure the distance between the phase points, but between the two phase curves (ellipses) it is clear that the motion is stable again. This type of stability is called *orbital stability*. Obviously, orbital stability is nothing else than the stability in the Poincaré mapping.

An important concept related to stability is that of an *attractor* ([84]). An attractor A, roughly speaking, is a set of trajectories in phase space to which all trajectories in its neighborhood are attracted for $t \to \infty$. Furthermore an attractor must contain a dense orbit. In Section 6.6.4, p. 277 an example of an attracting set of trajectories is given which is not an attractor because it does not contain a dense orbit. An attractor must also be an *invariant*, bounded set of the flow φ_t. By invariant we mean $\varphi_t A \subseteq A$. Finally, it must be asymptotically stable. That means that there exists a neighborhood V with $A \subseteq V$, such that for each $x \in V$ the relation

$$\lim_{t \to \infty} \varrho(\varphi_t(x), A) = 0$$

holds. The region V is called the domain of attraction of A. Further the flow on an attractor is *recurrent* and finally an attractor is *indecomposable*. Recurrency has been explained before (p. 27) and indecomposable refers to the fact that A cannot be decomposed into two separate pieces where the flow remains for $t \to \infty$. Examples of attractors are stable equilibrium positions, stable limit cycles, attractive tori, but also so-called *strange attractors* as they occur in connection with chaotic motions (Fig. 2.10). One possibility to characterize different types of attractors is

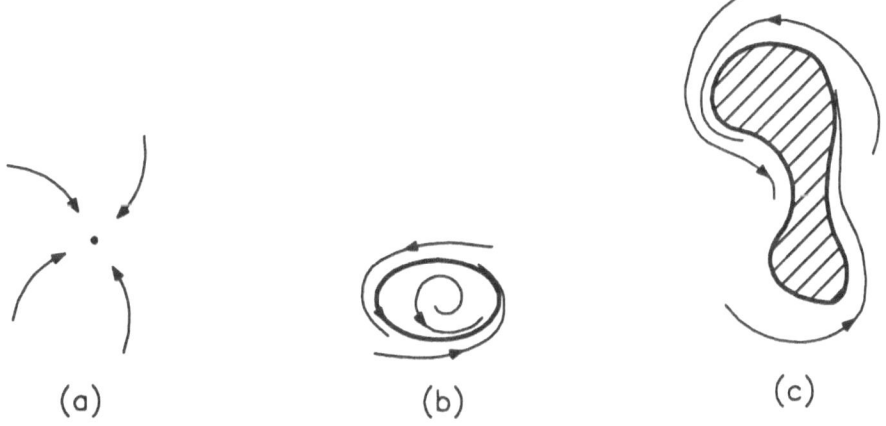

(a)   (b)   (c)

Figure 2.10. Three different types of attractors: (a) equilibrium, (b) limit cycle, (c) strange attractor in an at least three-dimensional space

by their dimension. Whereas points, limit cycles, and tori have integer values for their dimension, strange attractors have non-integer values. This type of dimension is called *fractal dimension* ([131], [100]).

Finite- versus infinite-dimensional systems

Theoretically, there exists – besides other points (see Section 3.2) – one fundamental difference in carrying-out a stability investigation for a given problem whether one considers a discrete mathematical model with a finite number of degrees of freedom or a continuous model with an infinite number of degrees of freedom. For finite-dimensional systems there exists the *principle of the equivalence of norms*, which states that any norm $\varrho(\boldsymbol{x})$ ($\boldsymbol{x} \in \mathbb{R}^n$) can be bounded from above and below by the *Euclidean norm*

$$\|\boldsymbol{x}\| = (x_1^2 + x_2^2 + \ldots + x_n^2)^{1/2} \tag{2.36}$$

using two constants $\alpha > 0$, $\beta > 0$ in the following form ([59] p. 77)

$$\alpha\|\boldsymbol{x}\| \leq \varrho(\boldsymbol{x}) \leq \beta\|\boldsymbol{x}\|. \tag{2.37}$$

From (2.37) follows for systems with finitely many degrees of freedom that stability in one specific norm guarantees stability in any other norm. For infinite-dimensional systems, a relation like (2.37) does not exist. Indeed, there exist examples, which may be considered academic, but for which one can show to have stability in one norm and instability in another ([74]). A physical example of norm dependence of stability ([17]) is given in [134] for an example in linear isotropic elastodynamics. There a spherically converging wave motion is considered. Prescribing an infinitesimal small disturbance on the surface of a sphere of radius a it is shown that the strains do not remain infinitesimal in a small region about the center of the sphere. The spherical wave propagating inward leads to a focussing effect, however, the energy remains constant. Hence, the state of rest would be unstable for a norm measuring the maximum strain but it would be stable if an energy norm ([18], [112]) were taken.

Another important question is whether the *energy criterion* ([168]), which has been used with great success in engineering calculations, gives stability in the sense of Ljapunov. Recall that the energy criterion states that an equilibrium position of a system is stable if its potential energy for this position is a minimum (see also Section 3.2.1). Here some open questions still exist for infinite-dimensional systems ([78]). This is easy to understand, if we select as norm (2.34) and study two- or three-dimensional structures (for example plates or shells). From the boundedness in the norm obviously one cannot necessarily conclude the boundedness of certain variables ([86] p. 199). For systems with a finite number of degrees of freedom the energy criterion guarantees stability in the sense of Ljapunov ([83]).

2.3.2 Structural stability (robustness, coarseness)

All those properties of a system which persist under small perturbations of the system are called robust (coarse). An important property of a technical system is the stability of its steady or periodic states. Thus one can ask the following question: what happens to an attractor of a system under small perturbations of the system? That is, if the system is replaced by a slightly different one, does the attractor persist? Here it is important to note that such changes of the system can occur, either by controlling parameters externally (for example increasing a load or the speed of a vehicle) or by a change of parameters which are independent of the behavior of the system (for example a change of the weather).

Let us consider an example from vehicle dynamics. From driving experience it is well-known that the tire profile depth has a decisive influence on the braking behavior of a car on a wet road. However, the profile depth of a tire varies between a maximum and a minimum value during the tire's lifetime. Therefore one requirement of driving safety is that the prescribed minimum value of the tire's profile depth must still guarantee a stable braking motion. In other words, no qualitative changes in the braking behavior of a car are allowed to occur over the range of admissible profile depth variation occurring in the lifetime of the tire.

The mathematical formulation of the robustness (coarseness) problem is given by the concept of *structural stability* of a system, which can be understood by studying a vector field $\mathbf{v}(\mathbf{x})$ and its flow on the phase space M.

We give the following *definition* of structural stability ([7] p. 88, [149]):

Definition 2.4 (Structural stability) *The dynamical system* $(\mathbf{v}, M)$ *is called structurally stable if for a sufficiently small perturbation* $\boldsymbol{\delta}(\mathbf{x})$ *of the vector field* $\mathbf{v}$, *the perturbed system* $(\mathbf{v} + \boldsymbol{\delta}, M)$ *is topologically equivalent to the unperturbed system.*

In other words: a structurally stable (robust) system of differential equations is one whose phase portrait retains its topological structure under any small perturbation of the system ([141]).

Two vector fields or their corresponding flows are *topologically equivalent* if there exists an one-to-one and continuous mapping h with continuous inverse, carrying the flow of one system into the flow of the other system.

A point in the parameter space of a family of systems at which a system loses structural stability, such that its solutions change their topological type, is called a bifurcation or a catastrophe point.

To illustrate this concept with a simple example, the harmonic oscillator is studied. The equation of motion is given by

$$\ddot{x} + x = 0. \tag{2.38}$$

The vector field $\mathbf{v}(\mathbf{x})$ of (2.38) is $(x_1 = x, x_2 = \dot{x})$

$$
\begin{aligned}
v_1 &= x_2 \\
v_2 &= -x_1.
\end{aligned}
\tag{2.39}
$$

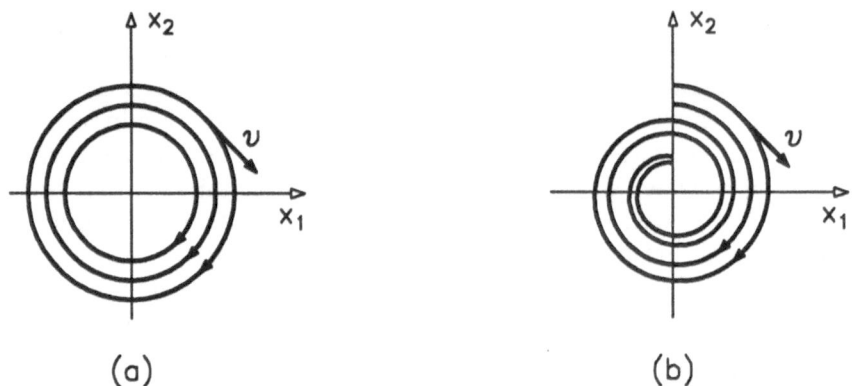

Figure 2.11. (a) Structurally unstable flow, (b) structurally stable flow

The corresponding phase curves are circles and shown in Fig. 2.11a. Now
$v(x)$ of (2.39) is perturbed with the small quantity $\delta(x)$, where a physically meaningful perturbation (linear damping)

$$\delta = \begin{pmatrix} 0 \\ -\varepsilon x_2 \end{pmatrix} \tag{2.40}$$

is chosen. The consequence of (2.40) on the structure of the trajectories
is shown in Fig. 2.11b. Instead of circles, spirals are obtained. No matter
how small $\varepsilon \neq 0$ is selected, a qualitative change in the orbit structure of
(2.38) is obtained, because there does not exist a one-to-one continuous
function h giving the equivalence between these two flows.
    If, however, the system

$$\ddot{x} + a\dot{x} + x = 0 \tag{2.41}$$

is studied, then it turns out that for $a \neq 0$ any small perturbation does
not lead to a qualitative change in the orbit structure in the phase plane.
However, if (2.41) is considered as a one-parameter family, with parameter a, then $a = 0$ is the critical parameter value, for which (2.41) is
a structurally unstable system. For negative values of a (2.41) is again
robust, though the equilibrium $x_1 = x_2 = 0$ is now unstable.
    An important question is whether the structurally stable systems in a
given phase space M form an open and dense set in the space of dynamical
systems ([59] p. 158). From an application oriented point of view one can
put this question also in the following form: is it possible to approximate
each system of differential equations (vector fields) on M by a structurally
stable one? The answer which may be disappointing at the first glance
is given by the following theorem ([7] p. 139, [149]):

Theorem 2.1 (Structural stability) *The structurally stable systems form an open and dense set in the space of vector fields, only if* dim $M \leq 2$ *(e. g. plane, sphere, torus).*

This means that for dim $M \leq 2$ almost all systems are robust and the non-robust systems are rare exceptions. However, the non-robust systems must not be disregarded, because they occur inevitably in systems in which parameters are varied. They are the structurally unstable systems and give a partition of the parameter space of dynamical systems in regions of structurally stable systems which possess qualitatively different properties. However, the restriction dim $M \leq 2$ from above does not exclude higher dimensional systems. This is already mentioned in the Introduction and will be explained in detail in Chapter 3. The reason is that due to the reduction process for many practically important cases only one- or two-dimensional systems of bifurcation equations will be obtained.

If dim $M \geq 3$, it is in general not true that the structurally stable systems form an open and dense set ([7] p. 139). However, there are still some classes of systems for which this is true. For example if we restrict to *gradient systems* like (2.19). These form an open and dense subset even for dim $M \geq 3$. These latter systems are those treated in *elementary catastrophe theory.*

Test for structural stability

In Section 6.6 the investigation of planar systems of the form

$$\dot{x} = P(x, y)$$
$$\dot{y} = Q(x, y) \tag{2.42}$$

will be of great importance in the calculation of *bifurcation diagrams.* Three conditions that must be fulfilled such that (2.42) is structurally stable (robust, coarse) are given in the following *theorem* ([98] p. 187):

Theorem 2.2 (Structural stability of planar systems)

1. *A robust system* (2.42) *cannot have an equilibrium position* (x_0, y_0) *for which*

$$\Delta = \begin{vmatrix} P_x(x_0, y_0) & P_y(x_0, y_0) \\ Q_x(x_0, y_0) & Q_y(x_0, y_0) \end{vmatrix} = 0 \tag{2.43}$$

or if $\Delta > 0$,

$$\sigma = P_x + Q_y = 0 . \tag{2.44}$$

2. *A robust system cannot have closed trajectories*
   $(\varphi(t), \chi(t))$ *for which*

$$\int_0^\tau [P_x(\varphi, \chi) + Q_y(\varphi, \chi)]dt = 0 .$$

3. *A robust system cannot have separatrices connecting saddle points.*

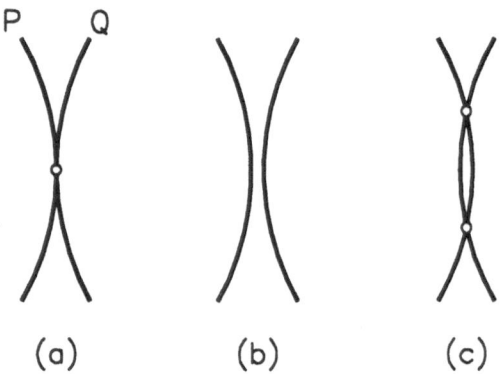

(a)   (b)   (c)

Figure 2.12. System (a) is structurally unstable because a small perturbation
leads to system (b) with no or to system (c) with two solutions

Condition (1) is equivalent to the requirement that all equilibria must
be hyperbolic. Relation (2.43) means that the curves $P(x_0, y_0) = 0$ and
$Q(x_0, y_0) = 0$ do not intersect transversally but have a point of contact.
Hence, a small perturbation of the system will lead to a change in the
number of equilibria (Fig. 2.12a,b,c). The meaning of condition (2) is that

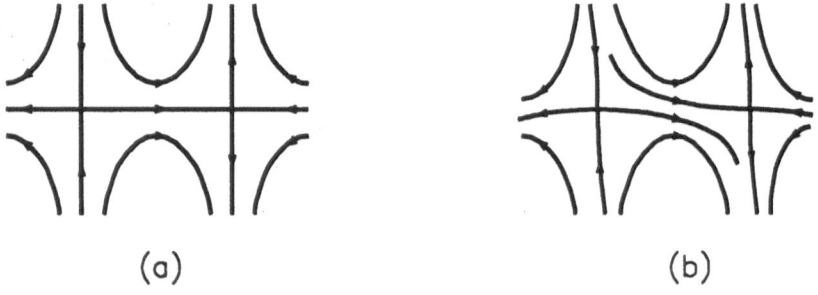

(a)   (b)

Figure 2.13. (a) System with a saddle connection, (b) a small perturbation of
the system of (a) results in a qualitatively different flow

each closed orbit is either a stable or an unstable limit cycle ([59] p. 314).

Hence, it excludes the occurrence of center type closed orbits, that is, orbits that are determined only by the initial conditions (see Fig. 6.28). In addition also the bifurcation of limit cycles is excluded (in Fig. 6.32 the parabola in the second quadrant is such a nonrobust system).

The third condition is easy to understand by breaking a saddle connection ([59] p. 315) which changes the flow qualitatively (Fig. 2.13).

Chapter 3

Reduction process, bifurcation equations

In this chapter the first, basic step in the treatment of nonlinear stability problems is presented, namely, the reduction of the given n- or infinite-dimensional system to a generally low-dimensional *bifurcation system*. In the neighborhood of the bifurcation point this bifurcation system has the same qualitative behavior as the original system, and therefore, completely describes the local stability problem.

We study four different cases of increasing complexity. First, stability of steady state solutions (equilibria) of finite-dimensional dynamical systems. Second, stability of periodic solutions of finite-dimensional dynamical systems, where we introduce a *point mapping*. Third, the special case of stability of equilibria of infinite-dimensional statical systems, whose behavior can be described with sufficient accuracy by potential functions. Finally, a short treatment of stability of equilibrium solutions of infinite-dimensional dynamical systems closes this chapter.

The methods used to perform the reduction are *center manifold theory* and the *Ljapunov-Schmidt method*.

3.1 Finite-dimensional dynamical systems

3.1.1 Steady states

A system, whose equations of motion are given by (2.1)

$$\dot{x} = \mathbf{F}(x, \overline{\lambda}), \tag{3.1}$$

is considered, where $x, \mathbf{F} \in \mathbb{R}^n$ and $\overline{\lambda} \in \mathbb{R}^\ell$. $\overline{\lambda}$ is the parameter vector, as it appears quite naturally in almost any stability problem (see also the examples in Chapter 4). The solution, whose stability will be investigated, is the equilibrium position x_0 of (3.1).

For the calculation of x_0 an important remark must be made concerning the parameter vector $\overline{\lambda}$. At this stage of the stability investigation we separate the components of $\overline{\lambda}$ into those which describe the perfect system and will be varied to calculate the loss of stability of x_0. These are the *main* or *bifurcation parameters* denoted by λ (without a bar) whereas all other components of $\overline{\lambda}$ are the imperfection parameters designated by e (see Section 4.1.4 and Fig. 4.8) and are set to zero now, even if for a practical problem they are never exactly equal to zero. That is, we can write $\overline{\lambda} = (\lambda, e)$. The imperfection parameters will be introduced in the stability analysis when the unfolding of the bifurcation equations in Sections 6.5 and 6.6 is performed. The motivation why we use this way of approach instead of trying to eliminate the imperfections by means of a variable transformation is given on p. 252. Hence, in the following we are treating the so-called *perfect* stability or bifurcation problem. From the vector of the main parameters we further select one component as *distinguished*. This distinguished component will be varied while all other main parameter components are kept fixed. Often the vector of main parameters λ has only one component, which is then the distinguished parameter designated by λ.

In the example of Fig. 1.2, x_0 is the straight line motion of the bogy. For x_0 being an equilibrium position of (3.1),

$$\mathbf{F}(x_0, \lambda) = \mathbf{0} \tag{3.2}$$

must hold. Introducing local coordinates ξ about x_0 by

$$x = x_0 + \xi , \tag{3.3}$$

equation (3.1) can be rewritten in the form

$$\dot{\xi} = \mathbf{A}(\lambda)\xi + \mathbf{f}(\xi, \lambda), \tag{3.4}$$

where

$$\mathbf{A}(\lambda) = \left. \frac{\partial F_i}{\partial x_j} \right|_{x=x_0} \tag{3.5}$$

and $\mathbf{f}(\xi, \lambda) = O(|\xi|^2)$ contains only terms of at least second or higher order. From the *stability theorems of Ljapunov* ([83] p. 48) follows that:

(a) x_0 is stable if all eigenvalues of (3.5) have negative real parts,

(b) x_0 is unstable if at least one eigenvalue of (3.5) has a positive real part,

(c) a critical case in the sense of Ljapunov occurs if the real part of one or several eigenvalues is zero while all the remaining eigenvalues have negative real parts.

If a system with fixed parameter values is investigated, then the critical case (c) is a very special situation, which may not have much practical significance. However, if parameters are varied, starting at a stable state, then this critical situation inevitably appears at a loss of stability of the state.

The type of loss of stability and the behavior of the system after loss of stability of the considered state is determined by the number and location of the eigenvalues with zero real part and the nonlinear terms appearing in the description of the system. These quantities are described by the concept of *codimension* (Section 6.2).

Hence, the task to be performed is to study the stability of a specific solution over a range of values of λ. In nearly all problems of interest there exists a range of parameter values λ, for which x_0 is asymptotically stable. That is, all eigenvalues of (3.5) have a negative real part. In the example of the railway bogy, this is certainly true for low speed. To perform the stability analysis we increase one distinguished component of the vector of main or bifurcation parameters λ *quasistatically*. That is, the stability problem is studied at continuously increasing, but constant values of the distinguished component of λ keeping the other components of λ fixed. Continuing in this manner, finally a critical value λ_c of λ is reached, at which for the first time one or several eigenvalues of (3.5) will have a zero real part. This type of approach results in the stability boundary in the space of main or bifurcation parameters (see the corresponding figures in Chapter 4).

Before we study the problem of loss of stability, it should be considered from another point of view. One could also ask, in connection with (3.2), whether there exists locally a unique solution of the form

$$x_0 = x_0(\lambda) \qquad (3.6)$$

passing through the solution tuple $(x_0(\lambda_0), \lambda_0)$ (Fig. 3.1). The answer is given by the *implicit function theorem* ([119]), (see also Section 3.2.1 where it is formulated in detail) which states:

If $\mathbf{A}(\lambda_0)$ in (3.5) is invertible, then the solution of (3.2) is locally a smooth curve $x_0(\lambda)$ passing through $(x_0(\lambda_0), \lambda_0)$.

Hence, we conclude that if for a certain $\lambda = \lambda_c$, $\mathbf{A}(\lambda_c)$ is not invertible, then bifurcation can occur. That is, there may exist several solutions in the neighborhood of the point $(x_0(\lambda_c), \lambda_c)$. A typical case of non-invertibility is the occurrence of zero eigenvalues. We show in Section 6.6.2 that the pitchfork bifurcation in Fig. 3.1 corresponds to a zero eigenvalue.

Here we indicate that at a turning point also a zero eigenvalue occurs. To show this, we first assume that $(x_0(s), \lambda(s))$ is a regular parametrization of the solution curve, for example by the arc length s. Second we

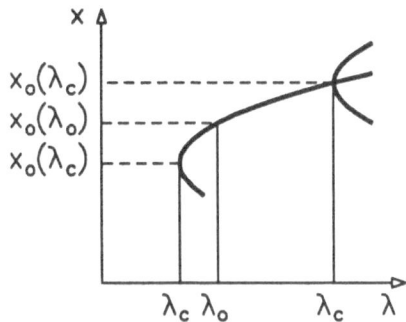

Figure 3.1. Turning point and pitchfork bifurcation as typical cases of non-invertibility of the linear operator (3.5)

calculate from (3.2)

$$\frac{\partial F}{\partial x}\frac{\partial x}{\partial s} + \frac{\partial F}{\partial \lambda}\frac{\partial \lambda}{\partial s} = 0 \ .$$

For the turning point we have $\partial \lambda / \partial s = 0$ and $\partial x / \partial s \neq 0$. Hence, the above relation yields that $\partial F / \partial x$ must be singular. In the simplest case, this can happen due to a zero eigenvalue. Thus we see that the stability and bifurcation problem are closely related (see the examples in Section 4.3).

We remark that for the occurrence of a purely imaginary pair of roots (3.5) does not become singular. Nevertheless a bifurcation occurs. In this case a flutter instability due to a Hopf bifurcation (Section 6.1) sets in. However, the steady state remains the unique equilibrium position. At the bifurcation point where the oscillation starts an *exchange of stability* from a stable to an unstable equilibrium occurs.

Now we proceed with the dynamic problem (3.4). First, the eigenvalues of (3.5) are calculated at the critical parameter value $\lambda = \lambda_c$ and the matrix $\mathbf{A}$ is transformed to *diagonal* form (with 2×2 blocks for complex eigenvalues) if n linearly independent eigenvectors exist. If n independent eigenvectors do not exist, then the *principal vectors* must be calculated and the corresponding *Jordan form* will be obtained (Appendix B). The transformation matrix $\mathbf{B}$ is composed by the eigenvectors (for a complex eigenvalue by the real and imaginary part of the complex eigenvector) and principal vectors, respectively. Thus, the transformation of variables is given by

$$\xi = \mathbf{B}y \ . \tag{3.7}$$

Inserting (3.7) into (3.4) results in

$$\dot{y} = \mathbf{B}^{-1}\mathbf{A}(\lambda_c)\mathbf{B}y + \mathbf{B}^{-1}\mathbf{f}(\mathbf{B}y, \lambda_c) = \mathbf{J}y + \mathbf{g}(y) \tag{3.8}$$

where $\mathbf{J} = \mathbf{B}^{-1}\mathbf{A}\mathbf{B}$ is either a *diagonal matrix* or contains *Jordan blocks*. In Appendix B two examples of the calculation of eigenvectors taken from [59] are given.

To explain the next step of the reduction process, we consider the simplest case of loss of stability that is, the occurrence of a simple zero eigenvalue. The matrix $\mathbf{J}$ in (3.8) then could have the following form

$$
\mathbf{J} = \begin{pmatrix}
0 & & & & & \\
& \mu_2 & & & 0 & \\
& & \ddots & & & \\
& & & \mu_m & & \\
& & & & \ddots & \\
& & & & & \varepsilon_{\frac{n-m}{2}} & -\omega_{\frac{n-m}{2}} \\
& 0 & & & & \omega_{\frac{n-m}{2}} & \varepsilon_{\frac{n-m}{2}}
\end{pmatrix}, \qquad (3.9)
$$

where the μ_k and ε_j are negative. For simplicity, but without loss of generality, it is assumed that in (3.9) no Jordan blocks appear. The system of differential equations (3.8) written in components has the form

$$
\begin{aligned}
\dot{y}_1 &= g_1(y_1, \ldots, y_n) \\
\dot{y}_2 &= \mu_2 y_2 + g_2(y_1, \ldots, y_n) \\
&\ \vdots \\
\dot{y}_{n-1} &= \varepsilon_{\frac{n-m}{2}} y_{n-1} - \omega_{\frac{n-m}{2}} y_n + g_{n-1}(y_1, \ldots, y_n) \\
\dot{y}_n &= \omega_{\frac{n-m}{2}} y_{n-1} + \varepsilon_{\frac{n-m}{2}} y_n + g_n(y_1, \ldots, y_n).
\end{aligned} \qquad (3.10)
$$

The nonlinear terms satisfy

$$
g_i(y_1, \ldots, y_n) = O(|y_1|^2 + \ldots + |y_n|^2), \qquad i = 1, \ldots, n.
$$

There is no linear term in the first equation of (3.10) because of the zero eigenvalue.

In the remaining $n - 1$ equations ($i = 2, \ldots, n$) all linear terms correspond to eigenvalues with negative real parts. Therefore, for the judgement of stability of the equilibrium position $y = 0$ one could plausibly, but incorrectly, argue that due to the negative linear terms for small perturbations of the system about the equilibrium the variables $y_2, \ldots, y_n$ will tend to zero as $t \to \infty$. Then the stability behavior would be governed by the equation

$$
\dot{y}_1 = \hat{g}_1(y_1) = g_1(y_1, 0, \ldots, 0),
$$

which depends only on the variable y_1. This reasoning is in error because it misses the fact that due to nonlinear couplings, $y_2, \ldots, y_n$ need not necessarily tend to zero. This can be seen from a simple two-dimensional example given below.

For the correct treatment of (3.10) we separate the variables into two sets. Namely in y_1, which is called the *active* or *critical variable* and in

$y_2, \ldots, y_n$ which are called the *passive* or *non-critical variables*. Accordingly the equations are also separated into equation $(3.10)_1$ which is called *bifurcation equation* and the remaining $n-1$ equations. Only the way the passive variables are eliminated from the bifurcation equation $(3.10)_1$ is, in general, incorrect if they are simply set to zero. But nevertheless the active variable obtained from a correctly determined bifurcation equation, that is where the passive variables are properly eliminated, will govern the stability problem.

We explain this point now with an example ([21] p. 5) where the stability of the equilibrium $y_0 = 0$ of the system

$$
\begin{aligned}
\dot{y}_1 &= y_1 y_2 + a y_1^3 + b y_1 y_2^2 \\
\dot{y}_2 &= -y_2 + c y_1^2 + d y_1^2 y_2
\end{aligned}
\tag{3.11}
$$

is investigated. The linear part of (3.11) is already given in diagonalized form. The eigenvalues are 0 and -1. Thus, y_1 is the critical and y_2 the non-critical variable. If we concluded from the second equation of (3.11) that $y_2 \to 0$ the first equation would take the form

$$
\dot{y}_1 = a y_1^3 \ .
\tag{3.12}
$$

The stability analysis of (3.12) yields the equilibrium position $y_0 = 0$ to be stable if $a < 0$ and unstable if $a > 0$. This follows from integration of (3.12) or making use of Ljapunov's second method. In order to present a slightly more general result we consider instead of (3.12)

$$
\dot{y} = a y^n
\tag{3.13}
$$

with $n \geq 2$ and where we have omitted the index. We explain briefly both approaches:

(α) Integration of (3.13) can be performed using separation of variables that is, we write (3.13) in the form

$$
\frac{dy}{y^n} = a\, dt
$$

and obtain with $y(0) = y_0$

$$
y^{n-1}(t) = \frac{y_0^{n-1}}{1 + (1-n)y_0^{n-1} a t} \ .
\tag{3.14}
$$

From (3.14) follows that we always obtain unbounded solutions if n is even because then the sign of the initial value has an influence on the boundedness of the solution. Hence, a necessary condition that the origin is (asymptotically) stable is that n is odd. In addition, $a < 0$ must be required, because then for $t \to \infty$ the solution is always smaller than y_0. However, if n is odd but $a > 0$ the solution becomes unbounded if t approaches $-1/((1-n)y_0^{n-1}a)$.

(β) According to *Ljapunov's second method* ([83]) a solution is stable if
   the following requirements are fulfilled: (i) the existence of a positive definite function $V(y)$ called *Ljapunov function* and (ii) that
   the derivative of V in the direction of the vectorfield is zero (stability) or strictly negative (asymptotic stability). That is,

$$\dot{V} = \frac{\partial V}{\partial y} \cdot \dot{y} \leq 0 .$$

For (3.13) we take $V = y^2$. Then we obtain for (3.13)

$$\dot{V} = 2yay^n = 2ay^{n+1} .\tag{3.15}$$

From (3.15) we see that $\dot{V}$ is strictily negative only if n is odd and
a negative.

The stability analysis of the equilibrium of (3.11) with (3.12) is incorrect,
as the nonlinear coupling term $y_1 y_2$ in $(3.11)_1$ has a decisive influence on
the stability behavior of $y_0 = 0$. The correct treatment of (3.10) and
(3.11) requires the application of *center manifold theory* ([21]).

Center manifold theory

First, we define the concept of an *invariant manifold* for the system (3.1).

Definition 3.1 (Invariant manifold) *A set $S \subset \mathbb{R}^n$ is called a locally
invariant manifold for (3.1), if for any initial value $x_0 \in S$ there exists
some $\tau > 0$, such that the solution $x(t)$ remains in S for $t < \tau$. If $\tau = \infty$,
S is called an invariant manifold.*

Consider now a system in the form (3.8). We rewrite (3.8) in the form
([50])

$$\begin{aligned}
\dot{y}_c &= \mathbf{J}_c y_c + \mathbf{g}_c(y_c, y_s) \\
\dot{y}_s &= \mathbf{J}_s y_s + \mathbf{g}_s(y_c, y_s).
\end{aligned}\tag{3.16}$$

The matrices $\mathbf{J}_c$ and $\mathbf{J}_s$ are of dimension $n_c \times n_c$ and $n_s \times n_s$ whose
eigenvalues have, respectively, zero and negative real parts ($n_c + n_s = n$).
   Note that if $\mathbf{g}_c = \mathbf{g}_s = \mathbf{0}$ all solutions would tend exponentially fast
to solutions of $\dot{y}_c = \mathbf{J}_c y_c$. That is, equation $(3.16)_1$ would determine the
asymptotic behavior of the entire system up to exponentially decaying
terms. The *center manifold theorems* stated below enable us to extend
this argument to the case where $\mathbf{g}_c$ and $\mathbf{g}_s$ are not equal to zero.
   The components of y_c are called *critical variables*. In the first set
of equations of (3.16), which will become the *bifurcation equations*, the
non-critical variables y_s are still present, and hence, must be eliminated.

For this purpose, the *non-critical variables* y_s are represented by a *local graph* of the critical variables y_c by means of a function h

$$\mathbf{h} : \mathbb{R}^{n_c} \to \mathbb{R}^{n_s} \qquad \text{or} \qquad y_s = \mathbf{h}(y_c) \qquad (3.17)$$

with

$$\mathbf{h}(0) = 0 \qquad \text{and} \qquad \mathbf{h}'(0) = 0. \qquad (3.18)$$

Inserting (3.17) into (3.16)$_1$, the system of bifurcation equations

$$\dot{y}_c = \mathbf{J}_c y_c + \mathbf{g}_c(y_c, \mathbf{h}(y_c)) \qquad (3.19)$$

follows, which depends only on the critical variables.

This procedure allows us to give a formal definition of the *center manifold* ([21]):

Definition 3.2 (Center manifold) *A smooth invariant manifold* $\mathbf{h}(y_c)$ *for* (3.16) *is called center manifold if* $\mathbf{h}(0) = \mathbf{h}'(0) = 0$.

The meaning of (3.18) is that the center manifold is tangent to the eigenspace E^c of the linearized system of dimension n_c. The procedure used above is justified by the following theorem ([7] p. 266):

Theorem 3.1 (Center manifold) *If* n_c, n_s, n_u *are the numbers of eigenvalues of* $\mathbf{A}$ *in* (3.4) *with zero, negative and positive real parts, respectively, where* $n = n_c + n_s + n_u$, *then system* (3.4) *is topologically equivalent to the following form*

$$\begin{aligned} \dot{y}_c &= \mathbf{g}_c(y_c, \lambda_c) \\ \dot{y}_s &= -y_s \\ \dot{y}_u &= +y_u \ . \end{aligned} \qquad (3.20)$$

where

$$y_c \in \mathbb{R}^{n_c}, \qquad y_s \in \mathbb{R}^{n_s} \qquad \text{and} \qquad y_u \in \mathbb{R}^{n_u}.$$

The first equation in (3.20) governs the motion on the center manifold, whereas the remaining equations yield the topological behavior of the trajectories in the neighborhood of the center manifold.

For physical or engineering applications we can always assume that $n_u = 0$, that is, the loss of stability of a stable equilibrium is considered. Thus, the third equation in (3.20) can be omitted. The second equation yields an exponential contraction towards the center manifold, and hence, locally, the stability behavior is governed by the reduced system of dimension n_c. This reduced set of equations of dimension n_c are the *bifurcation equations* realized on some smooth neutral submanifold of dimension n_c, the *center manifold*, which depends smoothly on $\lambda - \lambda_c$ in phase space. The smoothness of the center manifold can be finite and

the manifold need not be unique as simple examples show ([9]). Even if
the original system is analytic the center manifold will only be smooth.
Nevertheless, the local behavior of the trajectories, including the whole
picture of bifurcations, is determined for the full system (3.20) by what
takes place on the indicated center manifold, and in particular, does not
depend on the choice of the center manifold.

Let us return to the example (3.11). The linear system is

$$\dot{y}_1 = 0$$
$$\dot{y}_2 = -y_2. \tag{3.21}$$

This is a peculiar system since it has a continuum of equilibria. The
equilibrium of interest, which persists when nonlinear terms are added, is
$(0,0)$. There exist two invariant manifolds for $(0,0)$, the y_1-axis $(y_2 = 0)$
and the straight line $y_1 = 0$. The lines $y_1 = $ const. are the stable manifolds
for the points $(\text{const.},0)$. The line $y_2 = 0$ is the center manifold (Fig. 3.2a).
Since the center manifold of the nonlinear system must pass through the
origin and also be tangent to the eigenspace of the linearized system, the
conditions (3.18) follow. In Fig. 3.2b and c, sketches of the flows and the
center manifolds for the system

$$\dot{y}_1 = y_1 y_2 , \qquad \dot{y}_2 = -y_2 + a y_1^2 \tag{3.22}$$

are given. The stability of the origin of (3.22) depends on the coefficient
a, because according to (3.17) it follows from the second equation of
(3.22) that $y_2 = a y_1^2$. Inserting this expression into the first equation
yields

$$\dot{y}_1 = a y_1^3$$

which is (3.12). If $a < 0$ (Fig. 3.2b) the origin is stable and if $a > 0$
(Fig. 3.2c) it is unstable. Thus, we can conclude that in the critical
or bifurcation case, the nonlinear terms determine the stability of an
equilibrium position. We shall return to this example in Section 6.6.2
case (3).

The general procedure to compute the center manifold $\mathbf{h}$ is to insert
(3.17) into (3.16)$_2$ resulting in

$$\mathbf{h}'(\mathbf{y}_c)\dot{\mathbf{y}}_c = \mathbf{J}_s \mathbf{h}(\mathbf{y}_c) + \mathbf{g}_s(\mathbf{y}_c, \mathbf{h}(\mathbf{y}_c)). \tag{3.23}$$

Replacing $\dot{\mathbf{y}}_c$ in (3.23) from (3.16)$_1$ yields

$$\mathbf{h}'(\mathbf{y}_c)(\mathbf{J}_c \mathbf{y}_c + \mathbf{g}_c(\mathbf{y}_c, \mathbf{h}(\mathbf{y}_c))) = \mathbf{J}_s \mathbf{h}(\mathbf{y}_c) + \mathbf{g}_s(\mathbf{y}_c, \mathbf{h}(\mathbf{y}_c)) . \tag{3.24}$$

In principle, $\mathbf{h}$ can be calculated from equation (3.24) together with
(3.18). In general, an exact solution of (3.24) is impossible, since it would

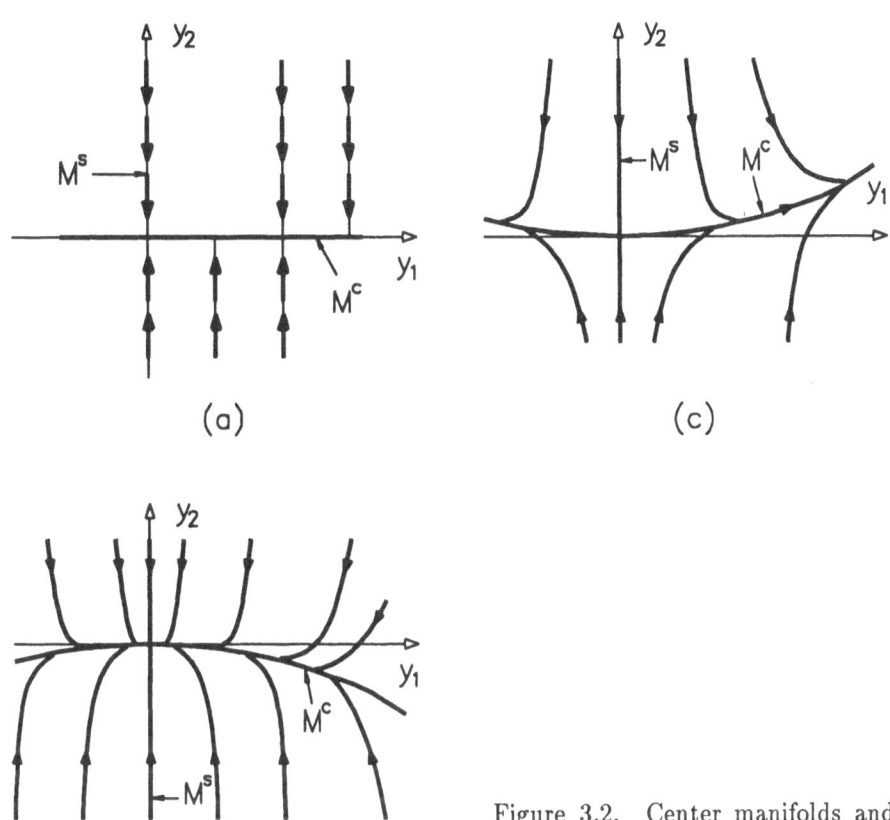

Figure 3.2. Center manifolds and flows (a) for the linear system (3.21) and the nonlinear system (3.22) for (b) $a < 0$ and (c) $a > 0$

be equivalent to solving the original system of nonlinear differential equations. However, an approximation $\mathbf{H}(\mathbf{y}_c)$ for $\mathbf{h}(\mathbf{y}_c)$ can be calculated to any desired degree of accuracy ([21]).

Rewriting (3.24) for $\mathbf{H}$ and satisfying the following relation up to the order $|\mathbf{y}_c|^q$ yields

$$\mathbf{H}'(\mathbf{y}_c)(\mathbf{J}_c\mathbf{y}_c + \mathbf{g}_c(\mathbf{y}_c, \mathbf{H}(\mathbf{y}_c))) - \mathbf{J}_s\mathbf{H}(\mathbf{y}_c) - \mathbf{g}_s(\mathbf{y}_c, \mathbf{H}(\mathbf{y}_c)) = O(|\mathbf{y}_c|^q) . \tag{3.25}$$

If (3.25) can be satisfied for $q > 1$, then as $|\mathbf{y}_c| \to \mathbf{0}$, $\mathbf{H}$ approximates the center manifold to order q: $|\mathbf{H}(\mathbf{y}_c) - \mathbf{h}(\mathbf{y}_c)| = O(|\mathbf{y}_c|^q)$ ([21]).

It seems convenient to sum up all important facts of center manifold theory in the following theorem ([21]):

Theorem 3.2 (Center manifold)   *1. There exists a center manifold $\mathbf{y}_s = \mathbf{h}(\mathbf{y}_c)$ for system (3.16) if $|\mathbf{y}_c|$ is sufficiently small. The behavior on the center manifold is governed by equation (3.19).*

2. *The zero solution of* (3.19) *has exactly the same stability properties as the zero solution of* (3.16).

3. *If* $\mathbf{H} : \mathbb{R}^{n_c} \to \mathbb{R}^{n_s}$ *is a smooth map with* $\mathbf{H}(0) = \mathbf{H}'(0) = 0$ *and* (3.25) *is satisfied for* $q > 1$ *as* $|\mathbf{y}_c| \to 0$, *then* $\mathbf{H}$ *approximates the center manifold* $\mathbf{h}$ *up to terms of* $O(|\mathbf{y}_c|^q)$.

We consider the example (3.11) for which we wish to calculate a sufficiently accurate approximation of the center manifold. What we mean by sufficiently accurate is made precise with the concept of *determinacy*, which is explained in detail in Section 6.3 and was already mentioned in the Introduction. We represent the non-critical variable y_2 as a series in the critical variable y_1 on the center manifold of the form

$$y_2 = H(y_1) = \alpha_2 y_1^2 + \alpha_3 y_1^3 + \alpha_4 y_1^4 + \dots \quad . \tag{3.26}$$

Because of (3.18), the series (3.26) starts with a quadratic term. Substituting (3.26) into (3.25) and setting to zero the coefficients up to the fourth order results in

$$\begin{aligned}
&(2\alpha_2 y_1 + 3\alpha_3 y_1^2 + \dots)[y_1(\alpha_2 y_1^2 + \alpha_3 y_1^3 + \dots) + ay_1^3 \\
&+ by_1(\alpha_2^2 y_1^4 + \dots)] + \alpha_2 y_1^2 + \alpha_3 y_1^3 + \alpha_4 y_1^4 + \dots - cy_1^2 \\
&- dy_1^2(\alpha_2 y_1^2 + \dots) = O(|y_1|^5)
\end{aligned}$$

$y_1^2 : \quad \alpha_2 - c = 0 \Longrightarrow \alpha_2 = c$

$y_1^3 : \quad \alpha_3 = 0$

$y_1^4 : \quad 2\alpha_2^2 + 2\alpha_2 a + \alpha_4 - d\alpha_2 = 0 \Longrightarrow \alpha_4 = dc - 2c(a + c) .$

Thus the center manifold is given by

$$y_2 = h(y_1) = cy_1^2 + (dc - 2c(a + c))y_1^4 + O(|y_1|^5). \tag{3.27}$$

Inserting (3.27) into the first equation of (3.11) yields

$$\dot{y}_1 = (a + c)y_1^3 + (dc - 2c(a + c) + bc^2)y_1^5 + O(|y_1|^6). \tag{3.28}$$

Contrary to the discussion following equation (3.11), the coefficient $(a+c)$ determines the stability of $y_0 = 0$, rather than the coefficient a in (3.12). If $a + c = 0$, then the term of fifth order in (3.28) determines the stability.

Let us now return to (3.10). To eliminate the non-critical variables $y_2, \dots, y_n$ from (3.10)$_1$, the functions $y_s = h_s(y_1)$, $s = 2, \dots, n$ must be calculated, recalling that by (3.18)

$$y_s = h_s(y_1) = O(|y_1|^2) . \tag{3.29}$$

The relation (3.29) often has the important consequence that, after inserting the y_s into the first equation of (3.10), the lowest order terms which determine the stability behavior will not be influenced by the y_s. Let us assume that in $(3.10)_1$

$$g_1(y_1, \ldots, y_n) = O(|y_1|^3 + \ldots + |y_n|^3) \ .$$

That is, terms $y_1^3, y_1^2 y_2, \ldots, y_1 y_n^2, \ldots, y_n^3 + h.o.t.$ are present. Because of (3.29) the term $y_1^2 y_2 = O(|y_1|^4)$ and $y_1 y_n^2 = O(|y_1|^5)$ and so on. Hence, the lowest order term is given by y_1^3 and all other terms can be neglected. That is, we can set $h_s(y_1) \equiv 0$ for $s = 2, \ldots, n$.

For example, if in (3.11) the term $y_1 y_2$ were absent, then in the bifurcation equation y_2 could be set to zero and the correct result would still be obtained. This is because the third order term does not vanish and we have a three-determinate system (Section 6.3). Thus, in many practical applications, where certain (reflectional) symmetry properties are present (Section 5.1), one can avoid the cumbersome calculation of the center manifold and still obtain correct bifurcation equations by setting the non-critical variables to zero.

We close this section with two examples and two comments:

Example 1: A picture of a three-dimensional system for which (3.5) has one eigenvalue equal to zero, one greater and one smaller than zero is shown in Fig. 3.3. In this figure, it can be seen that the eigenvectors spanning the eigenspaces E^u, E^c, E^s are tangent to the invariant manifolds M^u, M^c, M^s of the nonlinear system.

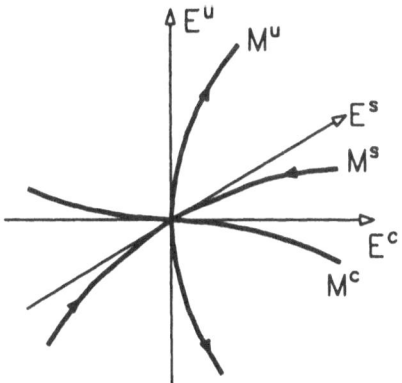

Figure 3.3. Linear eigenspaces E^u, E^c, E^s and invariant manifolds M^u, M^c, M^s of the nonlinear system corresponding to comment (i)

Example 2: In Fig. 3.4, the flow of a three-dimensional system is shown for which (3.5) has a purely imaginary pair of roots and one negative root. Here a so-called *Hopf bifurcation* occurs, which is explained in Sections 6.1 and 6.6.2. The corresponding center manifold is two-dimensional. Obviously the local stability behavior is completely determined by the flow on the center manifold.

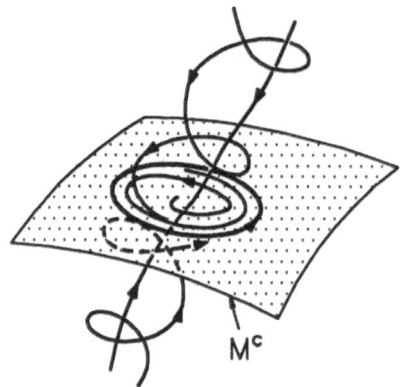

M^c

Figure 3.4. Flow nearby and on the center manifold for a three-dimensional system after a Hopf bifurcation

Comment 1: There exist simple examples with no unique center manifold ([50] p. 125, [9]). However, for practical applications, these cases are of minor significance. Furthermore, since the center manifold dynamics is equivalent on all center manifolds, finding just one and solving the problem with it is sufficient.

Comment 2: Strictly speaking we have shown so far that for (3.1) only for $\lambda = \lambda_c$ a center manifold exists. Hence, the existence of the center manifold for $\lambda \neq \lambda_c$ remains to be shown. For this a slightly different formulation, sometimes called "suspension trick", is used. In this case (3.1) is replaced by

$$\dot{\boldsymbol{x}} = \mathbf{F}(\boldsymbol{x}, \lambda)$$
$$\dot{\lambda} = 0 . \tag{3.30}$$

At $\lambda = \lambda_c$ a bifurcation occurs. This means that the flow changes qualitatively at $\lambda = \lambda_c$. In the case of (3.30) λ is considered as a trivial critical variable and the series expansion of the center manifold includes λ, and therefore, also values $\lambda \neq \lambda_c$. Similar to (3.16) we write (3.30)

$$\dot{\boldsymbol{y}}_c = \mathbf{J}_c \boldsymbol{y}_c + \mathbf{g}_c(\boldsymbol{y}_c, \boldsymbol{y}_s, \lambda)$$
$$\dot{\boldsymbol{y}}_s = \mathbf{J}_s \boldsymbol{y}_s + \mathbf{g}_s(\boldsymbol{y}_c, \boldsymbol{y}_s, \lambda) \tag{3.31}$$
$$\dot{\lambda} = 0$$

with $\mathbf{J}_c$ and $\mathbf{J}_s$ as in case of (3.16). The functions $\mathbf{g}_c$ and $\mathbf{g}_s$ vanish together with their derivatives at $(\boldsymbol{y}_c, \boldsymbol{y}_s, \lambda) = (0, 0, \lambda_c)$. For (3.31) exists a center manifold of the form

$$\boldsymbol{y}_s = \mathbf{h}(\boldsymbol{y}_c, \lambda) \tag{3.32}$$

for small $|\boldsymbol{y}_c|$ and $|\lambda - \lambda_c|$. Introducing (3.32) into $(3.31)_1$ we obtain for small solutions of (3.31)

$$\dot{\boldsymbol{y}}_c = \mathbf{J}_c \boldsymbol{y}_c + \mathbf{g}_c(\boldsymbol{y}_c, \mathbf{h}(\boldsymbol{y}_c, \lambda), \lambda)$$
$$\dot{\lambda} = \mathbf{0}.$$

This means that the center manifold depends smoothly on the parameters. This justifies our approach presented so far because one can perform the center manifold reduction at the critical parameter values $\lambda = \lambda_c$ and afterwards introduce the influence of the parameters involved by an unfolding of the bifurcation equations (Section 6.4).

Consider as example the system ([50])

$$\begin{aligned} \dot{y}_1 &= \lambda y_1 - y_1^3 \\ \dot{y}_2 &= -y_2 \\ \dot{\lambda} &= 0. \end{aligned} \tag{3.33}$$

In (3.33), the parameter λ plays the role of a trivial dependent variable. For this system linearized at $(y_1, y_2, \lambda) = (0, 0, 0)$, it follows that the y_2-axis is a stable subspace and the (y_1, λ)-plane is the center subspace, because at $(0, 0, 0)$ two zero eigenvalues appear, related to the coordinates y_1 and λ. The set of equilibria of (3.33) consists of the λ-axis and the parabola $\lambda = y_1^2$ in the (y_1, λ) plane (Fig. 3.5). Since $\dot{\lambda} \equiv 0$, the planes $\lambda =$

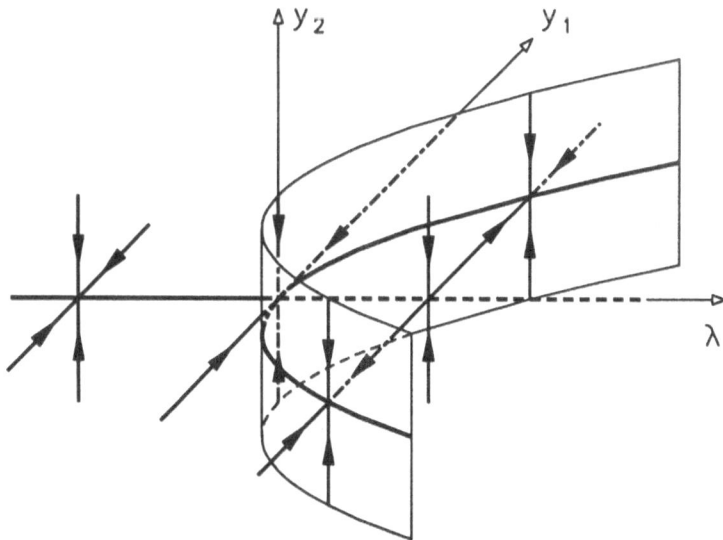

Figure 3.5. Suspended three-dimensional system (3.33)

const. are invariant under the flow of (3.33). In a plane $\lambda = \text{const.} \neq 0$ all equilibria are *hyperbolic*. That is, the linearization is non-degenerate or, in other words, no eigenvalues with zero real part occur. The equilibria on the λ-axis with $\lambda < 0$ are nodes, while those along the positive λ-axis are two stable nodes and one saddle. The center manifold is given by the (y_1, λ) coordinate plane. Geometrically, this means that the three-dimensional flow is everywhere attracted by this plane and the interesting stability behavior occurs only in this plane in its dependence on the parameter λ.

Applications of the suspension trick are given in Section 6.6.2 subsections (2) and (7).

3.1.2 Periodic motions

Now we study the stability of a periodic solution $x_0(t) = x_0(t + T)$, with period T. In the neighborhood of this solution a transformation of the time-continuous equations of motion to time-discrete equations of motion

$$x_{t+1} = P(x_t, \lambda) \qquad (3.34)$$

is possible by introducing a Poincaré mapping. The detailed description of the calculation of (3.34) for the periodic motion of a simple robot is given in Section 4.2. Let x_t and $P \in \mathbb{R}^n$ and $\lambda \in \mathbb{R}^\ell$. The periodic motion $x_0(t)$ of the time-continuous system corresponds to a fixed point x_{t0} (Fig. 2.3) of the time-discrete system (3.34). The equation for the calculation of the fixed point x_{t0} follows from (3.34) by setting $x_{t+1} = x_t$ to

$$x_{t0} = P(x_{t0}, \lambda). \qquad (3.35)$$

Similar to (3.1), the right-hand side of (3.34) is split into its linear part and the remaining nonlinear terms by setting $x_t = x_{t0} + \xi_t$. Hence,

$$\xi_{t+1} = A(\lambda)\xi_t + f(\xi_t, \lambda) \qquad (3.36)$$

is obtained, where

$$A(\lambda) = \left. \frac{\partial P_i}{\partial x_{tj}} \right|_{x_{t0}} \qquad (3.37)$$

and $f(\xi_t, \lambda) = O(|\xi_t|^2)$.

The stability problem of the fixed point $\xi_{t0} = 0$ is determined by the eigenvalues of (3.37):

(a) If all eigenvalues of (3.37) have modulus smaller than *one* then $\xi_{t0} = 0$ is stable.

(b) If at least one eigenvalue of A has a modulus larger than *one*, then $\xi_{t0} = 0$ is unstable.

(c) The critical case in the sense of Ljapunov is given if one or several eigenvalues have modulus equal to *one* and all the remaining eigenvalues are of modulus smaller than *one*.

Thus for mappings, the modulus *one* of the eigenvalues plays the same role as the *zero real part* for differential equations (see p. 47). This function of the modulus is easy to understand if we look at the linear part of (3.36)

$$\xi_{t+1} = A(\lambda)\xi_t. \tag{3.38}$$

Let us assume that all eigenvalues of (3.37) are distinct. In this case, the matrix A can be transformed into diagonal or block diagonal form by a linear coordinate transformation (Appendix B), and hence, (3.38) is divided into one- or two-dimensional subsystems, depending on whether the eigenvalues are real or complex conjugate. Writing one of these one-dimensional subsystems in the form

$$\xi_{t+1,i} = \mu_i \xi_{t,i} , \tag{3.39}$$

it is clear that for arbitrary initial conditions $\xi_{t,i}$ will decay to zero only if $|\mu_i| < 1$. This follows immediately from iterating (3.39).

Center manifold reduction

To reduce (3.34) to a lower-dimensional bifurcation system, one can proceed as in the time-continuous case. First, a parameter value λ is chosen for which all eigenvalues of (3.37) are of modulus smaller than *one*. Then one distinguished component of λ is picked and increased quasistatically, until the first time one or several eigenvalues have modulus equal to *one* at the critical parameter value $\lambda = \lambda_c$. In order to transform A into Jordan form $J = B^{-1}AB$ (Appendix B), new coordinates y_t are introduced by the linear transformation

$$\xi_t = By_t , \tag{3.40}$$

which is analogous to (3.7). Using (3.40), (3.36) may be written as

$$y_{t+1} = Jy_t + g(y_t, \lambda_c). \tag{3.41}$$

In (3.41), the equations are rearranged in such a way that the first n_c equations have eigenvalues with modulus equal to *one* and the remaining n_s equations have eigenvalues with modulus smaller than *one*. This allows us to rewrite (3.41) in the following form

$$\begin{aligned} y_{t+1,c} &= J_c y_{t,c} + g_c(y_{t,c}, y_{t,s}) \\ y_{t+1,s} &= J_s y_{t,s} + g_s(y_{t,c}, y_{t,s}) . \end{aligned} \tag{3.42}$$

We designate the active part of y_t by $y_{t,c} \in \mathbb{R}^{n_c}$ and the passive part by $y_{t,s} \in \mathbb{R}^{n_s}$ with $n_c + n_s = n$. The elimination of $y_{t,s}$ from the bifurcation equation $(3.42)_1$ can be performed by applying *center manifold theory* ([21] p. 33). This involves representing the passive variables $y_{t,s}$ as a function of the active variables $y_{t,c}$ in the form

$$y_{t,s} = \mathbf{h}(y_{t,c}), \qquad (3.43)$$

where $\mathbf{h} : \mathbb{R}^{n_c} \to \mathbb{R}^{n_s}$ and the properties (3.18) hold. We substitute (3.43) into $(3.42)_2$ and use $(3.42)_1$ to find

$$\mathbf{h}(\mathbf{J}_c y_{t,c} + \mathbf{g}_c(y_{t,c}, \mathbf{h}(y_{t,c}))) - \mathbf{J}_s \mathbf{h}(y_{t,c}) - \mathbf{g}_s(y_{t,c}, \mathbf{h}(y_{t,c})) = \mathbf{0}. \qquad (3.44)$$

Again, as in the time-continuous case, one looks for an approximation $\mathbf{H}(y_{t,c})$ of $\mathbf{h}(y_{t,c})$. For $\mathbf{H}(y_{t,c})$ we rewrite (3.44) to obtain

$$\mathbf{H}(\mathbf{J}_c y_{t,c} + \mathbf{g}_c(y_{t,c}, \mathbf{H}(y_{t,c}))) - \mathbf{J}_s \mathbf{H}(y_{t,c}) - \mathbf{g}_s(y_{t,c}, \mathbf{H}(y_{t,c})) = O(|y_{t,c}|^q). \qquad (3.45)$$

Center manifold theory ([21]) shows that the difference

$$\mathbf{h}(y_{t,c}) - \mathbf{H}(y_{t,c}) = O(|y_{t,c}|^q) \qquad (3.46)$$

as well. Inserting $\mathbf{H}(y_{t,c})$ into $(3.42)_1$, the bifurcation system becomes

$$y_{t+1,c} = \mathbf{J}_c y_{t,c} + \mathbf{g}_c(y_{t,c}, \mathbf{H}(y_{t,c})). \qquad (3.47)$$

An example where we calculate the function $\mathbf{H}(y_{t,c})$ explicitly up to terms of the necessary order is given in Section 4.2.3.

3.2 Infinite-dimensional statical and dynamical systems

In this section we distinguish between *statical* and *dynamical systems*.

We recall that we call a system *statical*, if its loading is conservative and varies quasistatically. Then after loss of stability of its equilibrium position a *divergent* transition into another equilibrium position occurs. Such a transition is mostly fast and strongly damped. This transition is generally not of practical interest, and hence, can be neglected in the description of the system. Then we speak of a case of *divergent instability* under statical loading. We further restrict to elastic behavior of the system. Then it can be described by a potential. *Elementary catastrophe theory* (Section 6.5) will be the proper way to treat stability or bifurcation problems of such systems. However, *catastrophe theory* can only be applied to finite-dimensional bifurcation equations. To obtain such a set of finite-dimensional bifurcation equations a method of reduction is needed which is called *Ljapunov-Schmidt* method.

In addition, at the end of this section a short extension of *center manifold theory* to infinite-dimensional dynamical systems is given.

3.2.1 Statical systems

Ljapunov-Schmidt method

The Ljapunov-Schmidt method will not be explained in its full generality. The interested reader will find a more general treatment and proofs of the statements that follow in [156], [48], [26]. Restriction is also made to cases that occur in the theory of elasticity. As a result of this restriction, it will be possible to make substantial simplifications.

We assume that a statical buckling problem can be described by an equation

$$G(u, \lambda) = 0. \tag{3.48}$$

For example consider the scalar equation (2.32). G is a differentiable mapping between two Hilbert spaces H and K, that is, $G : H \times \Lambda \to K$, where Λ is the finite-dimensional parameter space. We note that usually a less restrictive assumption concerning the function spaces is made in the literature ([156], [48], [26]). In fact, H and K are only required to be Banach spaces. However, the requirement of H and K to be Hilbert spaces simplifies some steps in the reduction process, for example, the decomposition introduced below, considerably. Suppose that there exists an equilibrium solution

$$(u_0, \lambda_0) \tag{3.49}$$

such that

$$G(u_0, \lambda_0) = 0. \tag{3.50}$$

Now, one may ask the question (see Section 3.1.1): does a unique solution of (3.48) in the form

$$u = u(\lambda) \tag{3.51}$$

exist in a small neighborhood of (3.49)? The answer to this question is given by the *implicit function theorem* ([119]):

Theorem 3.3 (Implicit function) *With the definition of the Hilbert spaces* H, K *and the operator* G *given above we assume that* G *is Fréchet differentiable (Section 2.2). The Fréchet derivative of* G *at* (u_0, λ_0), *designated by* $G_u(u_0, \lambda_0)$ *is a linear mapping* $G_u : H \to K$. *If* $G_u(u_0, \lambda_0)$ *possesses a bounded inverse, then locally for* $|\lambda - \lambda_0|$ *sufficiently small, there exists a differentiable mapping* $u(\lambda)$, *such that* $G(u(\lambda), \lambda) = 0$. *Furthermore, in a sufficiently small neighborhood of* (u_0, λ_0), $(u(\lambda), \lambda)$ *is the only solution of* $G = 0$.

From the theorem, one can conclude that the case of bifurcation can only occur if the linear mapping G_u, evaluated at a specific solution (u_0, λ_c), is singular and hence no inverse of the linear operator $A := G_u(u_0, \lambda_c) :$ $H \to K$ exists. Such a solution (u_0, λ_c) is a *bifurcation point*. In this case, the linear operator A has a nontrivial *kernel* $N(A) \neq \{0\}$ and a *range*

$R(A) \neq K$ (Appendix A). In those cases, which will be examined later on, it can be assumed that $N(A)$ and $R(A)$ possess closed complements ([97]) (Appendix A). Making use of the *adjoint operator* $A^\star$ (Appendix C), the two Hilbert spaces H and K can be decomposed in the following way

$$H = N(A) \oplus R(A^\star)$$
$$K = R(A) \oplus N(A^\star).$$

A is called a *Fredholm* operator if $N(A)$ and $N(A^\star)$ are finite-dimensional. This is the case in *elasticity* ([97]). The difference $\dim N(A) - \dim N(A^\star)$ is called the *index* of A. In the sequel we will assume that A is a Fredholm operator with index *zero* (Appendix A). Furthermore, we always can make a transformation of coordinates to obtain

$$u_0 = 0$$

and we retain u as variable. We define the *orthogonal projection* P of K onto $R(A)$, that is, $P : K \to R(A)$, (Appendix D) and decompose (3.48) by means of P and the *complementary projection* $(E - P)$, to obtain the two relations

$$PG(u, \lambda_c) = 0$$
$$(E - P)G(u, \lambda_c) = 0 . \qquad (3.52)$$

Furthermore, u can be decomposed into the form

$$u = u_c + u_s \qquad (3.53)$$

where $u_c \in N(A)$ and $u_s \in R(A^\star)$. We substitute (3.53) into $(3.52)_1$ to obtain

$$PG(u_c + u_s, \lambda_c) = 0. \qquad (3.54)$$

The mapping $PG(u_c + u_s, \lambda_c) : H \times \Lambda \to R(A)$ is regular, and hence, possesses an invertible *Fréchet derivative* at $(0, \lambda_c)$. According to the *implicit function theorem* it is possible to express u_s as a function of u_c

$$u_s = h(u_c, \lambda_c). \qquad (3.55)$$

Equation (3.55) exists only locally, in a neighborhood of $(0, \lambda_c)$, where h possesses the following properties

$$h(u_c, \lambda_c) = 0 \qquad h_u(u_c, \lambda_c) = 0 . \qquad (3.56)$$

That is, h is at least of second order in u_c

$$h = O(\|u_c\|^2). \qquad (3.57)$$

To obtain the bifurcation equations, we introduce (3.53) in $(3.52)_2$ and use (3.55) to find

$$(E - P)G(u_c + h(u_c, \lambda_c), \lambda_c) = 0. \qquad (3.58)$$

The bifurcation equations (3.58) are a set of $\dim N(\mathbf{A}^\star)$ equations in $\dim N(\mathbf{A})$ unknowns. As $\mathbf{A}$ is assumed to be a Fredholm operator with index $i = 0$, exactly n_c ($= \dim N(\mathbf{A}^\star)$) equations for n_c unknowns are obtained. In engineering terminology, n_c is the *multiplicity* of the critical eigenvalue and (3.58) are an n_c-dimensional system of algebraic equations, which are the *amplitude equations* of the critical modes. It is important to point out that the solution of (3.58) gives locally the solutions to the original problem (3.48).

Comparison of the Ljapunov-Schmidt method with the Rayleigh-Ritz-Galerkin method

As was mentioned in the Introduction, the engineer who is used to treat buckling problems with the Rayleigh-Ritz-Galerkin method, will have noticed that there is a strong similarity between the Galerkin and the Ljapunov-Schmidt method. However, there is also a significant difference between these two methods, which can now be explained. In the application of the Rayleigh-Ritz-Galerkin method to buckling problems, instead of (3.53) generally an ansatz (or assumed choice) for u only in the *critical variables* u_c is made. Hence, the operation of elimination of the *non-critical variables* u_s does not even occur in this method. Therefore, this approach is equivalent to setting $\mathbf{h} \equiv \mathbf{0}$ in (3.58). As already mentioned in the Introduction, for many practical applications, this may be correct, if only nonlinear terms of third order need to be retained in the bifurcation equations. Furthermore, this also requires the system to be reflectionally symmetric, that is, no quadratic terms must be present in the bifurcation system (Section 5.1). For example, this is not the case for a shallow arch ([153]). See also p. 239 where for the buckling problem of a spherical shell it is shown that even for the third order terms qualitatively incorrect bifurcation equations are obtained from the Rayleigh-Ritz-Galerkin reduction.

Applications of the method of Ljapunov-Schmidt are given in Chapter 4, where we treat several examples in detail.

In order to give the reader a feeling for the basic idea of the method of Ljapunov-Schmidt, we consider a simple finite-dimensional example. Let the potential V of a two-dimensional, one-parameter system be given by ([111] p. 294)

$$V(x_1, x_2, \lambda) = \frac{1}{2}(x_1^2 + \lambda x_2^2) + 4x_1 x_2^2 + 2x_2^4. \tag{3.59}$$

We want to investigate the stability of $x_1 = x_2 = 0$ by means of the *energy criterion* ([168]) (sometimes also called *criterion of Dirichlet* or *principle of minimum of potential energy*). This criterion states that for a stable equilibrium position, the potential energy must have a minimum.

For the equilibrium equations, we obtain from (3.59)

$$\frac{\partial V}{\partial x_1} = x_1 + 4x_2^2 = 0$$

$$\frac{\partial V}{\partial x_2} = \lambda x_2 + 8x_1 x_2 + 8x_2^3 = 0 .$$

(3.60)

It follows from (3.60) that $x_1 = x_2 = 0$ is indeed an equilibrium position. We also see that if $\lambda \neq 0$, the second derivative at the equilibrium is regular and we can determine its stability from the quadratic terms in (3.59). Obviously, the equilibrium is locally stable for $\lambda > 0$ and unstable for $\lambda < 0$.

However, we are interested in the case where $\lambda = \lambda_c = 0$. Then the quadratic part of the energy in (3.59) no longer determines stability, because the potential energy is not definite due to the occurrence of a zero eigenvalue. Hence, the system is at a bifurcation point and a nonlinear analysis must be performed.

As the linear part in (3.60) is already in diagonal form, the system (3.60) corresponds to the decomposed system (3.52). The decomposition (3.53) is trivially given by $u_c = x_2$ and $u_s = x_1$.

If we, incorrectly, ignored the passive variable x_1 in the bifurcation equation $(3.60)_2$, we would obtain the non-vanishing nonlinear term $8x_2^3$. However, this is an incorrect result, because according to (3.55) the passive variable x_1 must be calculated in its dependence on the active variable x_2 from equation $(3.60)_1$ which corresponds to $(3.52)_1$. This results in $x_1 = -4x_2^2$. Inserting this expression into $(3.60)_2$ yields, in lowest order, the nonlinear term: $-24x_2^3$. Obviously, this results in a qualitatively different behavior compared to the term $8x_2^3$ found before. The correct unfolded one-parameter bifurcation equation is

$$\lambda x_2 - 24x_2^3 = 0$$

instead of the incorrect one

$$\lambda x_2 + 8x_2^3 = 0.$$

A discussion of the qualitatively different solutions of these two equations is given in Section 6.2.1.

3.2.2 Dynamical systems

Center manifold theory

For the mathematical subtleties in the application of center manifold theory to *infinite-dimensional* problems we refer the reader to the literature

([21], [26], [64]). Formally we can write the equations of motion as an evolution equation in *Hilbert space* just as we did it in $\mathbb{R}^n$.

We assume the infinite-dimensional system to be given in the form

$$\dot{u} = \mathbf{A}(\lambda)u + \mathbf{g}(u, \lambda) , \qquad (3.61)$$

where u is an element of a Hilbert space $\mathbf{H}$ (Appendix A). We study the equilibrium state $u_0 = 0$. The nonlinear operator $\mathbf{g} : \mathbf{H} \to \mathbf{H}$ is smooth. In addition, $\mathbf{g}(0, \lambda) = 0$ and $\mathbf{g}_u(0, \lambda) = 0$.

However, for an infinite-dimensional system the spectrum of the operator $\mathbf{A}(\lambda)$ may be more complicated than for the finite-dimensional case. There may be an infinite number of eigenvalues and even a continuum of them. Even if the case of a continuum of eigenvalues is excluded, problems can still arise.

We state now some properties under which the existence of a center manifold is guaranteed:

(a) In [64] p. 94, necessary requirements for simple losses of stability (one zero root or a purely imaginary pair of roots) are given for which a reduction to bifurcation equations is possible. The linear evolution problem following from (3.61) must have a solution of the form $e^{\mu_1 t}x_1$. Here, x_1 is the eigenvector and μ_1 is an isolated eigenvalue possessing the following properties:

   (i)   No other eigenvalue μ_i of the linear eigenvalue problem should have a larger real part than μ_1.

   (ii) μ_1 must be of finite multiplicity n_c.

   (iii) The same properties (i) and (ii) are required for the eigenvalues of the adjoint operator $\mathbf{A}^\star$.

(b) Similar, but more general, requirements are formulated in [21] p. 117. Here it is assumed that:

   (i)   It must be possible to decompose $\mathbf{H}$ into the form

$$\mathbf{H} = \mathbf{H}_c \oplus \mathbf{H}_s \qquad (3.62)$$

   where $\mathbf{H}_c$ is finite- (n_c-) dimensional and $\mathbf{H}_s$ is closed.

   (ii) The space $\mathbf{H}_c$ is invariant under $\mathbf{A}$ and if $\mathbf{A}_c$ is the restriction of $\mathbf{A}$ to $\mathbf{H}_c$, then all eigenvalues of $\mathbf{A}_c$ have zero real parts.

   (iii) All solutions of the linear evolution equation $\dot{u}_s = \mathbf{A}_s u_s$, defined below, must decay strictly exponentially. That is, $\|u_s(t)\| \leq ae^{-bt}$, for some positive a, b.

In order to perform the reduction one defines a projection $(\mathbf{E} - \mathbf{P})$ onto $\mathbf{H}_c$ along $\mathbf{H}_s$, giving $\mathbf{A}_c = (\mathbf{E} - \mathbf{P})\mathbf{A}$. Let $\mathbf{A}_s = \mathbf{P}\mathbf{A}$ and with $u_c \in \mathbf{H}_c$, $u_s \in \mathbf{H}_s$ let

$$u = u_c + u_s \tag{3.63}$$

and

$$\mathbf{g}_c(u_c, u_s) = (\mathbf{E} - \mathbf{P})\mathbf{g}(u_c, u_s), \qquad \mathbf{g}_s(u_c, u_s) = \mathbf{P}\mathbf{g}(u_c, u_s). \tag{3.64}$$

Combining these definitions, (3.61) can be written in the form

$$
\begin{aligned}
\dot{u}_c &= \mathbf{A}_c u_c + \mathbf{g}_c(u_c, u_s) \\
\dot{u}_s &= \mathbf{A}_s u_s + \mathbf{g}_s(u_c, u_s) \ .
\end{aligned}
\tag{3.65}
$$

After a change of coordinates, (3.65) can be put into canonical form and, hence, is formally similar to (3.16) for the finite-dimensional case. For (3.65), there exists an invariant manifold ([21] p. 118), the center manifold, which is tangent to the $\mathbf{H}_c$ space at the origin. Hence, a mapping

$$u_s = \mathbf{h}(u_c) \tag{3.66}$$

exists locally and from (3.65) the finite-(n_c)-dimensional bifurcation system follows to

$$\dot{u}_c = \mathbf{A}_c u_c + \mathbf{g}_c(u_c, \mathbf{h}(u_c)) \ . \tag{3.67}$$

(3.67) are ordinary differential equations. The application of this concept to an example is given in Section 5.3.

Chapter 4

Application of the reduction process

4.1 Loss of stability of equilibrium positions of finite-dimensional dynamical systems

In this chapter we will first study three examples from mechanics. All are plane double pendula. These examples serve to explain some of the assertions made in the Introduction. These pendula are not intended to represent models of real systems, but they are considered because their dynamic behavior is easy to understand. However, they are important in order to become familiar with the possible types of behavior of a system after loss of stability of an equilibrium position.

4.1.1 Double pendulum with axially elastic rods and follower force loading

A double pendulum with axially elastic rods and follower force loading ([144], [146]) is shown in Fig. 4.1. This system serves as a simple example to demonstrate that it can be worthwhile to treat a bifurcation problem which is higher degenerate than it would follow from the given system. This higher degeneracy will be achieved by a proper choice of a second parameter. Such an approach can help to get a global rather than local understanding of a stability problem.

The double pendulum has four degrees of freedom, designated by φ_1, φ_2, ℓ_1 and ℓ_2. We assume that the hinges and the rods in the direction of their axis are viscoelastic, with the stiffnesses γ and c and the coefficients of viscoelasticity k_1 and k_2, respectively, as shown in Fig. 4.1. The load F is assumed to have always the direction of the second rod. The masses of the rods are concentrated at the end of each rod. The length ℓ_0 is that of the unloaded rods. The equations of motion can be obtained by means

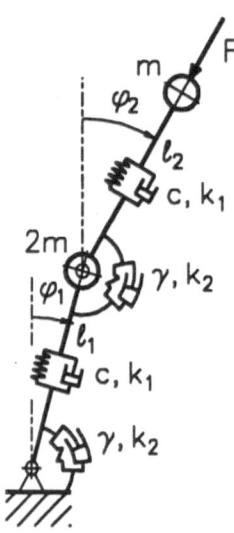

Figure 4.1. Double pendulum with four degrees of freedom $\varphi_1, \varphi_2, \ell_1$ and ℓ_2 loaded with a follower force F. The rods are axially and the hinges torsionally viscoelastic

of Lagrange's equation and are ([144])

$$
\begin{aligned}
& 3m(\ell_1^2\ddot{\varphi}_1 + 2\ell_1\dot{\ell}_1\dot{\varphi}_1)+ \\
& \quad + m(\ell_1\ell_2\ddot{\varphi}_2 + 2\ell_1\dot{\ell}_2\dot{\varphi}_2)\cos(\varphi_2 - \varphi_1)+ \\
& \quad + m(\ell_1\ddot{\ell}_2 - \ell_1\ell_2\dot{\varphi}_2^2)\sin(\varphi_2 - \varphi_1)+ \\
& \quad + k_2(2\dot{\varphi}_1 - \dot{\varphi}_2) + \gamma(2\varphi_1 - \varphi_2) - F\ell_1\sin(\varphi_1 - \varphi_2) = 0
\end{aligned}
$$

$$
\begin{aligned}
& m(\ell_2^2\ddot{\varphi}_2 + 2\ell_2\dot{\ell}_2\dot{\varphi}_2)+ \\
& \quad + m(\ell_1\ell_2\ddot{\varphi}_1 + 2\ell_2\dot{\ell}_1\dot{\varphi}_1)\cos(\varphi_2 - \varphi_1)- \\
& \quad - m(\ell_1\ddot{\ell}_1 - \ell_1\ell_2\dot{\varphi}_1^2)\sin(\varphi_2 - \varphi_1)+ \\
& \quad + k_2(\dot{\varphi}_2 - \dot{\varphi}_1) + \gamma(\varphi_2 - \varphi_1) = 0
\end{aligned}
$$

$$
\begin{aligned}
& 3m(\ddot{\ell}_1 - \ell_1\dot{\varphi}_1^2)+ \\
& \quad + m(\ddot{\ell}_2 - \ell_2\dot{\varphi}_2^2)\cos(\varphi_2 - \varphi_1)- \\
& \quad - m(\ell_2\ddot{\varphi}_2 + 2\dot{\ell}_2\dot{\varphi}_2)\sin(\varphi_2 - \varphi_1)+ \\
& \quad + k_1\ell_1 + c(\ell_1 - \ell_0) + F\cos(\varphi_2 - \varphi_1) = 0
\end{aligned}
\qquad (4.1)
$$

$$
\begin{aligned}
& m(\ddot{\ell}_2 - \ell_2\dot{\varphi}_2^2)+ \\
& \quad + m(\ddot{\ell}_1 - \ell_1\dot{\varphi}_1^2)\cos(\varphi_2 - \varphi_1)+ \\
& \quad + m(\ell_1\ddot{\varphi}_1 + 2\dot{\ell}_1\dot{\varphi}_1)\sin(\varphi_2 - \varphi_1)+ \\
& \quad + k_1\ell_2 + c(\ell_2 - \ell_0) + F = 0.
\end{aligned}
$$

We introduce non-dimensional quantities $\tau, \overline{F}$ and $\overline{\gamma}$ according to the following relations (g is the gravitational acceleration)

$$
\tau = \Omega t, \qquad \Omega = \sqrt{g/\ell_0}, \qquad \overline{\gamma} = \frac{\gamma}{c\ell_0^2}, \qquad \overline{F} = \frac{F}{c\ell_0}.
$$

In order to rewrite (4.1) as system of first order, we introduce

$$\boldsymbol{x} = (\varphi_1, \dot{\varphi}_1, \varphi_2, \dot{\varphi}_2, \xi_1, \dot{\xi}_1, \xi_2, \dot{\xi}_2)^T \in \mathbb{R}^8.$$

The ξ_i are defined as the deviations of the length ℓ_i of the rods from the length ℓ_{i0} in the vertical equilibrium position, that is, $\xi_i = \ell_{i0} - \ell_i$. This definition implies that $\xi_i \equiv 0$ as long as $\varphi_i \equiv 0$ and hence $\boldsymbol{x}_0 = \boldsymbol{0}$ is the equilibrium position which must be studied for its stability. The dot designates the derivative with respect to the dimensionless time τ. The first order system takes the form

$$\mathbf{R}(\boldsymbol{x})\dot{\boldsymbol{x}} = \mathbf{N}(\boldsymbol{x}), \tag{4.2}$$

where $\mathbf{R}(\boldsymbol{x}) \in \mathbf{L}(\mathbb{R}^8, \mathbb{R}^8)$ and $\boldsymbol{x}, \mathbf{N} \in \mathbb{R}^8$. Each element in $\mathbf{R}$ and $\mathbf{N}$ can be expanded into a power series of $\boldsymbol{x}$ up to the required order. Here and where appropriate in the following the Einstein summation assumption over repeated indices is adopted and all indices run from 1 to 8. The elements are

$$
\begin{aligned}
r_{ij}(\boldsymbol{x}) &= r_{ij} + r_{ijk}x_k + r_{ijk\ell}x_k x_\ell + r_{ijk\ell m}x_k x_\ell x_m + \\
&\quad + r_{ijk\ell mn}x_k x_\ell x_m x_n + O(|x|^5),
\end{aligned}
\tag{4.3}
$$

$$
\begin{aligned}
n_i(\boldsymbol{x}) &= n_{ij}x_j + n_{ijk}x_j x_k + n_{ijk\ell}x_j x_k x_\ell + n_{ijk\ell m}x_j x_k x_\ell x_m + \\
&\quad + n_{ijk\ell mn}x_j x_k x_\ell x_m x_n + O(|x|^6).
\end{aligned}
\tag{4.4}
$$

In order to transform (4.2) into the form (3.1), $\mathbf{R}(\boldsymbol{x})$ must be inverted. This is explained in Appendix O.1. Formally writing $\mathbf{R}^{-1}(\boldsymbol{x}) = \mathbf{T}(\boldsymbol{x})$, we obtain

$$
\begin{aligned}
t_{ij}(\boldsymbol{x}) &= t_{ij} + t_{ijk}x_k + t_{ijk\ell}x_k x_\ell + t_{ijk\ell m}x_k x_\ell x_m + \\
&\quad + t_{ijk\ell mn}x_k x_\ell x_m x_n + O(|x|^5).
\end{aligned}
\tag{4.5}
$$

From the relationship

$$r_{ik}(\boldsymbol{x})t_{kj}(\boldsymbol{x}) = \delta_{ij}$$

and comparing coefficients, we obtain

$$
\begin{aligned}
t_{ij} &= r_{ij}^{-1} \\
t_{ijk} &= -t_{ir}r_{rsk}t_{sj} \\
t_{ijk\ell} &= -t_{ir}(r_{rsk}t_{sj\ell} + r_{rsk\ell}t_{sj}) \\
t_{ijk\ell m} &= -t_{ir}(r_{rsk}t_{sj\ell m} + r_{rsk\ell}t_{sjm} + r_{rsk\ell m}t_{sj}]) \\
t_{ijk\ell mn} &= -t_{ir}(r_{rsk}t_{sj\ell mn} + r_{rsk\ell}t_{sjmn} + \\
&\qquad + r_{rsk\ell m}t_{sjn} + r_{rsk\ell mn}t_{sj}) \;.
\end{aligned}
\tag{4.6}
$$

From (4.6), the coefficients of the inverse matrix $\mathbf{T}(\boldsymbol{x})$ can be successively calculated. Here r_{ij}^{-1} designates the elements of the inverse matrix of the constant part of $\mathbf{R}$ given by the elements r_{ij}. Finally, if we write (3.1) in the form $\dot{\boldsymbol{x}} = \mathbf{f}(\boldsymbol{x}, \boldsymbol{\lambda})$, we obtain

$$f_i(\boldsymbol{x}) = f_{ij}x_j + f_{ijk}x_jx_k + f_{ijk\ell}x_jx_kx_\ell + f_{ijk\ell m}x_jx_kx_\ell x_m + \\ + f_{ijk\ell mn}x_jx_kx_\ell x_mx_n + O(|\boldsymbol{x}|^6) \tag{4.7}$$

where

$$\begin{aligned}
f_{ij} &= t_{ir}n_{rj} \\
f_{ijk} &= t_{ir}n_{rjk} + t_{irk}n_{rj} \\
f_{ijk\ell} &= t_{ir}n_{rjk\ell} + t_{irk}n_{rj\ell} + t_{irk\ell}n_{rj} \\
f_{ijk\ell m} &= t_{ir}n_{rjk\ell m} + t_{irk}n_{rj\ell m} + t_{irk\ell}n_{rjm} + t_{irk\ell m}n_{rj} \\
f_{ijk\ell mn} &= t_{ir}n_{rjk\ell mn} + t_{irk}n_{rj\ell mn} + t_{irk\ell}n_{rjmn} + t_{irk\ell m}n_{rjn} + t_{irk\ell mn}n_{rj} \, .
\end{aligned} \tag{4.8}$$

For the numerical calculations of (4.8), the coefficients in (4.3), (4.4) and (4.7) are ordered in such a way that each monomial in x_j appears only once, hence $k \leq \ell \leq m \leq n$ in (4.3) and $j \leq k \leq \ell \leq m \leq n$ in (4.4) and (4.7).

We end up with a system in the form (3.1). Clearly, $\boldsymbol{x}_0 = \boldsymbol{0}$ is the equilibrium position to be investigated for its stability. Introduction of local coordinates leads to (3.4) and (3.5) with the parameter vector $\boldsymbol{\lambda} = (\overline{F}, \overline{\gamma})^T$. For various but fixed values of $\overline{\gamma}$, the distinguished component $\overline{F}$ of $\boldsymbol{\lambda}$ is increased quasistatically until a critical value $\overline{F}_c$ is reached, for which one or several eigenvalues with zero real part occur for the first time. All the remaining eigenvalues have negative real parts. The parameter vector $\boldsymbol{\lambda}$ corresponding to F_c is called $\boldsymbol{\lambda}_c$. In Fig. 4.2, we give the result of the calculation, which is the *stability boundary* in parameter

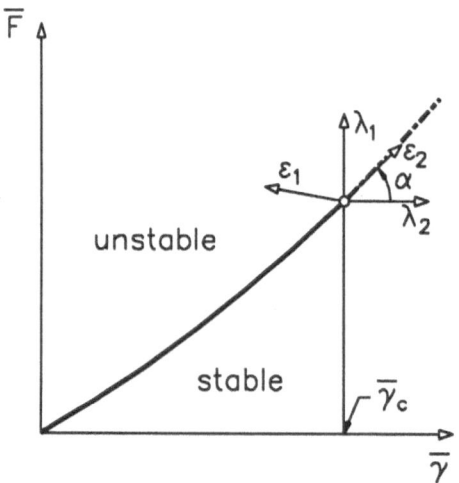

Figure 4.2. Stability boundary $\overline{F}_c$ in $\overline{F}, \overline{\gamma}$ parameter space. For $\overline{\gamma} = \overline{\gamma}_c$ the dependence between the physical parameters λ_1, λ_2 and the mathematical unfolding parameters $\varepsilon_1, \varepsilon_2$ used in (6.198) is shown (see p. 267)

space.

The form of the stability boundary gives no hint why we should use a two parameter family and why we need to calculate all expansions up to the required order such that we can derive bifurcation equations of fifth order. However the calculations show that by means of the second parameter $\bar{\gamma}$ we are able to annihilate the third order terms in the bifurcation equations. If such a case is met higher order terms must be retained. The considerable extra work of the calculation of the fifth order terms pays off because now we can perform a single global bifurcation analysis, instead of two local investigations, as it was indicated in the Introduction. Moreover, it is possible to study the stability of the limit cycle oscillation shown in Fig. 1.3 by still studying a steady state bifurcation problem (see p. 257).

Reaching the stability boundary for quasistatically increasing $\bar{F}$ in Fig. 4.2, always a pair of purely imaginary eigenvalues is found. Hence, making the change of variables (3.7), we obtain (3.8) or (3.16) where the matrices $\mathbf{J}_c$ and $\mathbf{J}_s$ of (3.16) have the following form

$$\mathbf{J}_c = \begin{pmatrix} 0 & -\omega_1 \\ \omega_1 & 0 \end{pmatrix},$$

$$\mathbf{J}_s = \begin{pmatrix} \sigma_2 & -\omega_2 & & & & & & \\ \omega_2 & \sigma_2 & & & & 0 & & \\ & & \sigma_3 & -\omega_3 & & & \\ & & \omega_3 & \sigma_3 & & & \\ & & & & \sigma_4 & -\omega_4 \\ & 0 & & & \omega_4 & \sigma_4 \end{pmatrix}, \tag{4.9}$$

with $\sigma_2, \sigma_3, \sigma_4$ all negative. It is interesting to note that also all the stable eigenvalues are complex. In order to obtain (3.19), the passive variables $\mathbf{y}_s = (y_3, \ldots, y_8)^T$ must be expressed in terms of the active variables $\mathbf{y}_c = (y_1, y_2)^T$. (3.17) takes the form

$$\mathbf{y}_s = \mathbf{H}(\mathbf{y}_c),$$

where we have already used the notation $\mathbf{H}$ for the approximation of h. We expand the components H_j into series in the critical variables and obtain

$$H_j(\mathbf{y}_c) = \sum_{i=2}^{4} H_j^i(\mathbf{y}_c), \qquad j = 3, \ldots, 8 \tag{4.10}$$

where $H_j^i(\mathbf{y}_c)$ is a monomial in the active variables of degree i. For example, we obtain terms like

$$H_3^2 = \alpha_{320} y_1^2 + \alpha_{311} y_1 y_2 + \alpha_{302} y_2^2.$$

Substituting (4.10) into (3.25) and setting $q = 6$ yields a system of algebraic equations for the determination of the coefficients in (4.10). These

operations can be performed numerically. The result are the equations
(3.19) given in the form

$$\dot{\boldsymbol{y}}_c = \mathbf{J}_c \boldsymbol{y}_c + \mathbf{M}_3(\boldsymbol{y}_c^3) + \mathbf{M}_5(\boldsymbol{y}_c^5) + O(|\boldsymbol{y}_c|^7) \qquad (4.11)$$

where $\boldsymbol{y}_c \in \mathbb{R}^2$ and $\mathbf{M}_3(\boldsymbol{y}_c^3)$ and $\mathbf{M}_5(\boldsymbol{y}_c^5)$ contain nonlinear terms of third
and fifth order, respectively. No terms of quadratic or fourth order are
present in (4.11). The absence of terms of second and fourth order is a
consequence of the elimination process of the non-critical variables from
the bifurcation equations. The bifurcation equations (4.11) are reflectional symmetric (Section 5.1). The two-dimensional system (4.11) replaces the original eight-dimensional system, and determines the local
stability behavior of the full eight-dimensional system. Equation (4.11)
appears again in (6.111) in the classification of Section 6.5.1 and is case
(4) in the treatment of the codimension two bifurcations in Section 6.6.2.

4.1.2   Double pendulum with elastic end support and follower force loading

The double pendulum with elastic end support and follower force loading
([126]) is shown in Fig. 4.3. The rods are rigid and the hinges are linearly

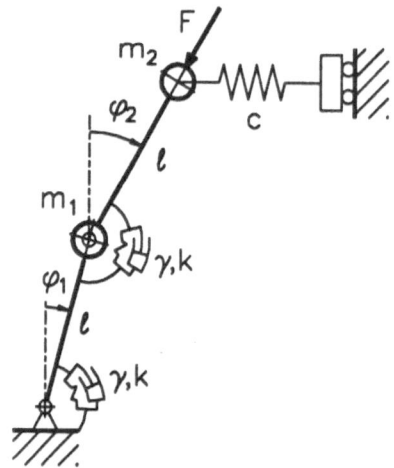

Figure 4.3. Double pendulum with
two degrees of freedom φ_1 and φ_2,
with an elastic end support and
loaded with a follower force F

viscoelastic, with torsional stiffness γ and coefficient of viscosity k. The
masses of the rods are concentrated in the two masses m_1 and m_2. Furthermore, there is an elastic end support with stiffness c. Additionally,
external damping is assumed, proportional to the speed of the masses m_i
with a coefficient κ. We introduce the damping ratio $d = \kappa/k$ defined
as the ratio of the coefficients of external to internal damping. In the
following, we keep this ratio fixed, but we increase the absolute value of
damping k. The system has two degrees of freedom denoted by φ_1 and

φ_2. As shown in [126], the equations of motion can be given in the form (3.1), where $\lambda = (F, c, k)^T \in \mathbb{R}^3$ and $x = (\varphi_1, \dot\varphi_1, \varphi_2, \dot\varphi_2)^T \in \mathbb{R}^4$. Expanding the equations of motion about the equilibrium position $x_0 = 0$, we obtain (3.4).

Keeping c and k fixed, we increase F quasistatically until we reach a critical value λ_c where for the first time an eigenvalue with zero real part appears. Obviously, keeping k fixed two extreme values of c exist. One extreme case we obtain for $c = 0$. Now no elastic constraint is present and a purely imaginary pair of eigenvalues is found at loss of stability. The other extreme value of c is $c = \infty$. That is, the upper end of the double pendulum is forced to move on the vertical axis through the lower end. In this case a zero root appears at loss of stability. If we vary c between 0 and ∞ the graphs shown in Fig. 4.4 are obtained for the stability boundary in the F, c parameter space for various but fixed

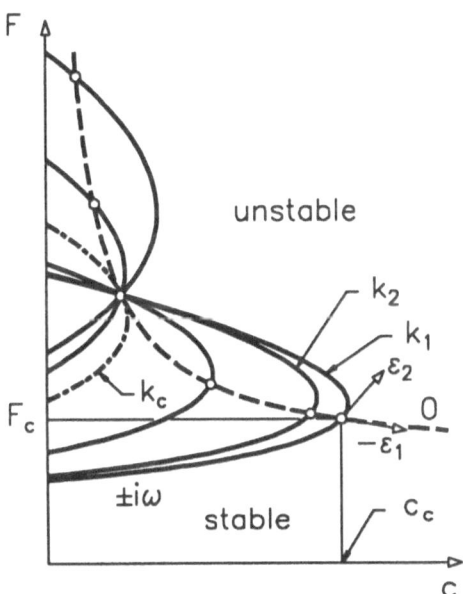

Figure 4.4. Stability boundary F_c in F, c parameter space. The parameters of the curves are the internal damping coefficients $k_1, k_2, \ldots,$ $(k_{i+1} > k_i)$. The critical eigenvalues are a purely imaginary pair for c small (full lines) and a zero root for c large (dashed line). The values of ω depend on c

values of damping k_i. Selecting a value $k = k_i$ we see that for all values of c except $c = c_c$ a simple loss of stability occurs. For the values $c = c_c$ (which still depend on k_i), a more degenerate situation occurs, because either a zero root and a purely imaginary pair of roots for small absolute values of damping $(k < k_c)$ or a double zero root in a Jordan block for large values of damping $(k > k_c)$ are obtained. The value of damping that separates these two different cases is $k = k_c$ and the corresponding stability boundary is depicted in Fig. 4.4.

If we transform (3.5) to (3.16), the five different cases for $\mathbf{J}_c$ are as follows:

(1)   If c is large $\mathbf{J}_c = (0)$

(2)   If c is small $\mathbf{J}_c = \begin{pmatrix} 0 & -\omega \\ \omega & 0 \end{pmatrix}$

(3)   If $c = c_c,\, k < k_c$ $\mathbf{J}_c = \begin{pmatrix} 0 & 0 & 0 \\ 0 & 0 & -\omega_c \\ 0 & \omega_c & 0 \end{pmatrix}$ (4.12)

(4)   If $c = c_c,\, k > k_c$ $\mathbf{J}_c = \begin{pmatrix} 0 & 1 \\ 0 & 0 \end{pmatrix}$

(5)   If $c = c_c,\, k = k_c$ $\mathbf{J}_c = \begin{pmatrix} 0 & 1 & 0 \\ 0 & 0 & 1 \\ 0 & 0 & 0 \end{pmatrix}$.

Cases (1) and (2) are generic cases both for the variation of stiffness c and damping k. Cases (3) and (4) are nongeneric with respect to c but generic concerning k. Whereas, case (5) is nongeneric both with respect to c and k. Case (5) obviously occurs if the two curves forming the stability boundary have a point of tangency as is shown by the dot-broken line in Fig. 4.4, corresponding to the special value of $k = k_c$. In Appendix H it is explained that the occurrence of points of tangencies are an indication that the system is not properly modelled and perhaps other parameters must be introduced. In fact this case is of codimension 3 (Section 6.2), and therefore, not generic in two-parameter families. We will not consider it further.

Following a representation in [160] we show in Fig. 4.5 how the different cases in (4.12) are obtained. The numbers of the cases in (4.12) correspond to the plate numbers in Fig. 4.5. The numbers 1 to 5 in each plate describe five successive positions of the motion of the two or three largest eigenvalues of interest in the complex plane under the variation of the distinguished parameter F. The bold-faced numbers designate the eigenvalue structure at the critical parameter value F_c as it is also given by (4.12). Moreover one can see from the plates how complex eigenvalues behave if they hit the real axis.

For the cases (1)–(4), either a one-, two- or three-dimensional bifurcation system is obtained.

Let us briefly consider case (3) of (4.12). The equations of the double pendulum written in the from (3.16) up to third order are

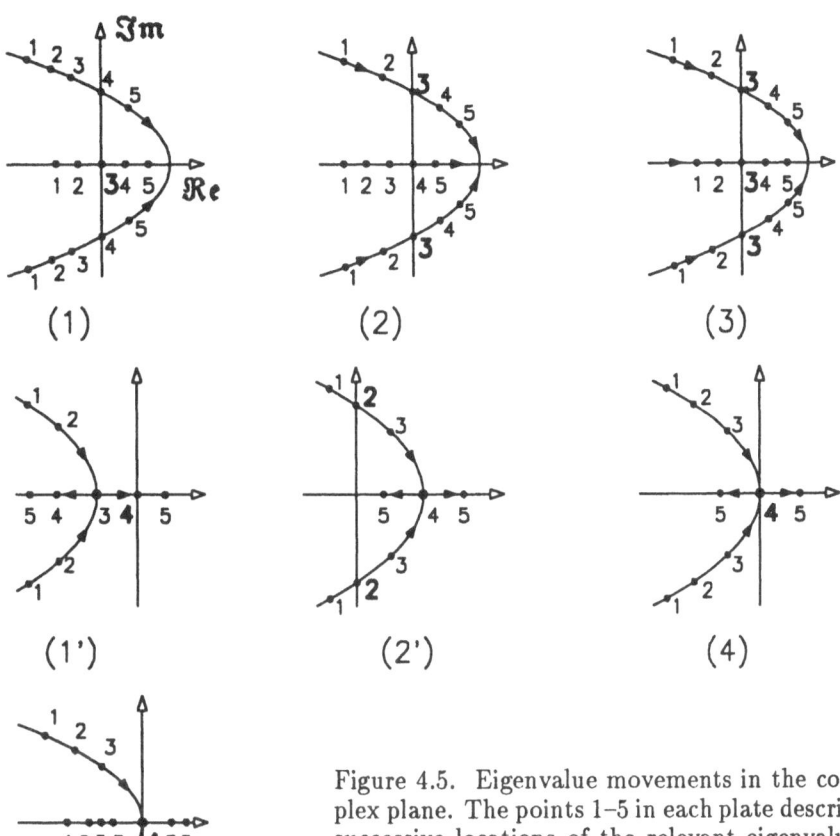

Figure 4.5. Eigenvalue movements in the complex plane. The points 1–5 in each plate describe successive locations of the relevant eigenvalues near the critical parametervalue $F = F_c$ under variation of F. The bold-faced numbers designate the eigenvalues at the stability boundaries. The plate numbers 1–5 correspond to the various cases of (4.12)

$$\dot{y}_1 = \sum_{i,j,k=1}^{4} a_{1ijk} y_i y_j y_k + O(|y|^5)$$

$$\dot{y}_2 = -\omega y_3 + \sum_{i,j,k=1}^{4} a_{2ijk} y_i y_j y_k + O(|y|^5)$$

$$\dot{y}_3 = \omega y_2 + \sum_{i,j,k=1}^{4} a_{3ijk} y_i y_j y_k + O(|y|^5)$$

$$\dot{y}_4 = \mu_4 y_4 + \sum_{i,j,k=1}^{4} a_{4ijk} y_i y_j y_k + O(|y|^5)$$

$$(4.13)$$

where y_1, y_2, y_3 are the active variables and y_4 is the passive variable. The eigenvalue μ_4 is negative. Due to the reflectional symmetry, no nonlinear terms of even, specifically second order, are present (Section 5.1). To eliminate y_4 from the first three equations of (4.13), y_4 can be expressed in the form $y_4 = h(y_1, y_2, y_3)$. However, we know that (3.18) holds for h. Hence, we obtain

$$y_4 = h(y_1, y_2, y_3) = O(|y_1|^2 + |y_2|^2 + |y_3|^2). \tag{4.14}$$

Substituting (4.14) into the first three equations of (4.13) for y_4 will, therefore, always result in terms of at least fourth order in the variables y_1, y_2, y_3. Hence, if third order terms in the bifurcation equations *determine* the behavior, then fourth or higher order terms can be neglected. Consequently, we can set $y_4 = 0$ in the first three equations resulting in

$$\dot{y}_1 = \sum_{i,j,k=1}^{3} a_{1ijk} y_i y_j y_k$$

$$\dot{y}_2 = -\omega y_3 + \sum_{i,j,k=1}^{3} a_{2ijk} y_i y_j y_k \tag{4.15}$$

$$\dot{y}_3 = \omega y_2 + \sum_{i,j,k=1}^{3} a_{3ijk} y_i y_j y_k.$$

Using the same line of reasoning, the bifurcation equations in the other three cases are obtained. For the loss of stability at the zero root (see also (3.9) and (3.10)), we obtain

$$\dot{y}_1 = a_{1111} y_1^3 . \tag{4.16}$$

For the loss of stability at the purely imaginary pair, we obtain

$$\dot{y}_1 = -\omega y_2 + \sum_{\substack{i,j=0 \\ i+j=3}}^{3} a_{1ij} y_1^i y_2^j$$

$$\dot{y}_2 = \omega y_1 + \sum_{\substack{i,j=0 \\ i+j=3}}^{3} a_{2ij} y_1^i y_2^j . \tag{4.17}$$

Finally, for the loss of stability at the double zero root, we obtain

$$\dot{y}_1 = y_2 + \sum_{\substack{i,j=0 \\ i+j=3}}^{3} a_{1ij} y_1^i y_2^j$$

(4.18)

$$\dot{y}_2 = \sum_{\substack{i,j=0 \\ i+j=3}}^{3} a_{2ij} y_1^i y_2^j \, .$$

Generically, for systems with arbitrarily chosen values of c only (4.16) or (4.17) are relevant. In a family of systems, where c and k are varied, (4.15) and (4.18) are also important. Engineers call the cases (4.16) and (4.17) *divergent* and *flutter instabilities*, respectively, while (4.15) and (4.18), are a coupling between divergence and flutter. The cases (4.11), (4.15), (4.16) and (4.18) are four of the five classified codimension *two* cases described in Section 6.5.1. In Section 4.1.3, we present another pendulum problem which gives a physical example for the last possible codimension *two* case. The two codimension *one* cases are (4.17) and (4.16) if the nonlinear term were quadratic.

In closing this Section, we make some comments concerning the physical relevance of systems at conicident eigenvalues as for example the three dimensional system (4.15). First, we mention the frequently observed experimental fact that not only for the case where the parameter c is exactly at c_c a coupling between divergence and flutter occurs, but that this phenomenon is also quite robust under perturbations of the critical parameter values. Second, we see from Fig. 4.4 that the system with $c = c_c$ has the biggest domain of stability. In some sense, one could call it an optimized system. On the other hand, one has to pay for such an optimal system. The price to be paid is that at loss of stability it behaves much more complicated than systems at a simple loss of stability. The third comment concerns the frequency of appearence of such more degenerate systems. Normally for technical systems, the cases with a simple loss of stability are most important. The higher the order of degeneracy of a loss of stability, the less likely is its appearence in a technical system provided that such a case is not systematically searched for by an optimization process or is not forced by symmetry constraints (Chapter 5).

4.1.3 Double pendulum under aerodynamic excitation

A double pendulum under aerodynamic excitation is sketched in Fig. 4.6. This system is similar to that of Fig. 4.3, but with an excitation due to a fluid flow vertical to the plane of motion. Such an excitation can lead

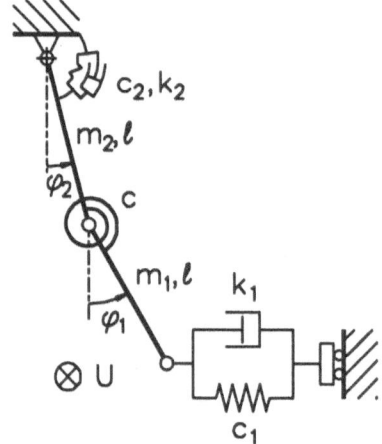

Figure 4.6. Double pendulum excited by a fluid flow force f acting in the plane of motion of the rod with mass m_1. The fluid flow with velocity U is directed vertical to the plane of motion of the rod

to a *galloping instability* ([16]). The fluid excitation is assumed to act only on the rod having mass m_1. The exciting mechanism is such that, due to the shape of the cross-section, an aerodynamic force is created in the plane of motion of the pendulum, which acts in the direction of the motion of the rod. Hence, a self-excitation is given as is explained in detail in [16]. Such an excitation follows from a steady fluid flow with velocity U and density ϱ acting on a rod the cross-section of which, as in this example, is of quadratic form with width d. If the cross-section were circular no excitation effect due to a galloping instability would be present. However, for a circular cross-section *vortex sheding* could occur, leading to an external periodic excitation which is not of interest here. The effect of the aerodynamic force yields a negative damping coefficient in the equations of motion. If this coefficient, which increases in absolute value with the velocity of the fluid flow, dominates the other dissipative effects then the equilibrium can lose its stability. Incidentally, this mechanism of self excitation may have led to the disastrous failure of the *Tacoma bridge*, mentioned in the Introduction. However, it is also thought that it could have been a combination of a galloping instability and vortex sheding ([150]).

The force f in the plane of motion of the pendulum acting vertical to a rod element of unit length is given by ([16])

$$f = a_1 v + a_3 |v|^2 v + \dots . \tag{4.19}$$

Here is v the in-plane component of the velocity of the unit element normal to the axis of the rod m_1. The coefficients a_1 and a_3 are given by

$$a_1 = \frac{b_1 \varrho d U}{2} , \qquad a_3 = \frac{b_3 \varrho d}{2U} .$$

For our example $b_1 = 2.7$ and $b_3 = -31.0$ ([16]).

The total flow exciting force acting on the rod m_1 follows from integration of (4.19) over the length of the rod.

The system of Fig. 4.6 has two degrees of freedom. Again, writing down the equations of motion in Lagrange's form, we end up with a system of the form (4.2). Inverting the matrix $\mathbf{R}(\boldsymbol{x})$, (4.2) can be transformed into (3.1). We proceed now analoguously to the previous two sections. We select the vector of the main parameters $\lambda = (U, c_2)^T \in \mathbb{R}^2$, where U is the speed of the fluid flow vertical to the plane of motion of the pendulum and c_2 is the stiffness of the upper torsion spring (Fig. 4.6).

The stability boundary is given in Fig. 4.7. The main difference of Fig. 4.7 compared to Fig. 4.4 is, that in Fig. 4.7 both curves correspond to a loss of stability at purely imaginary eigenvalues. The values of the frequencies ω_1 and ω_2 depend on c_2. In generic cases ($c_2 \neq c_{2c}$) a bifurcation system of the form (4.17) is obtained. However, the corresponding eigenvectors describing the vibration mode are qualitatively different for large or small values of c. If we are at the critical value $c_2 = c_{2c}$, both types of loss of stability coincide, however, $\omega_{1c} \neq \omega_{2c}$. In this case, the matrix corresponding to $\mathbf{J}_c$ in (3.19) is

$$
\mathbf{J}_c = \begin{pmatrix} 0 & -\omega_{1c} & & 0 \\ \omega_{1c} & 0 & & \\ & & 0 & -\omega_{2c} \\ 0 & & \omega_{2c} & 0 \end{pmatrix}. \tag{4.20}
$$

This leads to a four-dimensional bifurcation system with third order terms

$$
\begin{aligned}
\dot{y}_1 &= -\omega_{1c} y_2 + \sum_{i,j,k=1}^{4} a_{1ijk} y_i y_j y_k \\
\dot{y}_2 &= \omega_{1c} y_1 + \sum_{i,j,k=1}^{4} a_{2ijk} y_i y_j y_k \\
\dot{y}_3 &= -\omega_{2c} y_4 + \sum_{i,j,k=1}^{4} a_{3ijk} y_i y_j y_k \\
\dot{y}_4 &= \omega_{2c} y_3 + \sum_{i,j,k=1}^{4} a_{4ijk} y_i y_j y_k.
\end{aligned} \tag{4.21}
$$

We see from the classification given in Section 6.5.1 that the cases of loss of stability found for these three (academic) double pendula form the complete list of examples to the five different cases of loss of stability of codimension two.

Explanations of the physical behavior of these systems will be given

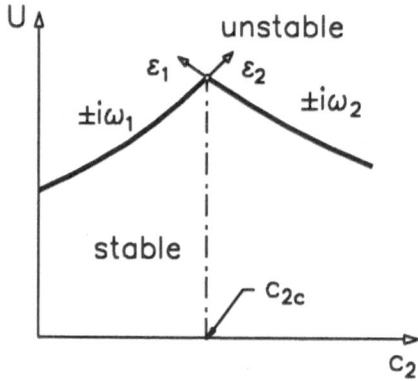

Figure 4.7.   Stability boundary U_c in U, c_2 parameter space.
Critical eigenvalues are always
purely imaginary pairs, but with
different frequencies and eigenmodes for small and large c_2, respectively

when the solutions of the bifurcation equations are discussed in Section 6.6.2.

In the next two sections we give two examples of mechanical models
of engineering systems.

4.1.4   Loss of stability of the straight line motion of a tractor-semitrailer

In [67] it is shown that the equations of motion of a driver controlled
tractor-semitrailer (Fig. 4.8) can be given in the form of equation (3.1),
where $x \in \mathbb{R}^9$ and the vector of the main parameters $\lambda = (V, d)^T \in \mathbb{R}^2$
has as its components the driving speed V and the length d measuring
the distance from the center of gravity of the trailer to the rear axle of
the trailer. Technically speaking, d is a measure of the loading condition
of the trailer.

We want to study the stability of the straight line motion of the
vehicle. This motion corresponds to $x_0 \equiv 0$ and satisfies (3.2) for any
values of λ. The first task is the calculation of the eigenvalues of (3.5).
Quite naturally, we select the speed V as the distinguished parameter.
That is, we keep d at a fixed value and evaluate the eigenvalues of (3.5)
for quasistatically increased values of V. It is natural to assume that
for small values of V, the motion is asymptotically stable, and therefore,
all eigenvalues will have a negative real part. However, it is also not
too difficult to understand that for increasing the speed the stability
boundary will be reached. In Fig. 4.9, the stability boundary is shown in
the V, d parameter plane for a motion with fixed steering angle δ, that
is, δ is kept at the value $\delta = 0$.

The form of the stability boundary corresponds to the requirements
postulated in Appendix H and in principle is the same as in Fig. 4.4 for the
double pendulum of Section 4.1.2. It consists of smooth arcs intersecting
in an acute angle pointing into the domain of instability. From the qual-

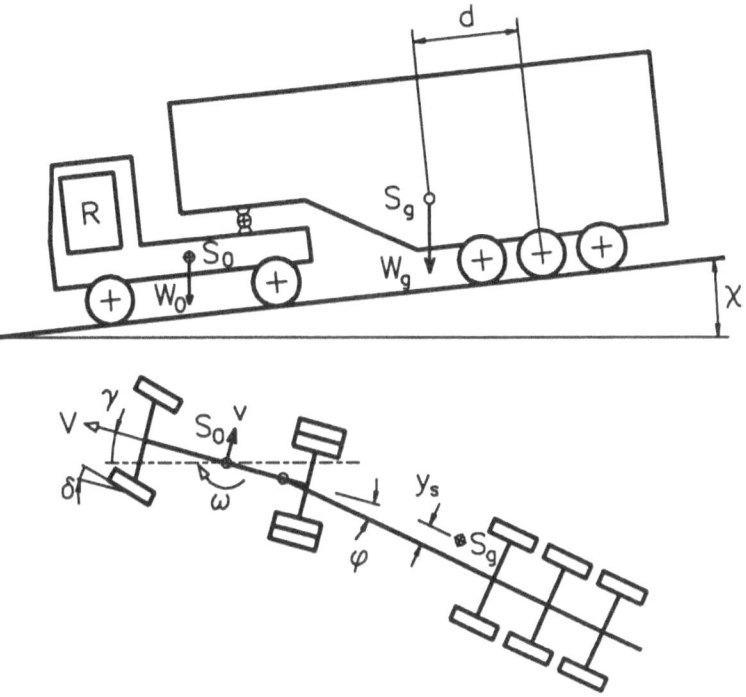

Figure 4.8. Tractor-semitrailer in downhill motion. Degrees of freedom are v, ω, φ. Main parameters are V, d. Imperfection parameters are δ and y_s

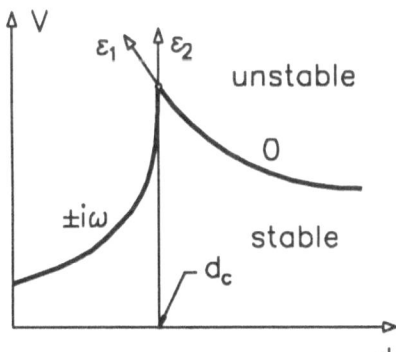

Figure 4.9. Stability boundary V_c in V, d parameter space for fixed steering angle $\delta = 0$. For d small a flutter instability and for d large a divergence stability occurs. The system corresponding to $d = d_c$ is a nongeneric case in a one parameter family

itative behavior of the stability boundary, several important conclusions, concerning the behavior of the system, can be drawn. Depending upon the loading condition of the trailer, which is expressed by the location of its center of mass d, either a zero root or a pair of purely imaginary roots occur at loss of stability. However, there exists also a special value $d = d_c$, where the zero root and the pair of purely imaginary roots occur simultaneously. This case, although it occurs only for a very special loading condition, is nevertheless important from a practical point of view for two reasons. First, for systems close to the system corresponding to $d = d_c$ a coupling between the two instability modes will occur. Second, the special loading condition $d = d_c$ is the one which yields the largest critical speed. Hence, this system can be considered as the system with an optimized load distribution. As has already been mentioned in the Introduction, and in Section 4.1.2, this results in a more complicated problem with multiple eigenvalues at the loss of stability.

From the linear stability analysis and examination of the stability boundary, we can conclude that, depending on the parameter d, either a one, two or three dimensional bifurcation system can be obtained.

This leads to exactly the same matrices $\mathbf{J}_c$ as given in the cases (1)–(3) of (4.12). The corresponding bifurcation equations are (4.16), (4.17) and (4.15). This is the case because the same simplification for the elimination of the passive variables as could be made in Section 4.1.2 can be made here. That is, no influence on the third order terms in the bifurcation equations results from the passive variables. Because this step is so important in applications, we explain it again for the case of one zero root. The equations (3.16) in diagonalized form are

$$\dot{y}_1 = \sum_{i,j,k=1}^{n} a_{1ijk} y_i y_j y_k + O(|y|^4)$$

$$(4.22)$$

$$\dot{y}_s = \mu_s y_s + \sum_{i,j,k=1}^{n} a_{sijk} y_i y_j y_k + O(|y|^4) \,, \qquad s = 2,\ldots,n \,.$$

To simplify the notation, we assume that in (4.22) all eigenvalues of $\mathbf{J}_s$ are simple. This assumption has no qualitative influence on the analysis. According to (3.17) the y_s must be expressed as functions of y_1 in the form

$$y_s = h_s(y_1) \,, \qquad s = 2,\ldots,n \,. \tag{4.23}$$

However, (3.29) holds. Therefore, if we insert (4.23) into (4.22)$_1$, we obtain at least fourth order terms, because no quadratic terms appear in (4.22)$_1$. Hence, we obtain the important result that if the term a_{1111} is not equal to zero, (4.22) is *three-determinate* (Section 6.3). This means that for this problem the y_s, $s = 2,\ldots,n$; can be set to zero in (4.22)$_1$ and we still obtain a correct bifurcation equation up to and including terms of third order. Therefore, the calculation of the center manifold becomes trivial and the bifurcation equation is

$$\dot{y}_1 = a_{1111} y_1^3. \tag{4.24}$$

The above reasoning applies also to the other two cases of tractor-semi-trailer dynamics, namely to the cases (2) and (3) of (4.12) and the corresponding bifurcation equations are exactly equations (4.17) and (4.15).

Thus in (4.24) as in (4.16) one coefficient, in (4.17) eight coefficients and in (4.15) thirty coefficients must be determined. In Section 6.1, it will be shown that the use of *normal form theory* leads to a substantial reduction of the number of coefficients needed in the further discussion of the bifurcation equations.

Finally, we discuss the case of a tractor-semitrailer with a driver who does not keep the steering wheel at a fixed value $\delta = 0$ but according to a certain control law makes appropriate steering corrections ([67]). In this case, the stability boundary takes the form given in Fig. 4.10. Now, for any loading condition, at the critical speed $V = V_c$ a purely imaginary

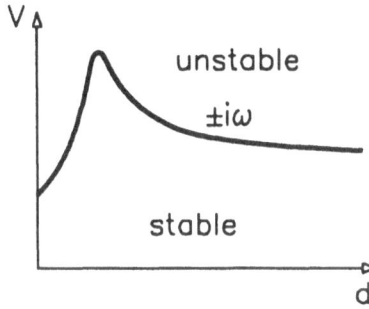

Figure 4.10. Stability boundary V_c in V, d space for the tractor-semi-trailer with automatic controller (driver). Only purely imaginary pairs of eigenvalues appear at loss of stability

pair of eigenvalues occurs. Hence, no divergent instability, as it occurs for fixed steering, is possible in this case. Also, contrary to the example in Section 4.1.3, the frequency ω at loss of stability is a continuous function of the parameter d. Hence, the bifurcation equations in this case will always have the form of (4.17).

4.1.5 Loss of stability of the straight line motion of a railway vehicle

We consider the motion of a railway bogy (in Fig. 4.11 a bogy with four axles as it is used for special railway cars to transport trucks is shown) on a straight track ([162]) under the action of the forces at rail–wheel contact and the forces acting from the car to which this bogy belongs. We treat this example because it possesses a stability behavior that is qualitatively similar to that of a whole railway car. Moreover it is completely analogous to the behavior of the double pendulum in Section 4.1.1.

Experimental results ([99]) show that the behavior of a railway bogy (Fig. 1.2) is qualitatively of the form drawn in Fig. 1.3. We want to show that proceeding – as explained in the Introduction – by introducing an additional parameter besides the speed V, it will be possible to calculate the complete amplitude curve of the limit cycles of Fig. 1.3 in a simple manner. That is, we will be able to calculate the turning point of the amplitude curve of a periodic solution by treating only a steady state bifurcation problem.

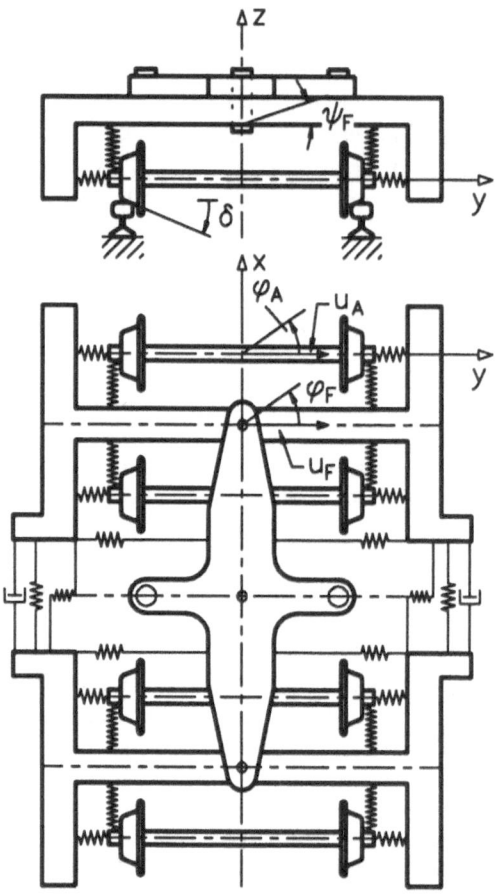

Figure 4.11. Railway bogy with 14 degrees of freedom. Each frame has the three degrees of freedom u_F, φ_F and ψ_F and each axle the two degrees u_A and φ_A. δ is the conicity angle of the wheel

The equations of motion of the bogy, which possesses 14 degrees of freedom, are derived in [162]. They are a system of 14 second order nonlinear ordinary differential equations. The nonlinearity follows from the nonlinear relations due to the rail wheel contact. In order to be able to calculate the turning point shown in Fig. 1.3 we must perform a center manifold reduction up to and including terms of fifth order. Transforming the equations of motion into a first order system we obtain the dynamical system (3.1) with the vector of main parameters $\lambda = (V, \delta) \in \mathbb{R}^2$, where δ is the angle of conicity of the wheel (Fig. 4.11). The system has 14 degrees of freedom, therefore $x \in \mathbb{R}^{28}$ in (3.1). For technically meaningful values of δ, we increase the velocity V, which is the distinguished component of λ, quasistatically starting with values of V for which the equilibrium position $\boldsymbol{x}_0 = \boldsymbol{0}$ is asymptotically stable. At the critical speed, we always find a purely imaginary pair of eigenvalues. Hence, from the linear analysis no indication of higher degeneracies results. However, the nonlinear investigation reveals that by variation of the parameter δ it is possible to annihilate the third order terms in the bifurcation equations. Therefore, from the outset, all terms contributing to those in the bifurcation equations up to fifth order must be included. The analysis is completely analogous to that given in Section 4.1.1 and the final result is the two-dimensional system of bifurcation equations (4.11).

4.1.6 Summary of Section 4.1

We repeat the steps necessary to obtain from the original equations (3.1) the reduced set of bifurcation equations (4.11), (4.15), (4.16), (4.17) and (4.21):

(i) Calculate the manifold of equilibrium solutions $\boldsymbol{x}_0(\lambda)$ for a quasistatic variation of the main parameters. The imperfection parameters (see p. 47) are set to zero. The relationship $\boldsymbol{x}_0(\lambda)$ may fail to be unique. (See, for example, the saddle node bifurcation in Section 6.6.2.)

(ii) Perform a linearization of (3.1) about the steady state (equilibrium) and perform a Taylor series expansion of the nonlinear terms up to the required order (mostly third but sometimes also fifth order).

(iii) Compute the eigenvalues of $\mathbf{A}(\lambda)$ (defined in (3.5)) for a quasistatic variation of the main parameters. These calculations yield the stability boundary in parameter space.

(iv) Perform a linear transformation of coordinates, transforming $\mathbf{A}$ into Jordan form $\mathbf{J}$ at a critical parameter value. The critical part of $\mathbf{J}$ called $\mathbf{J}_c$ in (3.16) is given by (4.12) and (4.20), respectively. Substitute the transformation into the nonlinear terms.

(v) Calculate the dependence of the non-critical variables on the critical variables (3.17) and eliminate the non-critical variables from the bifurcation equations (3.19).

(vi) If, as in (4.13) and (4.22), no quadratic terms appear in the equations for the active or critical variables and, for example, the coefficient a_{1111} in (4.22) is not equal to zero that is, third order terms determine the behavior then the calculation of the center manifold (point (v)) is trivial.

(vii) The determinacy of bifurcation equations in several variables, for example, equations (4.15) and (4.18) is not always easy to decide if some of the coefficients of third order are zero (Section 6.3).

4.2   Loss of stability of periodic motions: an example from robotics

In this section, we explain an important step in the stability analysis of periodic motions, namely, the explicit calculation of the lowest order nonlinear terms of the corresponding Poincaré mapping.

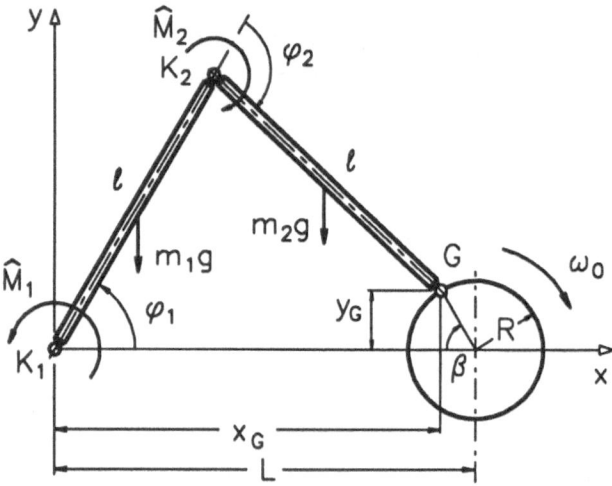

Figure 4.12.   Mechanical model of a simple planar two bar robot under the action of two drive moments $\hat{M}_1, \hat{M}_2$ yielding a prescribed circular motion of the endpoint G

We perform these calculations when investigating the dynamics of the simple robot sketched in Fig. 4.12. For example, such robots are used in the automobile manufacturing industry for painting. During such operations, the automobile is passing along the robot endpoint G with constant speed, while the robot endpoint G performs a prescribed motion. At the robot endpoint a painting device (spray pistol) may be mounted. For such a painting process, stability problems can become important, especially when parameters are varied. First, if the speed of the assembly line is increased, the endpoint of the robot must move faster and, hence, a loss of stability of the basic, periodic motion can occur if the parameters of the controller are kept fixed. Second, a change of the spray pistol to one with different mass can also result in a loss of stability of the prescribed motion.

4.2.1 Mechanical model and equations of motion

The mechanical model is a double pendulum with two drive moments $\hat{M}_1$ and $\hat{M}_2$ acting at its hinges. The motion of the endpoint G is assumed to follow a circle with constant angular velocity ω_0. Making use of Lagrange's equations we obtain ([87])

$$
\begin{pmatrix} \frac{\varrho}{3} + \frac{4}{3} + \cos\varphi_2 & -\left(\frac{1}{3} + \frac{1}{2}\cos\varphi_2\right) \\ -\left(\frac{1}{3} + \frac{1}{2}\cos\varphi_2\right) & \frac{1}{3} \end{pmatrix} \begin{pmatrix} \ddot{\varphi}_1 \\ \ddot{\varphi}_2 \end{pmatrix} +
$$

$$
+ \begin{pmatrix} k_1 & 0 \\ 0 & k_2 \end{pmatrix} \begin{pmatrix} \dot{\varphi}_1 \\ \dot{\varphi}_2 \end{pmatrix} + \begin{pmatrix} \frac{1}{2}\dot{\varphi}_2(\dot{\varphi}_2 - 2\dot{\varphi}_1)\sin\varphi_2 \\ \frac{1}{2}\dot{\varphi}_1^2 \sin\varphi_2 \end{pmatrix} + \qquad (4.25)
$$

$$
+ k_G \begin{pmatrix} \left(\frac{\varrho}{2}+1\right)\cos\varphi_1 + \frac{1}{2}\cos(\varphi_2 - \varphi_1) \\ -\frac{1}{2}\cos(\varphi_2 - \varphi_1) \end{pmatrix} = \begin{pmatrix} M_1 \\ M_2 \end{pmatrix},
$$

where the following dimensionless quantities (Fig. 4.12)

$$
\varrho = \frac{m_1}{m_2}, \quad t = \tau\omega_0, \quad M_i = \frac{\hat{M}_i}{m_2\ell^2\omega_0^2}, \quad k_i = \frac{K_i}{m_2\ell^2\omega_0}, \quad k_G = \frac{g}{\omega_0^2\ell}
$$

are used.

The components φ_1, φ_2 of the vector $\varphi = (\varphi_1, \varphi_2)^T$ are the two degrees of freedom and $\mathbf{M} = (M_1, M_2)^T$ is the vector of the drive moments, acting at the hinges of the robot. k_i are the viscous friction coefficients in the joints and k_G is a gravitational constant.

To obtain the prescribed circular motion of the endpoint G, the first thing to do is the calculation of the moment $\mathbf{M}(t) = \mathbf{M}_0(t)$. This requires the solution of the so-called inverse problem of robotics and is done in [87] for this special problem. More generally it is treated in [34]. The goal of this analysis is to express the left-hand side of (4.25) by the coordinates

$$
\begin{aligned}
x_G &= L - r\cos\beta \\
y_G &= r\sin\beta
\end{aligned} \qquad (4.26)
$$

and their derivatives under the condition

$$
\dot{\beta} = \omega_0 = \text{const.} \qquad (4.27)
$$

One starts with $x_G(t)$ and $y_G(t)$ in the form

$$
\begin{aligned}
x_G &= \ell(\cos\varphi_1 + \cos(\varphi_2 - \varphi_1)) \\
y_G &= \ell(\sin\varphi_1 - \sin(\varphi_2 - \varphi_1)),
\end{aligned} \qquad (4.28)
$$

which follow from Fig. 4.12. The dependence of all quantities $(\ddot{\varphi}_1, \ddot{\varphi}_2, \dot{\varphi}_1,$ $\dot{\varphi}_2, \cos\varphi_1, \cos\varphi_2, \cos(\varphi_2 - \varphi_1), \sin\varphi_2)$ appearing on the left-hand side of (4.25) on $(x_G, y_G, \dot{x}_G, \dot{y}_G, \ddot{x}_G$ and $\ddot{y}_G)$ can be calculated by lengthy, but straightforward calculations ([87], [34]). For example, one obtains for $\dot{\varphi}_2$ the following expression

$$\dot{\varphi}_2 = -2\frac{x_G\dot{x}_G + y_G\dot{y}_G}{(x_G^2 + y_G^2)(4\ell^2 - x_G^2 - y_G^2)} . \tag{4.29}$$

If we introduce these quantities into the left-hand side of (4.25), the two components of $\mathbf{M}_0(t)$ are obtained in their dependence on the motion of the endpoint G. These give rise to a periodic motion $\varphi_0(t)$, which is the basic state to be investigated for its stability. In order to formulate the stability problem, a control loop (Fig. 4.13) must be superposed to

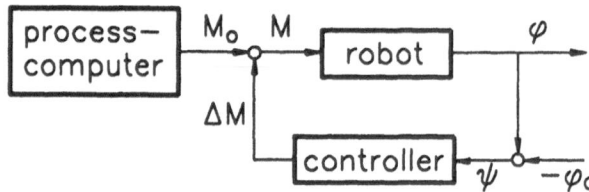

Figure 4.13. Control loop for the robot

the system. The task of the controller is to correct deviations from the prescribed motion. Mathematically this can be done by introducing new coordinates $\psi = (\psi_1, \psi_2)^T$ in the form

$$\varphi_i = \varphi_{0i} + \psi_i \qquad i = 1, 2 . \tag{4.30}$$

These new coordinates can be interpreted as deviations or perturbations from the fundamental motion. To force the robot back to its required motion the drive moment $\mathbf{M}_0$ must be corrected. We express this in the following way

$$M_i = M_{0i} + \Delta M_i \qquad i = 1, 2. \tag{4.31}$$

Now we transform to a system of first order equations by setting

$$x_1 = \psi_1, \qquad x_2 = \dot{\psi}_1, \qquad x_3 = \psi_2, \qquad x_4 = \dot{\psi}_2. \tag{4.32}$$

The resulting equations of motion can be written in the form

$$\dot{x} = \overline{\mathbf{A}}(t, \lambda)x + \mathbf{C}(t, \lambda)u + \mathbf{f}(t, x, \dot{x}, \lambda) \tag{4.33}$$

where $x, \mathbf{f} \in \mathbb{R}^4$, $\overline{\mathbf{A}}$ is a 4×4 matrix, $u = (\Delta M_1, \Delta M_2)^T$ and $\mathbf{C}$ is a 4×2 matrix. The vector $\mathbf{f}$ contains terms up to third order which have been obtained from a series expansion of the nonlinear terms in (4.25).

To eliminate the vector $\boldsymbol{u}$ from (4.33), a control law must be introduced. A simple compensation of the deviation from the prescribed motion is stipulated. Hence, the following relations are introduced:

$$u_1 = \Delta M_1 = Rx_1, \qquad u_2 = \Delta M_2 = \frac{1}{3}Rx_3 . \qquad (4.34)$$

To keep the number of parameters as small as possible, only one controller constant R is used which, of course, must have a negative value. To obtain a first order system, we substitute (4.34) into (4.33) and express the terms containing $\dot{\boldsymbol{x}}$ in $\mathbf{f}$ in (4.33) by means of its linear and quadratic terms, as it is explained in Appendix O.2. This yields

$$\dot{\boldsymbol{x}} = \mathbf{A}(t,\lambda)\boldsymbol{x} + \mathbf{f}_2(t,\boldsymbol{x},\lambda) + \mathbf{f}_3(t,\boldsymbol{x},\lambda) + O(|\boldsymbol{x}|^4) \qquad (4.35)$$

where $\mathbf{A}(t,\lambda)$, $\mathbf{f}_2(t,\boldsymbol{x},\lambda)$ and $\mathbf{f}_3(t,\boldsymbol{x},\lambda)$ are periodic in t with the period T of the given fundamental motion of the robot and the vectors $\mathbf{f}_2$ and $\mathbf{f}_3$ contain nonlinear functions of the variables of second and third order, respectively. The parameter vector λ is two-dimensional and given by

$$\lambda = \begin{pmatrix} \omega_0 \\ R \end{pmatrix} \qquad (4.36)$$

with ω_0 defined by (4.27) and R by (4.34).

4.2.2 Calculation of the power series expansion of the Poincaré mapping

The stability analysis of the prescribed periodic motion $\boldsymbol{\varphi}_0(t) = \boldsymbol{\varphi}_0(t+T)$ reduces in the formulation of (4.35) to the investigation of the trivial solution

$$\boldsymbol{x}_0(t) = \mathbf{0}. \qquad (4.37)$$

For a linearized stability analysis of $\boldsymbol{x}_0$ *Floquet theory* ([103]) could be used. However, we will soon see that the investigation of the Poincaré mapping includes the linear stability analysis supplied by Floquet's theory ([7] p. 282). In the case under investigation, the calculation of the Poincaré mapping, that is, the calculation of the points of intersection with a transversal surface is not very difficult. Because of the T-periodicity of (4.35) we only have to calculate the solution at time $T, 2T, \ldots$ and so on. The result of these calculations is sometimes called a *time-T* mapping. From Fig. 4.14 follows that the trajectory $\boldsymbol{x}(\boldsymbol{x}_1, t)$ leaving $\boldsymbol{x}_1$, in the neighborhood of the periodic solution $\boldsymbol{x}_0$, hits the surface again at $\boldsymbol{x}_2$. Hence, the following equation results

$$\boldsymbol{x}_2 = \mathbf{P}(\boldsymbol{x}_1) = \boldsymbol{x}(\boldsymbol{x}_1, T), \qquad (4.38)$$

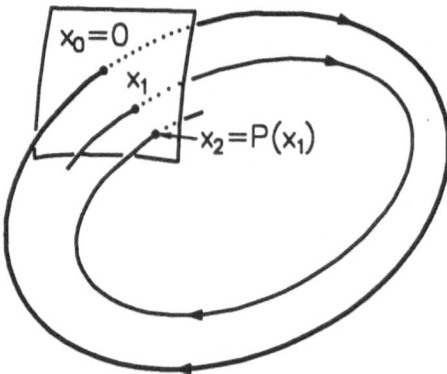

Figure 4.14. Poincaré mapping in the neighborhood of the periodic solution x_0

where in the notation of Section 2.1, $\mathbf{P}_T$ instead of $\mathbf{P}$ should have been used. However, since T is fixed, we omit the index.

To calculate the Poincaré mapping in the neighborhood of the periodic solution, which is given by the fixed point x_0 in the surface of section, the following formal power series expansion

$$\mathbf{P}(x_0 + \xi) = \mathbf{P}(x_0) + \mathbf{P}'(x_0)\xi + \frac{1}{2}\mathbf{P}''(x_0)(\xi, \xi) + \frac{1}{6}\mathbf{P}'''(x_0)(\xi, \xi, \xi) + \cdots$$
(4.39)

is introduced, where $\mathbf{P}(x_0) = x_0$, because $x(x_0, T) = x_0$. $\mathbf{P}'(x_0)$, $\mathbf{P}''(x_0)$, $\mathbf{P}'''(x_0)$,... are obtained from the solution of a series of initial value problems. This is easy to understand if (4.38) is written in the form

$$\mathbf{P}(x_0 + \xi) = x(x_0 + \xi, T) = x(x_0, T) + \frac{\partial x}{\partial x_0}(x_0, T)\xi + \quad (4.40)$$

$$+ \frac{1}{2}\frac{\partial^2 x}{\partial x_0^2}(x_0, T)(\xi, \xi) + \frac{1}{6}\frac{\partial^3 x}{\partial x_0^3}(x_0, T)(\xi, \xi, \xi) + \cdots$$

where $x_1 = x_0 + \xi$ is used. Comparing (4.39) and (4.40) yields

$$\mathbf{P}'(x_0) = \frac{\partial x}{\partial x_0}(x_0, T), \qquad \mathbf{P}''(x_0) = \frac{\partial^2 x}{\partial x_0^2}(x_0, T), \qquad \cdots \cdot \quad (4.41)$$

To calculate $\partial x / \partial x_0$ the differential equation (4.35) is written in the form

$$\dot{x} = \mathbf{F}(x, t). \quad (4.42)$$

Taking the derivative of both sides of (4.42) with respect to x_0, we obtain a differential equation for $\partial x / \partial x_0$ in the form

$$\left(\frac{\partial x}{\partial x_0}\right)^{\cdot} = \frac{\partial \mathbf{F}}{\partial x}\frac{\partial x}{\partial x_0}. \quad (4.43)$$

To obtain the first term in (4.41), the system (4.43) of 4×4 linear homogeneous differential equations must be integrated from $t = 0$ until $t = T$, with initial conditions $\partial \boldsymbol{x} / \partial \boldsymbol{x}_0 \, (0) = \mathbf{E}$, where $\mathbf{E}$ is the unit matrix. To calculate the next term in (4.41), the derivative of (4.43) with respect to $\boldsymbol{x}_0$ is taken, resulting in

$$\left(\frac{\partial^2 \boldsymbol{x}}{\partial \boldsymbol{x}_0^2} \right)^{\cdot} = \frac{\partial \mathbf{F}}{\partial \boldsymbol{x}} \frac{\partial^2 \boldsymbol{x}}{\partial \boldsymbol{x}_0^2} + \frac{\partial^2 \mathbf{F}}{\partial \boldsymbol{x}^2} \left(\frac{\partial \boldsymbol{x}}{\partial \boldsymbol{x}_0} \right)^2 . \tag{4.44}$$

This system of $4 \times 4 \times 4$ linear inhomogeneous differential equations must be integrated from $t = 0$ until $t = T$, with initial conditions $\partial^2 \boldsymbol{x} / \partial \boldsymbol{x}_0^2 \, (0) = \mathbf{0}$. Proceeding in this way, the coefficients of the third and higher order terms can be calculated. Of course, this process of numerical calculation of the higher order terms becomes computationally intense, if the dimension of (4.35) is large.

Introducing the notation $\boldsymbol{x}_2 = \boldsymbol{x}_0 + \boldsymbol{\xi}_{t+1}$ and $\boldsymbol{x}_1 = \boldsymbol{x}_0 + \boldsymbol{\xi}_t$ we obtain from (4.39) as result of these calculations the discrete dynamical system

$$\boldsymbol{\xi}_{t+1} = \mathbf{L}(\lambda)\boldsymbol{\xi}_t + \mathbf{Q}_2(\boldsymbol{\xi}_t, \boldsymbol{\xi}_t, \lambda) + \mathbf{Q}_3(\boldsymbol{\xi}_t, \boldsymbol{\xi}_t, \boldsymbol{\xi}_t, \lambda) + h.o.t. \tag{4.45}$$

in the form (3.34), where $\boldsymbol{\xi}_t \in \mathbb{R}^4$ and $\lambda \in \mathbb{R}^2$. $\mathbf{Q}_2, \mathbf{Q}_3, \ldots \in \mathbb{R}^4$ are vectors, the components of which are quadratic, cubic and higher order terms in the variables, respectively. For example, the first component of $\mathbf{Q}_2$ is of the form: $\sum_{i,j=1}^{4} \alpha_{1ij}(\lambda) \xi_{ti} \xi_{tj}$.

4.2.3 Stability boundary in parameter space

In order to investigate the stability of the fixed point $\boldsymbol{\xi}_t = \mathbf{0}$ which corresponds to the periodic solution $\boldsymbol{\varphi}_0(t)$ given by (4.37), we study the linear part of (4.45). The corresponding matrix

$$\mathbf{L}(\lambda) = \mathbf{P}'(\boldsymbol{x}_0) = \frac{\partial \boldsymbol{x}}{\partial \boldsymbol{x}_0}(\boldsymbol{x}_0, T) \tag{4.46}$$

is exactly the matrix also obtained using Floquet's theory. For a quasistatic variation of the distinguished parameter ω_0 (given by (4.27)) with the second parameter R (given by (4.34)) kept at a fixed value, we calculate the eigenvalues of (4.46). From the conditions given in Section 3.1.2, we know that the critical eigenvalues are those with modulus equal to one. Hence, we pick a parameter value of λ according to (4.36) such that all eigenvalues have modulus smaller than one. Now, ω_0 is increased quasistatically until the first time an eigenvalue μ with modulus equal to one is reached. Typically, in one parameter families only three different cases can occur ([7] p. 284 and Fig. 4.15):

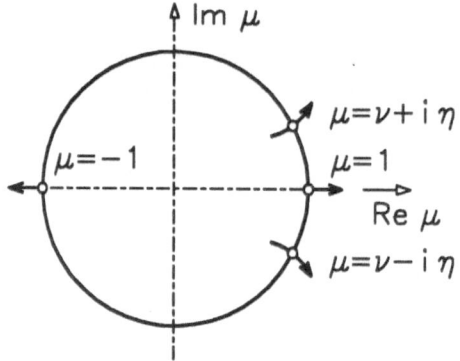

Figure 4.15. The typical cases of loss of stability in one parameter families of mappings are the three cases: $\mu = 1$; $\mu = -1$; $\mu = \nu \pm i\eta$, $|\mu| = 1$

$$
\begin{align}
(1) \qquad\qquad \mu &= 1 \\
(2) \qquad\qquad \mu &= -1 \qquad\qquad\qquad (4.47)\\
(3) \qquad\qquad \mu &= \nu \pm i\eta \qquad |\mu| = 1.
\end{align}
$$

(1), (2) and (3) are called *saddle-node*, *flip* and *Hopf* bifurcation, respectively. These names will become clear from the analysis of the nonlinear cases in Section 6.6.3. In Fig. 4.16, the stability boundary in the

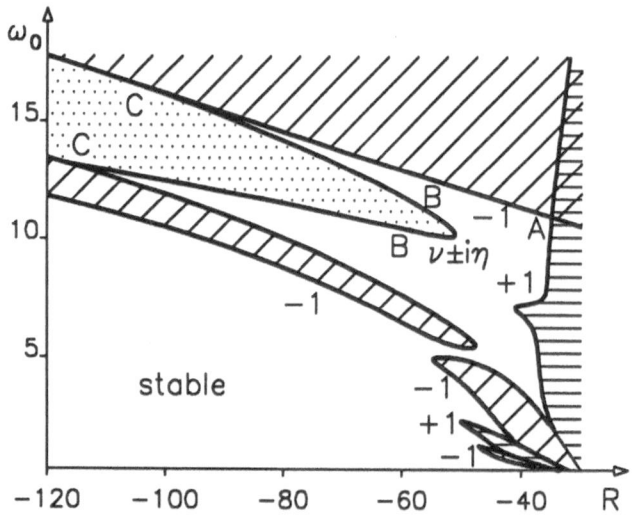

Figure 4.16. Stability boundary ω_{0c} in ω_0, R parameter space. Different types of loss of stability occur according to the critical eigenvalues 1, -1 and $|\nu \pm i\eta| = 1$

(ω_0, R) parameter plane is shown. In the range of R shown in Fig. 4.16 $(-120 \leq R \leq -30)$, the first instability obtained by increasing ω_0 is almost always due to a flip bifurcation ($\mu = -1$). However, saddle-node bifurcations can also occur. If the system is operated at values of ω_0 bigger than those leading to the first instability region also Hopf bifurcations can occur. Above the first instability region parameter domains exist where the original periodic motion is again stable. This behavior can be explained from the fact that these bifurcations result from resonance phenomenona (parametric resonances).

4.2.4 Center manifold reduction

By the transformation of coordinates (3.40), the discrete dynamical system (4.45) takes the form (3.41) and after a proper rearrangement the form of (3.42).

We give now the explicit calculations of one and two dimensional bifurcation equations for the cases $\mu = 1$ and $\mu = \nu \pm i\eta$.

$\mu_1 = 1$: Saddle-node bifurcation

In this case, $n_c = 1$ and (3.42) takes the form

$$
\begin{aligned}
y_{t+1,1} &= y_{t,1} + g_1(y_{t,1}, y_{t,i}) = y_{t,1} + \sum_{j,k=1}^{4} a_{1jk} y_{t,j} y_{t,k} + \\
&\quad + \sum_{j,k,\ell=1}^{4} a_{1jk\ell} y_{t,j} y_{t,k} y_{t,\ell} + \cdots
\end{aligned}
\tag{4.48}
$$

$$
y_{t+1,i} = \mu_i y_{t,i} + g_i(y_{t,1}, y_{t,i}), \qquad i = 2, 3, 4
$$

where for simplicity it has been assumed that the remaining eigenvalues μ_i $(i = 2, 3, 4)$ are distinct and real. From (3.43) follows

$$
y_{t,i} = h_i(y_{t,1}) \qquad i = 2, 3, 4.
$$

For the calculation, we use the approximation $\mathbf{H}$ of h in the form

$$
y_{t,i} = H_i(y_{t,1}) = \alpha_{i,2} y_{t,1}^2 + \alpha_{i,3} y_{t,1}^3 + \cdots ,
\tag{4.49}
$$

which starts with at least quadratic terms considering the general property (3.18). To calculate the coefficients $\alpha_{i,j}$, we substitute (4.49) according to (3.45) into (4.48)$_2$ where use is made of (4.48)$_1$ to find

$$
\begin{aligned}
&\alpha_{i,2}[y_{t,1} + g_1(y_{t,1}, H_i(y_{t,1}))]^2 + \alpha_{i,3}[y_{t,1} + g_1(y_{t,1}, H_i(y_{t,1}))]^3 + \cdots \\
&- \mu_i(\alpha_{i,2} y_{t,1}^2 + \alpha_{i,3} y_{t,1}^3 + \cdots) - g_i(y_{t,1}, H_i(y_{t,1})) = O(|y_{t,1}|^3).
\end{aligned}
\tag{4.50}
$$

We calculate the approximation $\mathbf{H}$ in (4.50) only up to quadratic terms. This approximation will give us bifurcation equations, correct up to and including third order terms as is explained in detail in Section 3.1.1. The quadratic terms follow from (4.50) to

$$\alpha_{i,2} y_{t,1}^2 - \mu_i \alpha_{i,2} y_{t,1}^2 - g_{i,2} y_{t,1}^2 = 0, \tag{4.51}$$

where $g_{i,2}$ are the coefficients of the quadratic terms in the series expansion of the nonlinear function g_i. From (4.51), we obtain the coefficients

$$\alpha_{i,2} = \frac{g_{i,2}}{1 - \mu_i} \qquad i = 2, 3, 4. \tag{4.52}$$

Inserting (4.52) into (4.48)$_1$ gives the one-dimensional, third order bifurcation equation

$$y_{t+1,1} = y_{t,1} + a_{111} y_{t,1}^2 + (a_{1111} + a_{112}\alpha_{2,2} + a_{113}\alpha_{3,2} + a_{114}\alpha_{4,2}) y_{t,1}^3 \cdots . \tag{4.53}$$

$\mu_{1,2} = \nu \pm i\eta$, $|\mu_{1,2}| = 1$: Hopf bifurcation

Now, $n_c = 2$. The most convenient way to handle this case is to use complex variables. Then system (3.41) takes the form

$$
\begin{aligned}
w_{t+1,1} &= \mu_1 w_{t,1} + g_1(w_{t,1}, \overline{w}_{t,1}, w_{t,j}, \overline{w}_{t,j}) \\
w_{t+1,j} &= \mu_j w_{t,j} + g_j(w_{t,1}, \overline{w}_{t,1}, w_{t,j}, \overline{w}_{t,j})
\end{aligned}
\tag{4.54}
$$

where in the robot example $j = 2$. In order to keep the treatment more general, $j = 2, \ldots, N$ can be assumed. Here

$$w_{t,1} = y_{t,1} + i y_{t,2} \qquad \text{and} \qquad \overline{w}_{t,1} = y_{t,1} - i y_{t,2}$$

are complex conjugate. Analogous to before, we set

$$w_{t,j} = H(w_{t,1}, \overline{w}_{t,1}) = \alpha_{11,j} w_{t,1}^2 + \alpha_{12,j} w_{t,1}\overline{w}_{t,1} + \alpha_{22,j}\overline{w}_{t,1}^2, \tag{4.55}$$

where from the outset, we look only for bifurcation equations (4.54)$_1$ of third order in $w_{t,1}, \overline{w}_{t,1}$. First, we write (4.55) for $(t + 1)$

$$w_{t+1,j} = \alpha_{11,j} w_{t+1,1}^2 + \alpha_{12,j}|w_{t+1,1}|^2 + \alpha_{22,j}\overline{w}_{t+1,1}^2$$

and substitute for each $w_{t+1,1}$ from (4.54)$_1$ to find

$$w_{t+1,j} = \alpha_{11,j}\mu_1^2 w_{t,1}^2 + \alpha_{12,j}\mu_1\overline{\mu}_1 w_{t,1}\overline{w}_{t,1} + \alpha_{22,j}\overline{\mu}_1^2\overline{w}_{t,1}^2 + O(|w_1, \overline{w}_1|^3). \tag{4.56}$$

Second, we consider (4.54)$_2$ where we substitute (4.55) on the right-hand side to obtain

$$
\begin{aligned}
w_{t+1,j} &= \mu_j(\alpha_{11,j} w_{t,1}^2 + \alpha_{12,j} w_{t,1}\overline{w}_{t,1} + \alpha_{22,j}\overline{w}_{t,1}^2) + \\
&\quad + g_{11,j} w_{t,1}^2 + g_{12,j} w_{t,1}\overline{w}_{t,1} + g_{22,j}\overline{w}_{t,1}^2 + O(|w_1|^3). \tag{4.57}
\end{aligned}
$$

The $g_{k\ell,m}$ are the coefficients of the quadratic terms in the series expansion of g_m, $(m = 1, 2, \ldots, N)$. Comparing the second order coefficients of (4.56) and (4.57), which corresponds to relation (3.45) with $q = 3$, we find

$$w_{t,1}^2 : \qquad \alpha_{11,j}(\mu_1^2 - \mu_j) = g_{11,j}$$

$$w_{t,1}\overline{w}_{t,1} : \qquad \alpha_{12,j}(\mu_1\overline{\mu}_1 - \mu_j) = g_{12,j} \qquad (4.58)$$

$$\overline{w}_{t,1}^2 : \qquad \alpha_{22,j}(\overline{\mu}_1^2 - \mu_j) = g_{22,j} \ .$$

Inserting the coefficients $\alpha_{11,j}, \alpha_{12,j}, \alpha_{22,j}$ obtained from (4.58) into (4.55) and then (4.55) into (4.54)$_1$ results in third order bifurcation equations for the robot $(j = 2)$ in the form

$$w_{t+1,1} = \mu_1 w_{t,1} + g_{11,1} w_{t,1}^2 + g_{12,1} w_{t,1}\overline{w}_{t,1} + g_{22,1}\overline{w}_{t,1}^2 +$$

$$+ \sum_{k=0}^{3} a_{3-k,k,1} w_{t,1}^{3-k}\overline{w}_{t,1}^k$$

or

$$w_{t+1} = \mu_1 w_t + a_{20}w_t^2 + a_{11}w_t\overline{w}_t + a_{02}\overline{w}_t^2 +$$
$$+ a_{30}w_t^3 + a_{21}w_t^2\overline{w} + a_{12}w_t\overline{w}_t^2 + a_{03}w_t^3 + O(|w_t, \overline{w}_t|^4) \qquad (4.59)$$

where we have omitted the index one.

The calculation of the coefficients of third order is straightforward. We do not write down their explicit form because in the further treatment of (4.59) it will turn out that in general not all coefficients are needed. In fact only one will be needed which follows from the application of the *normal form* theorem. This is explained in Section 6.1.

4.3 Loss of stability of finite- and infinite-dimensional statical systems

4.3.1 Buckling of a rod: discrete model

To explain the method of Ljapunov-Schmidt, we start with the double pendulum of Fig. 2.5a. This is one of the simplest examples of a model problem for buckling with more than one degree of freedom. It is a conservatively loaded system and if the load is applied quasistatically, the equilibrium equations can be derived from the potential function V (2.17). The bars are assumed to be massless. The torsional stiffnesses are $c_1 = c_2 = c$ and we divide V by c. Then from (2.17), we obtain the potential

$$V_1 = \frac{V}{c} = \frac{\psi_1^2}{2} + \frac{1}{2}(\psi_2 - \psi_1)^2 - F(2 - \cos\psi_1 - \cos\psi_2) \qquad (4.60)$$

expressed in dimensionless quantities with the dimensionless loading parameter $F = P\ell/c$. Taking the gradient of (4.60) the equilibrium equations are

$$
\begin{aligned}
G_1 &= \frac{\partial V_1}{\partial \psi_1} = 2\psi_1 - \psi_2 - F \sin \psi_1 = 0 \\
G_2 &= \frac{\partial V_1}{\partial \psi_2} = -\psi_1 + \psi_2 - F \sin \psi_2 = 0 .
\end{aligned}
\tag{4.61}
$$

We note that (4.61) are of the form $\mathbf{G}(\boldsymbol{\psi}, F) = \mathbf{0}$. Further, the state $\boldsymbol{\psi}_0 = (0,0)^T$ to be analyzed for stability is a solution of (4.61). According to the *implicit function theorem*, we have to calculate the Fréchet derivative of (4.61) at $\boldsymbol{\psi}_0$, which is

$$
\mathbf{A} = \mathbf{G}_\psi(\boldsymbol{\psi}_0, F) = \begin{pmatrix} 2 - F & -1 \\ -1 & 1 - F \end{pmatrix} .
\tag{4.62}
$$

For those values of F for which $\mathbf{A}$ is a regular linear operator a unique solution curve passes locally through $(\boldsymbol{\psi}_0, F_0)$ and no bifurcation can occur. To locate bifurcations we must find those values $F = F_c$ for which $\mathbf{A}$ is singular. These values F_c are given by the roots of the characteristic equation

$$
\det \mathbf{A} = (2 - F)(1 - F) - 1 = F^2 - 3F + 1 = 0 .
\tag{4.63}
$$

The roots are

$$
F_{1,2} = \frac{3}{2} \pm \sqrt{\frac{9}{4} - 1} = \frac{3 \pm \sqrt{5}}{2} .
\tag{4.64}
$$

The physically relevant value is the smaller one, which is

$$
F_c = \frac{3 - \sqrt{5}}{2} .
\tag{4.65}
$$

Next, we calculate the eigenvector of (4.62) corresponding to the eigenvalue (4.65) which spans the kernel $N(A)$ (Appendix A)

$$
N(A) = \text{span} \left\{ \begin{pmatrix} 1 \\ 2 - F_c \end{pmatrix} \right\} .
\tag{4.66}
$$

The operators $\mathbf{G}$ and $\mathbf{A}$ define a mapping: $\mathbb{R}^2 \times \mathbb{R}^1 \to \mathbb{R}^2$. The matrix $\mathbf{A}$ is self-adjoint (Appendix C) because it is symmetric. Hence, $N(\mathbf{A}^\star) = N(\mathbf{A})$ and the range $R(\mathbf{A})$ is spanned by a vector $\boldsymbol{b}$ defined by

$$
\boldsymbol{b}^T \cdot \boldsymbol{a} = 0
\tag{4.67}
$$

where $a \in N(\mathbf{A})$. Clearly, $\mathbf{A}$ is a *Fredholm operator* (Appendix A). From (4.67) it follows

$$R(\mathbf{A}) = \text{span} \left\{ \begin{pmatrix} 2 - F_c \\ -1 \end{pmatrix} \right\} . \qquad (4.68)$$

For the calculation to follow we assume that a and b are normalized to unit vectors $e_a = a/\|a\|$ and $e_b = b/\|b\|$ where a and b are given by (4.66) and (4.68), respectively, and $\|a\| = \|b\| = \|\cdot\| = \sqrt{5 - 4F_c + F_c^2}$.

Now we define the orthogonal projection P of an element x of the image space on $R(\mathbf{A})$ to

$$P\boldsymbol{x} = (\boldsymbol{x} \cdot e_b)e_b \qquad (4.69)$$

and the complementary projection $(E - P)$ on $N(\mathbf{A}^\star)$ to

$$(E - P)\boldsymbol{x} = (\boldsymbol{x} \cdot e_a)e_a . \qquad (4.70)$$

Finally, according to (3.53), ψ is decomposed into the form

$$\psi = y_c e_a + y_s e_b . \qquad (4.71)$$

Next, we write (4.61) in the form of the right-hand side of (3.4), that is,

$$\begin{pmatrix} 2 - F_c & -1 \\ -1 & 1 - F_c \end{pmatrix} \begin{pmatrix} \psi_1 \\ \psi_2 \end{pmatrix} + F_c \begin{pmatrix} \dfrac{\psi_1^3}{3!} - \dfrac{\psi_1^5}{5!} + \cdots \\ \dfrac{\psi_2^3}{3!} - \dfrac{\psi_2^5}{5!} + \cdots \end{pmatrix} = 0 \qquad (4.72)$$

and perform the two projections (4.69) and (4.70). Let us calculate (4.70) first and retain only third order nonlinear terms. This yields

$$\begin{aligned}
(E - P)\mathbf{G} &= \left\{ \left[(2 - F_c)\psi_1 - \psi_2 + F_c \frac{\psi_1^3}{6} \right] + \right. \\
&\qquad + \left. \left[-\psi_1 + (1 - F_c)\psi_2 + F_c \frac{\psi_2^3}{6} \right] (2 - F_c) \right\} e_a \qquad (4.73) \\
&= \frac{F_c}{6}(\psi_1^3 + (2 - F_c)\psi_2^3)e_a = 0 .
\end{aligned}$$

The complementary projection $P\mathbf{G}$ yields

$$\begin{aligned}
P\mathbf{G} &= \left\{ \left[(2 - F_c)\psi_1 - \psi_2 + F_c \frac{\psi_1^3}{6} \right] (2 - F_c) - \right. \\
&\qquad - \left. \left[-\psi_1 + (1 - F_c)\psi_2 + F_c \frac{\psi_2^3}{6} \right] \right\} e_b \\
&= \left\{ [(2 - F_c)^2 + 1]\psi_1 - (3 - 2F_c)\psi_2 + \right. \\
&\qquad + \left. (2 - F_c)F_c \frac{\psi_1^3}{6} - F_c \frac{\psi_2^3}{6} \right\} e_b = 0 .
\end{aligned} \qquad (4.74)$$

Substituting (4.71) into (4.74) and taking the inner product with e_b we find

$$(-1 + 3F_c - F_c^2)y_c + [(2 - F_c)^3 + 5 - 3F_c]y_s +$$
$$+ \frac{1}{6}F_c\Big\{(2 - F_c)[y_c + y_s(2 - F_c)]^3 - \qquad (4.75)$$
$$- [(2 - F_c)y_c - y_s]^3\Big\}\frac{1}{\|\cdot\|^2} = 0 \; .$$

The first term in (4.75) vanishes because of (4.63). Thus, it is possible to calculate from (4.75) the non-critical variable $y_s = h(y_c)$ as a function of the critical variable, as assumed in (3.55). Furthermore, we see that $h(y_c) = O(|y_c|^3)$. From the bifurcation equation which follows from the projection $(E - P)\mathbf{G} = 0$, (4.73), we can explicitly explain the influence of the non-critical variables. Inserting (4.71) into (4.73), we find

$$\frac{F_c}{6}\{[y_c + y_s(2 - F_c)]^3 + (2 - F_c)[y_c(2 - F_c) - y_s]^3\}e_a = 0 \; . \qquad (4.76)$$

Since $y_s = O(|y_c|^3)$ and assuming that the coefficient of y_c^3 is not equal to zero we can set $y_s = 0$ in (4.76) without effecting the third order terms. This follows from the fact that all terms which contain y_s (for example: $y_c^2 y_s, y_c y_s^2, y_s^3$) are at least of fifth order in y_c. Taking the inner product of (4.76) with e_a we obtain the bifurcation equation of third order

$$\frac{F_c}{6}(y_c^3 + (2 - F_c)(2 - F_c)^3 y_c^3) + O(|y_c|^5) = 0 \; . \qquad (4.77)$$

From (4.77), we see that the third order term in the one-dimensional bifurcation equation does not vanish, and therefore, the previous assumption that we do not have to retain higher order terms was correct. The problem is *three-determinate* (Section 6.3).

Finally, it would be convenient to include also the loading F in (4.77). This can be achieved by a one-parameter unfolding and yields the so-called one-parameter bifurcation equation. For this purpose, we replace F_c by $F_c + \lambda$ in (4.73) and obtain the following expression

$$(-3 + 2F_c)\lambda\psi_2 + \frac{F_c + \lambda}{6}[\psi_1^3 + (2 - F_c - \lambda)\psi_2^3] = 0 \; .$$

Inserting from (4.71) yields

$$(-3 + 2F_c)(2 - F_c)\lambda y_c + \frac{1}{6}F_c[1 + (2 - F_c)^4]y_c^3\frac{1}{\|\cdot\|^2} = 0 \qquad (4.78)$$

where, as is explained in Section 6.4 and 6.5.3, higher order terms in λ (for example $\lambda^2 y_c$ or λy_c^3) can be neglected. Thus the one-parameter bifurcation equation is

$$y_c^3 + a y_c = 0 \; , \qquad (4.79)$$

where a follows immediately from (4.78). From (4.79) also the order of λ

$$\lambda \approx a = O(|y_c|^2) \tag{4.80}$$

is obtained.

We mention one important point. After solving the nonlinear one-dimensional equation (4.79) for y_c, which is explained in Section 6.2.1, we substitute y_c into (4.71) to obtain the physical quantities ψ_1 and ψ_2. The second term in (4.71) y_s is at least of second order in y_c $(O(|y_c|^2))$ and can be neglected in a local analysis. For this example y_s is even $O(|y_c|^3)$.

4.3.2 Buckling of a rod: continuous model

In Section 2.2, we derived the differential equation (2.32) describing the statical buckling behavior of a continuous, axially inextensible rod. Dividing (2.32) by EJ and setting $p = P/EJ$, we obtain

$$G(\psi, p) = \frac{d^2\psi}{ds^2} + p\sin\psi = 0. \tag{4.81}$$

As boundary conditions we stipulate $\psi'(0) = \psi'(L) = 0$, that is, the rod is assumed to be simply supported at both ends (Fig. 2.7).

G is a nonlinear differential operator: $H \times \Lambda \rightarrow K$ where $\Lambda \subset R^1$, and the function spaces H, K are defined as (Appendix A, p. 291)

$$H = \{\psi \in C^2[0, L] \quad \text{and} \quad \psi'(0) = \psi'(L) = 0\} \quad \text{and} \quad K = C[0, L].$$

The solution ψ_0 to be investigated for its stability is given by $\psi_0 \equiv 0$ and satisfies (4.81) for any value of p. Now we ask whether values of p exist, for which the linear mapping $A = G_\psi(0, p)$ does not possess an invertible inverse. To answer this question, we must look for nontrivial solutions of the corresponding eigenvalue problem. The linear operator A is defined by taking the Fréchet derivative of (4.81)

$$A\chi = G_\psi(\psi_0, p)\chi = \chi'' + p\chi \cos\psi_0 .$$

Inserting $\psi_0 \equiv 0$ into this equation, we obtain the linear eigenvalue problem

$$G_\psi(0, p)\chi = A\chi = \chi'' + p\chi = 0 \tag{4.82}$$

with the boundary conditions $\chi'(0) = \chi'(L) = 0$.

From the example in Appendix C.2 follows that the eigenvalue problem (4.82) is self-adjoint, that is, $A = A^*$. The solution of (4.82) is $\chi(s) = B\cos(\sqrt{p}s) + D\sin(\sqrt{p}s)$. The boundary condition $\chi'(0) = 0$ implies $D = 0$. The boundary condition $\chi'(L) = 0$ yields $-\sqrt{p}B\sin(\sqrt{p}L) =$

0. For a non-trivial solution $\chi(s)$, $B \neq 0$ and hence $\sqrt{p}L = n\pi$. Thus, an infinite sequence of eigenvalues

$$p = p_n = \left(\frac{n\pi}{L}\right)^2 \tag{4.83}$$

with the corresponding eigenfunctions

$$\chi = \chi_n = B_n \cos\frac{n\pi s}{L} \qquad n = 1, 2, \ldots \tag{4.84}$$

is obtained. In (4.83), the eigenvalue corresponding to $n = 0$ can be excluded for two reasons. First, for $n = 0$ follows from (4.83) that $p = 0$, which means that the rod is not loaded. Second follows from (4.84) that $\chi = \chi_0 = B_0$. This is a rigid body rotation. However, such a rigid body motion is only compatible with the boundary conditions if $B_0 \equiv 0$.

The nullspace is spanned by the eigenfunction $\cos(n\pi s/L)$, which we express in the form

$$N(A) = \text{span} \left\{ \cos\frac{n\pi s}{L} \right\} \tag{4.85}$$

and, hence, is one-dimensional for a given n. Physically only the smallest buckling load is relevant, therefore, $n = 1$. However, we do not specify n at the moment. Since $A = A^*$, we have

$$\dim N(A) = \dim N(A^*) = 1 \qquad \text{and} \qquad N(A) = N(A^*).$$

The range $R(A)$ is given by those functions $g(s)$ which are orthogonal to $N(A^*)$, that is

$$R(A) = N(A^*)^\perp = \{g(s) \in C[0, L] : g \perp N(A^*)\}. \tag{4.86}$$

From (4.85) and (4.86) follows with (A.8) and (A.13)

$$\int_0^L g(s) \cos\frac{n\pi s}{L} ds = 0.$$

Next, we explicitly calculate the projections (Appendix D). We decompose the image space

$$K = R(A) \oplus N(A^*),$$

by representing $R(A)$ and $N(A^*)$ by means of the projection P. This results in

$$\begin{aligned} R(A) &= PK \\ N(A^*) &= (E - P)K. \end{aligned}$$

The projection P of K onto $R(A)$ is obtained in the following manner. For any $f(s) \in K$, the projection $(E - P)f(s)$ must be in $N(A^\star)$. That is, it must be proportional to $\cos(n\pi s/L)$, having the form

$$(E - P)f(s) = C \cos \frac{n\pi s}{L} \qquad (4.87)$$

with the constant C to be determined. The complementary projection $E - (E - P) = P$, which projects K on $R(A)$, is then given by (D.5)

$$Pf(s) = f(s) - C \cos \frac{n\pi s}{L}. \qquad (4.88)$$

The constant C can be calculated from the condition that each element in the range $R(A)$ is orthogonal to the nullspace $N(A^\star)$. This gives

$$\int_0^L \left(f(s) - C \cos \frac{n\pi s}{L} \right) \cos \frac{n\pi s}{L} ds = 0$$

$$\int_0^L f(s) \cos \frac{n\pi s}{L} ds = C \int_0^L \cos^2 \frac{n\pi s}{L} ds = C \frac{L}{2} \qquad (4.89)$$

$$C = \frac{2}{L} \int_0^L f(s) \cos \frac{n\pi s}{L} ds \ .$$

Thus, the projection of any $f(s) \in K$ onto $N(A^\star)$ is given by (4.87)

$$(E - P)f(s) = \left(\frac{2}{L} \int_0^L f(s) \cos \frac{n\pi s}{L} ds \right) \cos \frac{n\pi s}{L}.$$

The final step is the decomposition (3.53) of ψ in the form

$$\psi = \psi_c + \psi_s \qquad (4.90)$$

where

$$\psi_c = q \cos(n\pi s/L) \in N(A) \qquad \text{and} \qquad \psi_s = h(q, s) \in R(A^\star). \qquad (4.91)$$

In (4.91), q is the amplitude of the buckling mode. It is the active or critical variable in the one-dimensional algebraic bifurcation equation.

The bifurcation equation (3.58) follows by projecting (4.81) onto $N(A^\star)$, to

$$\int_0^L \left[\psi'' + p \left(\psi - \frac{\psi^3}{6} + \dots \right) \right] \cos \frac{n\pi s}{L} ds = 0. \qquad (4.92)$$

We substitute (4.90) into (4.92) and we make use of (4.91) to find

$$
\int_0^L \left[q \left(-\frac{n^2\pi^2}{L^2} + p \right) \cos \frac{n\pi s}{L} + h'' + ph - \right.
$$

$$
- \frac{p}{6} \left(q^3 \cos^3 \frac{n\pi s}{L} + 3q^2 h \cos^2 \frac{n\pi s}{L} + \right. \tag{4.93}
$$

$$
\left. \left. + 3qh^2 \cos \frac{n\pi s}{L} + h^3 \right) + \ldots \right] \cos \frac{n\pi s}{L} ds = 0 .
$$

If we substitute $p = p_n$ from (4.83), the linear terms in q in (4.93) vanish. Furthermore, since h is orthogonal to $N(A^*)$ the integral

$$
\int_0^L (h'' + ph) \cos \frac{n\pi s}{L} ds = 0 . \tag{4.94}
$$

Hence, the lowest order nonlinear terms in the bifurcation equation can be determined from

$$
\int_0^L \left\{ -\frac{p}{6} \left[q^3 \cos^4 \frac{n\pi s}{L} + 3q^2 h \cos^3 \frac{n\pi s}{L} + \right. \right.
$$

$$
\left. \left. + 3qh^2 \cos^2 \frac{n\pi s}{L} + h^3 \cos \frac{n\pi s}{L} \right] \right\} ds = 0. \tag{4.95}
$$

The function h in (4.95) must be calculated in its dependence on the critical variable from (3.54). We only indicate this here, because due to (3.57) follows that $h = O(|q|^2)$. Therefore, in (4.95) h does not influence the third order term, which has a non zero coefficient, because $p = p_n \neq 0$ and also the corresponding integral does not vanish. Hence, for reasons of determinacy (Section 6.3), $h \equiv 0$ can be introduced and a correct bifurcation equation of third order is obtained. Thus, similar to some of the problems in Section 4.1, application of the Ljapunov-Schmidt method becomes trivial for this problem. That is, it reduces to the Galerkin method.

However, we note that to calculate h, (4.81) must be projected onto $R(A)$. Using (4.88) and (4.89), we obtain

$$
PG(\psi, p_n) = \psi'' + p_n \sin \psi - \tag{4.96}
$$

$$
- \left[\frac{2}{L} \int_0^L (\psi'' + p_n \sin \psi) \cos \frac{n\pi s}{L} ds \right] \cos \frac{n\pi s}{L} = 0 .
$$

Expanding $\sin \psi$ in its Taylor series and using (4.91), we substitute (4.90) into (4.96) to find a differential equation for h of the form

$$
h'' + p_n h = N(q, h, s) \tag{4.97}
$$

where $N(q, h, s)$ is a nonlinear function of q and h. Since h is orthogonal to the eigenfunction $\cos(n\pi s/L)$ for which p_n is the eigenvalue, p_n is not an eigenvalue of the linear operator in (4.97), restricted to $N(A)^\perp$. Its inverse exists and according to the *implicit function theorem*, a smooth unique solution

$$h = h(q, s) \tag{4.98}$$

can be calculated. The practical calculation is done by series expansions ([156]) and will be explicitly given for an example in Section 4.3.3.

Finally, we present the one-parameter (load P) bifurcation equation. For this purpose, we set $p = p_n + \lambda$ in (4.93). This results in

$$\int_0^L \left[q\left(-\frac{n^2\pi^2}{L^2} + p_n + \lambda \right) \cos\frac{n\pi s}{L} - \right.$$
$$\left. -\frac{p_n + \lambda}{6} \left(q^3 \cos^3\frac{n\pi s}{L} + O(q^4) \right) \right] \cos\frac{n\pi s}{L} ds = 0.$$

After integrating and neglecting terms of $O(q^4)$, we find

$$\lambda q \frac{L}{2} - \frac{p_n + \lambda}{6} q^3 \frac{3L}{8} = 0. \tag{4.99}$$

The parameter $\lambda = p - p_n$ measures the small deviation from the bifurcation point. This gives $\lambda = O(q^2)$, which follows immediately from the solution of (4.99) for λ (Sections 4.3.1 and 6.4). Therefore, the term λq^3 can be neglected in comparison to $p_n q^3$. Considering this ordering, we obtain

$$\frac{8L^2\lambda}{\pi^2 n^2} q - q^3 = 0 \tag{4.100}$$

as the one-parameter bifurcation equation analogous to (4.79).

It must be pointed out, that the application of the Ljapunov-Schmidt method, in general, requires the solution of two nonlinear problems. First, (4.97) must be solved to obtain $h(q, s)$, and second (4.100) must be solved for q. However, the solution (4.98) of (4.97) has to be calculated only up to such an order that after substitution of (4.98) into (4.95) only those terms are affected that determine the local nonlinear behavior. In the case of the rod buckling problem, the calculation of the solution (4.98) was not necessary at all, because the substitution of (4.98) into (4.95) led to at least fourth order terms. Second, the solution of the nonlinear bifurcation equation (4.100) must be determined. This is examined in detail in Chapter 6.

4.3.3 Buckling of a circular ring

This problem is treated in ([81], [92]).

In Appendix I, the governing equation of a circular, elastic, and in-extensional ring under uniform compression p is derived. This equation may be written in the form

$$G(w,p) = w''(s) + pw(s) + \beta + \frac{1}{2}w(s)^2(w(s) + 3) = 0 \qquad (4.101)$$

with periodic boundary conditions

$$w(0) = w(2\pi), \qquad w'(0) = w'(2\pi) \qquad (4.102)$$

and the supplementary condition

$$\int_0^{2\pi} w(s)ds = 0. \qquad (4.103)$$

In (4.101) is p the loading parameter, $s \in [0, 2\pi]$ the dimensionless inde-pendent coordinate (arc length) and β a constant which has to be selected in such a way that the solution defines a closed ring without discontinu-ities in its slope. The function $w(s)$ is the difference of the curvature of the undeformed and the deformed ring and is also dimensionless.

From (4.101) follows that $w_0 \equiv 0, \beta_0 = 0$ is the fundamental solution to be studied for its stability under a quasistatically increasing load p. Setting $w = w_0 + \varepsilon$ and $\beta = \beta_0 + \eta$ the linear operator $A : H \to K$ with $H = \{(\varepsilon, \eta) \in C^2[0, 2\pi]$ and (4.102), (4.103) to hold$\}$ is given by

$$A\begin{pmatrix} \varepsilon \\ \eta \end{pmatrix} = \varepsilon'' + p\varepsilon + \eta$$

from which the linear eigenvalue problem

$$\varepsilon'' + p\varepsilon = -\eta \qquad (4.104)$$

with the boundary conditions (4.102) and (4.103) follows. The solution to (4.104) is

$$\varepsilon(s) = U_n \sin ns + V_n \cos ns + W$$

where $W = -\eta/p$ is a constant.

From (4.103) follows $\eta = 0$. We mention that this problem possesses a special symmetry, namely the $O(2)$-symmetry, which we will find again in Section 5.2 in connection with buckling of an annular plate. There it is explained that both $\sin ns$ and $\cos ns$ are buckling modes. Hence,

$$p = p_n = n^2 \qquad (4.105)$$

is a double eigenvalue with the corresponding eigenfunctions

$$\varepsilon_n = U_n \sin ns , \qquad \varepsilon_n = V_n \cos ns . \qquad (4.106)$$

We can pick either one, because as long as an imperfection is not present the location of the buckling pattern is not fixed on the ring. This is further explained in Section 5.2 for the annular plate. We proceed with $\sin ns$. Bifurcation can only occur if $n \geq 2$, because the solution corresponding to $n = 0$ is trivial and the solution to $n = 1$ corresponds to a rigid body displacement of the ring. Hence, the physically relevant value is $n = 2$.

The kernel of the operator A is one-dimensional and is spanned by the function $\sin ns$. Since A is self-adjoint, the following relationships hold

$$N(A) = N(A^\star) \qquad \text{and} \qquad \dim N(A) = \dim N(A^\star) = 1.$$

The range of A is defined by

$$R(A) = N(A^\star)^\perp = \{g(s) \in C[0, 2\pi] : g(s) \perp N(A^\star)\}. \qquad (4.107)$$

With the projection P defined analogously to (4.87) and (4.88) the image space (4.107) is decomposed into one part in $N(A^\star)$

$$(E - P)f(s) = C \sin ns \qquad (4.108)$$

and another part in $R(A)$

$$Pf(s) = f(s) - C \sin ns \ . \qquad (4.109)$$

Here, as in the example 4.3.2 is $f(s) \in K = C[0, 2\pi]$ an element of the image space.

The constant C can be calculated similarly to (4.89) to obtain

$$C = \frac{1}{\pi} \int_0^{2\pi} f(s) \sin nsds \ . \qquad (4.110)$$

Thus, the projection of any $f(s) \in K$ onto $N(A^\star)$ is given by

$$(E - P)f(s) = \left(\frac{1}{\pi} \int_0^{2\pi} f(s) \sin nsds \right) \sin ns \ . \qquad (4.111)$$

According to (4.89) and (4.90), $w(s)$ with $n = 2$ is represented as

$$w(s) = q \sin 2s + h(q, s), \qquad (4.112)$$

with $q \sin 2s \in N(A)$ and $h(q, s) \in R(A^\star)$. The variable q is the amplitude of the buckling mode, and hence, the critical or active variable. In order to obtain the bifurcation equation, (4.101) must be projected on $N(A^\star)$. From (4.111) follows

$$(E - P)G(w, p) = \left\{ \frac{1}{\pi} \int_0^{2\pi} \left[w'' + pw + \beta + \right. \right.$$

$$\left. \left. + \frac{1}{2} w^2(w + 3) \right] \sin 2s\, ds \right\} \sin 2s = 0. \qquad (4.113)$$

Substituting (4.112) into (4.113) results in

$$\int_0^{2\pi} [-4q \sin 2s + h'' + pq \sin 2s + ph + \beta +$$

$$+ \frac{1}{2}(q^3 \sin^3 2s + 3q^2 h \sin^2 2s + 3qh^2 \sin 2s + h^3) + \qquad (4.114)$$

$$+ \frac{3}{2} (q^2 \sin^2 2s + 2qh \sin 2s + h^2)] \sin 2s\, ds = 0.$$

The introduction of a local parameter λ with respect to (4.105) $(n = 2)$ gives

$$p = n^2 + \lambda = 4 + \lambda \qquad (4.115)$$

and analogously to (4.94)

$$\int_0^{2\pi} (h'' + (4 + \lambda)h) \sin 2s\, ds = 0$$

because h is orthogonal to $N(A)$. From (4.114) follows the one-dimensional one-parameter bifurcation equation

$$\int_0^{2\pi} \left(\lambda q \sin^2 2s + \frac{1}{2} q^3 \sin^4 2s + 3qh \sin^2 2s + \frac{3}{2} q^2 h \sin^3 2s + \right.$$

$$\left. + \frac{3}{2} h^2 \sin 2s + \frac{3}{2} qh^2 \sin^2 2s + \frac{1}{2} h^3 \sin 2s \right) ds = 0. \qquad (4.116)$$

If we set $\lambda = 0$ in (4.116) then the lowest order non-vanishing term decides the determinacy. The lowest order nonlinear terms are one term of $O(q^3)$ and one term $O(qh)$. From (3.57) it is known that $h = 0(|q|^2)$. Thus, we find that the term which is proportional to qh may be of third order. Hence, quite contrary to the buckling problem of the straight rod in Section 4.3.2, one cannot proceed by setting $h = 0$, but has to calculate h from (3.54), to at least second order by projection of (4.101) onto $R(A)$. With $\lambda = 0$, we obtain

$$PG(w, p_n) = G(w, p_n) - \left(\frac{1}{\pi} \int_0^{2\pi} G(w, p_n) \sin 2s\, ds \right) \sin 2s = 0. \qquad (4.117)$$

From (4.117), the equation

$$h'' + 4h + \beta + \frac{1}{2}(q \sin 2s + h)^2 (q \sin 2s + h + 3) = 0 \qquad (4.118)$$

immediately follows, because the second term in (4.117) vanishes due to (4.116). Rewriting (4.118) gives

$$\begin{aligned}
-(h'' + 4h) &= \beta + \frac{3}{2}(q^2 \sin^2 2s + 2qh \sin 2s + h^2) + \\
&\quad + \frac{1}{2}(q^3 \sin^3 2s + 3q^2 h \sin^2 2s + 3qh^2 \sin 2s + h^3).
\end{aligned} \qquad (4.119)$$

All functions h are orthogonal to the functions obtained from the linear eigenvalue problem (4.104). Hence, $p_n = n^2 = 4$ is not an eigenvalue for the linear operator on the left-hand side of (4.119) restricted to $N(A)^{\perp}$, and hence, this operator can be inverted. Therefore, an ansatz can be made for the solution of the form

$$h(s, q) = h_0(s) + h_1(s)q + h_2(s)q^2 + \dots \qquad (4.120)$$

From (3.55) and (3.56), follows that $h(s, 0) = 0$, $h'(s, 0) = 0$, and hence, $h = O(|q|^2)$. Thus in (4.120), we obtain

$$h_0(s) = h_1(s) = 0. \qquad (4.121)$$

Next, from (4.116) follows that only in the case that h is of second order in q, a contribution will be obtained which can enter in the third order bifurcation equation. Thus we concentrate on the calculation of $h_2(s)$. In order to be able to satisfy (4.103), we must also represent β in a power series in q:

$$\beta(q) = \beta_0 + \beta_1 q + \beta_2 q^2 + \dots \qquad (4.122)$$

For $q = 0$, β must be zero, and hence, $\beta_0 = 0$. Substituting (4.120) and (4.122) into (4.119) and equating coefficients of the same power in q yields the following conditions for the β_i

$$q \quad : \quad \beta_1 = 0$$
$$q^2 \quad : \quad -(h_2'' + 4h_2) = \beta_2 + \frac{3}{4} \sin^2 2s \ .$$

The last expression can be written as

$$h_2'' + 4h_2 = -\beta_2 - \frac{3}{4}(1 - \cos 4s). \qquad (4.123)$$

The solution of (4.123) is

$$h_2(s) = C_1 \sin 2s + C_2 \cos 2s - \frac{\beta_2}{4} - \frac{3}{16} - \frac{1}{16} \cos 4s. \qquad (4.124)$$

Since $h_2(s)$ must be orthogonal to $N(A)$, we obtain $C_1 = C_2 = 0$, because both functions are buckling modes. Substituting (4.124) into (4.103) results in $\beta_2 = -3/4$. Hence, the solutions (4.120) and (4.122) are

$$h(s,q) = -\frac{1}{16}q^2 \cos 4s + O(|q|^3) \qquad (4.125)$$

and

$$\beta = -\frac{3}{4}q^2 + O(|q|^3) , \qquad (4.126)$$

respectively. Substituting (4.125) into (4.116) yields the one-parameter bifurcation equation up to third order terms

$$q\left(\lambda + \frac{6}{16}q^2 + \frac{3}{32}q^2\right) = 0 \qquad (4.127)$$

where the third term results from inserting h into (4.116). If the Galerkin method had been used, instead of the Ljapunov-Schmidt method, this term would not appear in (4.127). However, although this term is of third order, it does not have a qualitative influence on the local bifurcation behavior of the ring because its inclusion does not change the sign of the third order term.

Equation (4.127) is of the same type as (4.100) and is discussed in Section 6.2.1.

4.3.4 Buckling at a double eigenvalue: rectangular plate

Formulation of the boundary value problem ([12])

In Appendix J the von Karman plate equations (J.23)

$$\overline{\Delta}^2 F \;=\; -\frac{1}{2}Eh[W,W]$$

$$\qquad (4.128)$$

$$K\overline{\Delta}^2 W \;=\; [F,W] + \bar{t}$$

with the physical parameters in the usual engineering notations are derived, giving the relationship between the displacement $W(X,Y)$ of the plate's midplane into Z direction, the stress function $F(X,Y)$ and the distributed transversal loading $\bar{t}(X,Y)$ in Z-direction. The bracket operator is defined in (J.22). E is Young's modulus, K is given by (J.12) and h is the thickness of the plate. The independent variables X and Y are the coordinates of the middle surface of the undeformed plate. The plate occupies the region $\Omega : 0 \leq X \leq a, 0 \leq Y \leq b, |Z| \leq h/2$ (Fig. 4.17). $\overline{\Delta} = \partial^2/(\partial X^2) + \partial^2/(\partial Y^2)$ is the Laplacian with respect to X, Y.

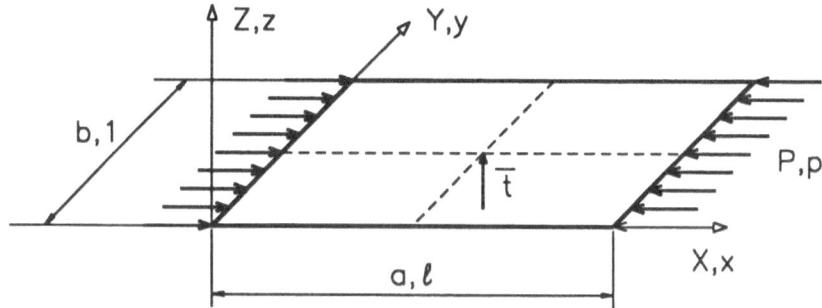

Figure 4.17. Rectangular plate under inplane loading P and distributed transversal loading indicated by $\bar{t}$

A compressive thrust of magnitude P which will be the distinguished parameter is applied normal to the edges $X = 0, a$. In addition we will use the ratio a/b as second parameter. Hence, P and a/b are the main parameters and $\bar{t}$ will be related to imperfection parameters. For the derivation of the one parameter bifurcation equations we set $\bar{t} \equiv 0$. $U(X, Y)$ and $V(X, Y)$ are the midplane displacements in the X and Y directions, respectively. To formulate the boundary conditions for (4.128), we assume that the boundary $\delta\Omega$ of the plate is simply supported. This implies that

$$W = \overline{\Delta}W = 0 \qquad \text{on } \delta\Omega. \qquad (4.129)$$

The condition $\overline{\Delta}W = 0$ follows from (J.11) and (J.13). This is easy to understand because for $X = 0$ and $X = a$ we have $M_x = 0$, and hence, from (J.11) we obtain

$$W_{,XX} + \nu W_{,YY} = 0 .$$

However, $W_{,Y} \equiv 0$ for $X = 0, a$. Hence, we have $W_{,YY} = 0$ and therefore $\overline{\Delta}W = 0$ holds at the boundary. Further, for the inplane forces and the deformation at the boundaries the relations

$$N_X = -P, V = \frac{\nu PY}{hE} \qquad \text{for} \qquad X = 0, a; \ 0 \le Y \le b \quad (4.130)$$

$$N_Y = 0, U = -\frac{PX}{hE} \qquad \text{for} \qquad 0 \le X \le a; \ Y = 0, b \quad (4.131)$$

hold. The relations for V and U in (4.130) and (4.131) are immediate consequences of (J.10). From (J.18) and $N_X = -P$ follows the stress function F_0 in the flat plate to $F_0 = -PY^2/2$ where we do not care about linear and constant terms, because F appears at least in its second derivative in all expressions. Thus,

$$F_0 = -\frac{PY^2}{2}, \qquad W_0 \equiv 0, \qquad U_0 = -\frac{PX}{hE}, \qquad V_0 = \frac{\nu PY}{hE} \qquad (4.132)$$

is a solution of the boundary value problem (4.128), (4.129), (4.130) and (4.131) for all values of the parameters a, b and P. Hence, (4.132) is the solution which we are going to investigate for its stability under quasistatically increasing values of the load P.

We reformulate the boundary value problem (4.128)–(4.131) as a buckling problem in dimensionless variables and parameters. First, we introduce a new stress function $\tilde{F}$ defined by

$$\tilde{F} = F - F_0 \tag{4.133}$$

with F_0 given by (4.132). Thus, $\tilde{F}$ is related to the buckled configuration. Substituting (4.133) into (4.128), we find

$$\overline{\Delta}^2 \tilde{F} \; = \; \frac{1}{2} Eh[W, W]$$

$$\tag{4.134}$$

$$K\overline{\Delta}^2 W \; = \; -PW_{,XX} + [\tilde{F}, W] \; .$$

Analogous to (4.133), we define

$$\tilde{U} \; = \; U - U_0$$
$$\tilde{V} \; = \; V - V_0 \; . \tag{4.135}$$

Second, dimensionless variables and parameters

$$x = \frac{X}{b}, \qquad y = \frac{Y}{b}, \qquad \ell \equiv \frac{a}{b} \tag{4.136}$$

are introduced. The derivatives with respect to the new variables are

$$\frac{\partial}{\partial X} = \frac{\partial x}{\partial X}\frac{\partial}{\partial x} = \frac{1}{b}\frac{\partial}{\partial x}, \qquad \frac{\partial}{\partial Y} = \frac{1}{b}\frac{\partial}{\partial y} \; .$$

From (4.134) follows

$$\frac{1}{b^4}\Delta^2 \tilde{F} \; = \; -\frac{Eh}{2b^4}[W, W]$$

$$\tag{4.137}$$

$$\frac{K}{b^4}\Delta^2 W + \frac{P}{b^2}W_{,xx} \; = \; \frac{1}{b^4}[\tilde{F}, W].$$

Using two constants c_1 and c_2, which still have to be determined, two dimensionless variables w and f are defined by

$$W = c_1 w(x, y) \qquad \text{and} \qquad \tilde{F} = c_2 f(x, y). \tag{4.138}$$

If we substitute (4.138) into (4.137), we find

$$\Delta^2 f = -\frac{Ehc_1^2}{2c_2}[w, w]$$

$$\Delta^2 w + \frac{Pb^2}{K} w_{,xx} = \frac{c_2}{K}[f, w].$$

(4.139)

Setting

$$\frac{Ehc_1^2}{c_2} = 1, \qquad \frac{c_2}{K} = 1, \qquad p = \frac{Pb^2}{K},$$

(4.140)

results in

$$c_1^2 = \frac{h^2}{12(1 - \nu^2)}, \qquad c_2 = \frac{Eh^3}{12(1 - \nu^2)}.$$

(4.141)

To transfer $\tilde{U}$ and $\tilde{V}$ into dimensionless quantities we insert $\tilde{U} = c_3 u$ and $\tilde{V} = c_3 v$ into (J.10), where for N_x and N_y expressions corresponding to $(4.140)_3$ must be inserted. This results in

$$c_3 = \frac{h^2}{12(1 - \nu^2)b} = \frac{c_1^2}{b}$$

and

$$u(x, y) = \frac{b}{c_1^2}\tilde{U}$$

$$v(x, y) = \frac{b}{c_1^2}\tilde{V}.$$

(4.142)

Using the dimensionless variables $w(x, y)$ and $f(x, y)$, the von Karman plate equations (4.139) are

$$\Delta^2 f = -\frac{1}{2}[w, w]$$

$$\Delta^2 w + pw_{,xx} = [f, w] .$$

(4.143)

In (4.143) $\Delta = \partial^2/(\partial x^2) + \partial^2/(\partial y^2)$ is the Laplacian with respect to x and y. The boundary conditions (4.129)–(4.131) are

$$w = \Delta w = 0 \qquad \text{on} \qquad \partial\Omega$$

(4.144)

$$\begin{array}{llll} f_{,yy} = 0, & v = 0 & \text{for} & x = 0, \ell; \quad 0 \le y \le 1 \\ f_{,xx} = 0, & u = 0 & \text{for} & 0 \le x \le \ell, \quad y = 0, 1 . \end{array}$$

(4.145)

Using (J.10) and (J.18), the conditions (4.145) imply

$$f_{,xx} = f_{,yy} = 0 \qquad \text{on} \quad \delta\Omega \ . \tag{4.146}$$

The conditions (4.146) can be further simplified ([12]) to yield

$$f = \Delta f = 0 \qquad \text{on} \qquad \partial\Omega. \tag{4.147}$$

To show that (4.147) follows from (4.145), the tangential derivatives $f_{,xx}$ and $f_{,yy}$ are integrated along each edge of $\partial\Omega$. Of the eight constants of integration, four can be eliminated, because f is required to be continuous at the corners of the plate. Since in problem (4.143)–(4.145) at least the second derivative of $f(x,y)$ appears, $f(x,y)$ can only be determined to within an arbitrary linear function in x and y. Hence, three more constants are eliminated and only one remains. Thus, for example, for f on $\partial\Omega$

$$f(x,0) = f(\ell, y) = 0, \qquad f(0, y) = Cy, \qquad f(x, 1) = -\frac{C}{\ell}(x - \ell)$$

is obtained, where C is an arbitrary constant (Fig. 4.18). Now we show

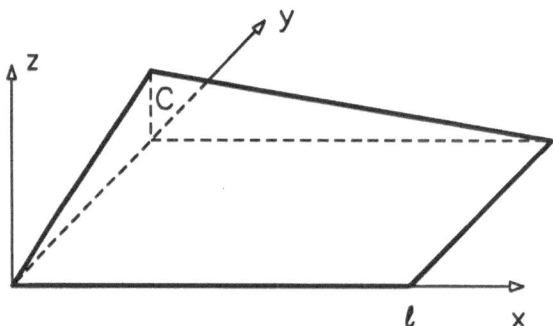

Figure 4.18. The stress function (4.148)

that if C is not equal to zero, the two boundary conditions for u and v in (4.145) are violated. If $C \neq 0$

$$w \equiv 0, \qquad f(x,y) = -\frac{C}{\ell}(x - \ell)y \tag{4.148}$$

is a solution of (4.143), (4.144) and of $f_{,yy} = f_{,xx} = 0$. However, (4.148) violates the conditions $u = 0, v = 0$ on $\partial\Omega$. This is an immediate consequence from (J.10)$_3$ in connection with (J.18) for N_{xy}, because the left hand side of (J.10)$_3$ vanishes but the right hand side does not. Therefore, C must be zero and, thus, $f = 0$ in (4.147).

The linear eigenvalue problem

The first step in the solution of the nonlinear buckling problem is the solution of the corresponding linear eigenvalue problem. We formulate it using the von Karman equations (4.143) with boundary conditions (4.144) and (4.147). To do so, (4.143) is written in a compact form as a functional equation $\mathbf{G}(\psi, p) = \mathbf{0}$ with $\psi = (w, f)^T, \mathbf{G} = (G_1, G_2)^T$ where

$$
\begin{aligned}
G_1(\psi, p) &= \Delta^2 f + \frac{1}{2}[w, w] = 0 \\
G_2(\psi, p) &= \Delta^2 w - [f, w] + p w_{,xx} = 0.
\end{aligned}
\tag{4.149}
$$

The Fréchet derivative of (4.149) at the equilibrium solution $\psi_0 = (w_0, f_0)^T$ applied to $\hat{\psi}$ is given by

$$
\mathbf{G}_{,\psi}(\psi_0, p)\hat{\psi} = \begin{pmatrix} \Delta^2 \hat{f} + [\hat{w}, w_0] \\ \Delta^2 \hat{w} - [f_0, \hat{w}] - [\hat{f}, w_0] + p \hat{w}_{,xx} \end{pmatrix}
\tag{4.150}
$$

where the infinitesimal variables are designated by $\hat{\psi} = (\hat{w}, \hat{f})^T$.

The expressions in (4.150) can be easily checked by writing out the nonlinear operator [,] in (4.149) according to (J.22). Let us do this for the first equation in (4.149). We substitute $w_0 + \hat{w}$ and $f_0 + \hat{f}$ into

$$
G_1(\psi, p) = \Delta^2 f + w_{,xx} w_{,yy} - w_{,xy}^2
$$

and retain only terms linear in $\hat{w}, \hat{f}$ to find

$$
G_{1,\psi}(\psi_0, p)\hat{\psi} = \Delta^2 \hat{f} + \hat{w}_{,xx} w_{0,yy} + w_{0,xx} \hat{w}_{,yy} - 2 w_{0,xy} \hat{w}_{,xy} \doteq \Delta^2 \hat{f} + [\hat{w}, w_0].
$$

The second expression in (4.150) can be similarly verified.

The undeflected state of the plate

$$
w_0 \equiv 0, \qquad f_0 \equiv 0
\tag{4.151}
$$

is an equilibrium solution of (4.149) for any value of p. Inserting (4.151) into (4.150) yields the following linear eigenvalue problem

$$
\Delta^2 \hat{f} = 0
\tag{4.152}
$$

$$
A\hat{w} = \Delta^2 \hat{w} + p \hat{w}_{,xx} = 0
\tag{4.153}
$$

with the boundary conditions on $\partial \Omega$

$$
\hat{w} = \Delta \hat{w} = 0
\tag{4.154}
$$

$$
\hat{f} = \Delta \hat{f} = 0.
\tag{4.155}
$$

From (4.152) and (4.155) follows that $\hat{f} \equiv 0$ and, hence, only (4.153) with (4.154) must be solved. This problem has nontrivial eigenfunctions

$$\hat{w} = \hat{w}_{mn}(x,y) = C_{mn} \sin \frac{m\pi x}{\ell} \sin n\pi y \qquad m,n = 1,2,\ldots \qquad (4.156)$$

if and only if

$$p = p_{mn} = \left(\frac{\pi}{\ell}\right)^2 \left(m + \frac{n^2\ell^2}{m}\right)^2, \qquad m,n = 1,2,\ldots \qquad (4.157)$$

which can be verified by inserting (4.156) into (4.153).

The C_{mn} are arbitrary constants and p_{mn} are the eigenvalues. From (4.157) the physically relevant minimum value for p_{mn} is obtained in any case for $n = 1$. This means that the plate always buckles with one half of a sine wave in the y direction (Fig. 4.20).

The corresponding number of waves in the x direction depends on the length ℓ of the plate. Rewriting (4.157) for $n = 1$ yields (the second index 1 for p_{m1} is omitted)

$$p = p_m = \frac{\pi^2(m^2 + \ell^2)^2}{m^2\ell^2}. \qquad (4.158)$$

In Fig. 4.19, p_m is drawn as a function of ℓ for different values of m. It

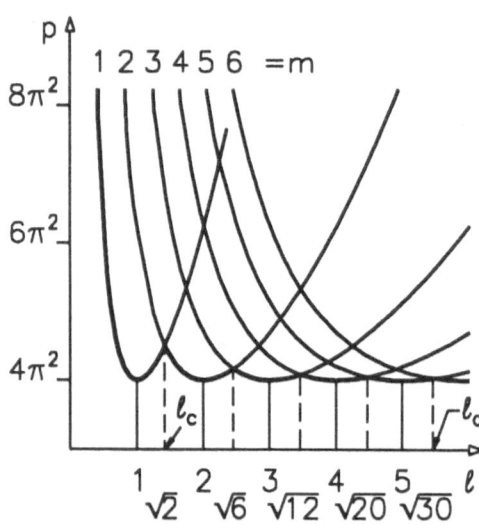

Figure 4.19. Eigenvalue curves p_m of the plate of Fig. 4.17 drawn as a function of plate length ℓ for different values of m. The envelope forms the stability boundary in the p, ℓ parameter plane

is obvious that for almost all values of ℓ, the plate buckles in a unique way at a simple eigenvalue. However, there exists a discrete, yet infinite set of special values of ℓ, designated by ℓ_c, for which double eigenvalues occur. These values ℓ_c are obtained from setting

$$p_m = p_{m+1}. \qquad (4.159)$$

Substituting (4.158) into (4.159) and setting $\ell = \ell_c$ in this equation, we find

$$\ell_c = \sqrt{m(m+1)} \,. \tag{4.160}$$

From (4.160), the first double eigenvalue is given for $m = 1$ at the length

$$\ell_c = \sqrt{2}. \tag{4.161}$$

The corresponding two buckling modes are those with one and two half waves in x direction, respectively (Fig. 4.20). In Fig. 4.21, the eigenvalue

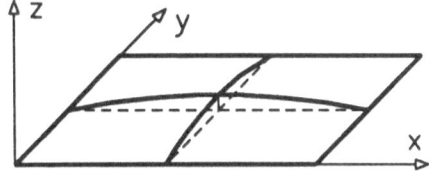

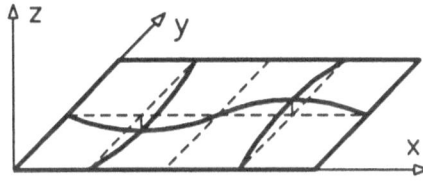

Figure 4.20. The two buckling modes $m = 1$ and $m = 2$ corresponding to the first double eigenvalue $\ell_c = \sqrt{2}$ of Fig. 4.19

curves p_m (4.158) are drawn again, but now with $1/\ell$ as parameter instead of ℓ as in Fig. 4.19. Fig. 4.21 will be discussed later with regard to shell buckling problems.

Ljapunov-Schmidt reduction to bifurcation equations at $\ell = \ell_c$

Before applying the method of Ljapunov-Schmidt to derive the bifurcation equations which are the amplitude equations of the critical modes at a double eigenvalue, a special peculiarity of the von Karman plate equations should be pointed out. This peculiarity is that it is possible to reduce the system of two equations (4.143) in two variables w, f to only one equation for w. From (4.143)$_1$, f can be formally expressed in the form

$$f = -\frac{1}{2}\Delta^{-2}[w, w]. \tag{4.162}$$

Substituting (4.162) into (4.143)$_2$

$$\Delta^2 w + p w_{,xx} - \left[-\frac{1}{2}\Delta^{-2}[w, w], w \right] = 0 \tag{4.163}$$

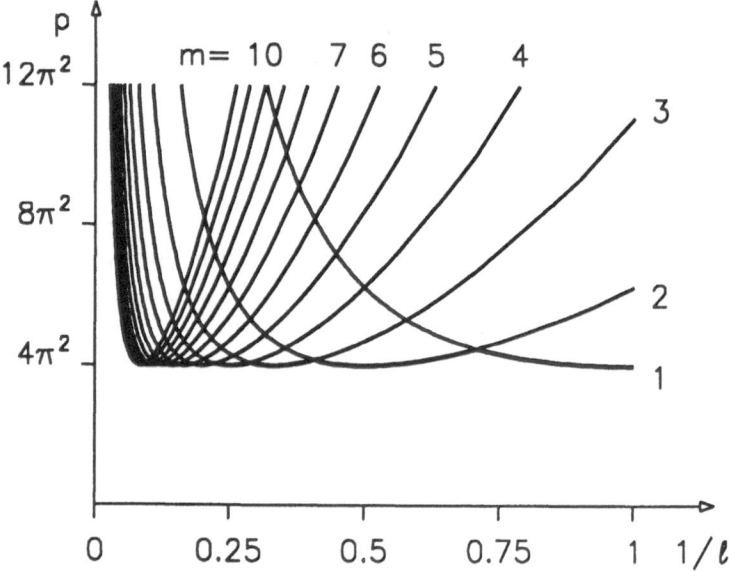

Figure 4.21. Stability boundary for the rectangular plate given by the envelope of p_m in the p, $1/\ell$ parameter space

is obtained. To arrive at (4.163), the inverse Δ^{-2} of the biharmonic operator Δ^2 must be calculated. In Appendix E it is indicated how to determine Δ^{-2}.

In carrying out the Ljapunov-Schmidt reduction we decompose w into $w = w_c + w_s$ according to (3.53), where $w_c \in N(A)$ and $w_s \in R(A^\star)$. Of course, A in (4.153) is a self-adjoint linear operator and we have $A^\star = A$ (Appendix C). The eigenspace of A at $\ell = \ell_c$ is two-dimensional. The two adjacent eigenmodes, spanning the kernel $N(A)$, are designated by w_m and w_{m+1}, according to (4.156) with $n = 1$ and the appropriate m from (4.160). Again, the index n is omitted in this notation. Thus at ℓ_c, we have

$$w(x, y) = rw_m(x, y) + sw_{m+1}(x, y) + w_s(x, y, r, s) . \qquad (4.164)$$

Here, r and s will be the variables (amplitudes of the critical modes) in the bifurcation equations. Before substituting (4.164) into (4.163) we recall from (3.57) that

$$w_s = O(|r|^2 + |s|^2). \qquad (4.165)$$

It will turn out that no second order terms appear in the bifurcation equations (this follows from the reflectional symmetry of this problem (Section 5.2), because downward and upward deflections of the plate are equivalent) and, further that third order terms will determine the behavior of the plate in the neighborhood of the bifurcation point. Hence, only the third order terms must be determined. The third order terms in (4.163), after inserting (4.164), are: r^3, rs^2, r^2s, s^3, rw_s^2, r^2w_s, sw_s^2, s^2w_s, rsw_s and w_s^3. It is immediately obvious that in this listing all terms including w_s are at least of fourth order because of (4.165). In fact, they are of fifth order as the bifurcation equations must satisfy the same symmetry requirements as the plate equations, which is shown below for the associated potential. Thus, we will obtain correct bifurcation equations with third order nonlinear terms even if (4.164) is simplified to

$$w(x,y) = rw_m(x,y) + sw_{m+1}(x,y) . \qquad (4.166)$$

Using (4.166) instead of (4.164) is nothing else than using the method of Ritz-Galerkin with the buckling modes as *ansatz-functions* instead of the method of Ljapunov-Schmidt. However, it must be pointed out that we have verified that up to third order terms both methods yield the same bifurcation equations.

It is convenient to introduce a new loading parameter λ measured from the buckling value p_m by

$$p = p_m + \tilde{\lambda} . \qquad (4.167)$$

Now we substitute (4.166) into (4.163). Making use of

$$
\begin{aligned}
[w, w] &= [rw_m + sw_{m+1}, rw_m + sw_{m+1}] \\
&= r^2[w_m, w_m] + s^2[w_{m+1}, w_{m+1}] + 2rs[w_m, w_{m+1}]
\end{aligned}
$$

the following expression for the bracket operator

$$
\begin{aligned}
[w, \Delta^{-2}[w, w]] &= [rw_m + sw_{m+1}, \\
&\quad \Delta^{-2}[rw_m + sw_{m+1}, rw_m + sw_{m+1}]] \\
&= r^3[w_m, \Delta^{-2}[w_m, w_m]] + \\
&\quad + r^2s\{[w_{m+1}, \Delta^{-2}[w_m, w_m]] + \\
&\quad\quad + 2[w_m, \Delta^{-2}[w_m, w_{m+1}]]\} + \qquad (4.168)\\
&\quad + rs^2\{[w_m, \Delta^{-2}[w_{m+1}, w_{m+1}]] + \\
&\quad\quad + 2[w_{m+1}, \Delta^{-2}[w_m, w_{m+1}]]\} + \\
&\quad + s^3[w_{m+1}, \Delta^{-2}[w_{m+1}, w_{m+1}]]
\end{aligned}
$$

is obtained. Moreover, with (4.167), we find

$$pw_{,xx} = (p_m + \tilde{\lambda})(rw_{m,xx} + sw_{m+1,xx} + w_{s,xx}) . \qquad (4.169)$$

From (4.163) we obtain the two bifurcation equations by projection onto the eigenspace. The projections are denoted by $(\cdot, w_m)$ and $(\cdot, w_{m+1})$ where

$$(u, v) = \int_\Omega uv d\Omega. \qquad (4.170)$$

Before writing down the bifurcation equations up to and including third order terms, we make use of some simple symmetry arguments in order to reduce the number of terms in the bifurcation equations from the outset (see Chapter 5 for more details). Here it is essential to note that, since a conservative system is treated, it must be possible to derive the bifurcation equations from the potential energy $V(r, s)$ of the plate. The potential, however, must satisfy certain symmetry requirements, namely ([122])

$$V(r, s) = V(-r, s) = V(r, -s) = V(-r, -s), \qquad (4.171)$$

which are obvious, since the energy must be invariant no matter whether the deflection is upward or downward or whether a reflexion of the deflection surface with respect to the line $x = \ell/2$ is made. A potential of fourth order obeying (4.171) has the following form

$$V(r, s) = \frac{1}{4}Br^4 + \frac{1}{2}Cr^2s^2 + \frac{1}{4}Ds^4 - \frac{1}{2}\alpha\tilde{\lambda}r^2 - \frac{1}{2}\beta\tilde{\lambda}s^2 . \qquad (4.172)$$

Here, B, C, D, α, β are constants and $\tilde{\lambda}$ is defined by (4.167). From (4.172), the so-called one-parameter ($\tilde{\lambda}$) bifurcation equations or the bifurcation equations of the perfect system follow to

$$\frac{\partial V}{\partial r} = Br^3 + Crs^2 - \alpha\tilde{\lambda}r = 0$$

$$\qquad (4.173)$$

$$\frac{\partial V}{\partial s} = Cr^2s + Ds^3 - \beta\tilde{\lambda}s = 0.$$

Thus, only the coefficients B, C, D, α and β must be calculated, which follow according to $(3.52)_2$ from (4.168) and (4.169) by projection onto the eigenspace spanned by w_m and w_{m+1}. Hence, we obtain

$$B = \frac{1}{2}([w_m, \Delta^{-2}[w_m, w_m]], w_m)$$

$$C = \left(\left\{\frac{1}{2}[w_{m+1}, \Delta^{-2}[w_m, w_m]] + [w_m, \Delta^{-2}[w_m, w_{m+1}]]\right\}, w_m\right)$$

$$= \left(\left\{\frac{1}{2}[w_m, \Delta^{-2}[w_{m+1}, w_{m+1}]] + [w_{m+1}, \Delta^{-2}[w_m, w_{m+1}]]\right\}, w_{m+1}\right)$$

$$\qquad (4.174)$$

$$D = \frac{1}{2}([w_{m+1}, \Delta^{-2}[w_{m+1}, w_{m+1}]], w_{m+1})$$

$$\alpha = -(w_{m,xx}, w_m)$$

$$\beta = -(w_{m+1,xx}, w_{m+1}).$$

We note that the passive variable w_s in (4.169) does not contribute to the expressions for α and β, because w_s is orthogonal to w_m and w_{m+1}, and hence, the inner product vanishes. For the numerical calculations of (4.174), the following two identities

$$(\Delta^{-2}u, v) \;=\; (\Delta^{-1}u, \Delta^{-1}v)$$

(4.175)

$$([u, v], w) \;=\; ([u, w], v)$$

are useful.

To verify (4.175)$_1$, use is made of Green's formula for Δ^2 ([38] p. 543)

$$\iint_{\Omega}(u\Delta^2 v - \Delta u \Delta v)dx\,dy = \int_{\partial\Omega}\left(u\frac{\partial \Delta v}{\partial n} - \Delta v\frac{\partial u}{\partial n}\right)ds.$$

(4.176)

Using the boundary conditions (4.144) it follows from (4.176)

$$\iint_{\Omega} v\Delta^2 u\,dx\,dy = \iint_{\Omega}\Delta u\Delta v\,dx\,dy = \iint_{\Omega} u\Delta^2 v\,dx\,dy.$$

(4.177)

Replacing u and v in (4.177) by $\Delta^{-2}u$ and $\Delta^{-2}v$, respectively, gives (4.175)$_1$.

To verify (4.175)$_2$, the identity ([14] p. 98)

$$(u_{,yy}v_{,x} - u_{,xy}v_{,y})_{,x} + (u_{,xx}v_{,y} - u_{,xy}v_{,x})_{,y}$$

(4.178)

$$= u_{,yy}v_{,xx} - u_{,xy}v_{,xy} + u_{,xx}v_{,yy} - u_{,xy}v_{,xy} = [u, v]$$

is used. Multiplication of the first line of (4.178) by w, integration over Ω, integration by parts and making use of the boundary conditions results in

$$\iint_{\Omega}\{(u_{,yy}v_{,x} - u_{,xy}v_{,y})w_{,x} + (u_{,xx}v_{,y} - u_{,xy}v_{,x})w_{,y}\}dx\,dy$$

$$= \iint_{\Omega}\{(u_{,yy}w_{,x} - u_{,xy}w_{,y})v_{,x} + (u_{,xx}w_{,y} - u_{,xy}w_{,x})v_{,y}\}dx\,dy\;,$$

from which (4.175)$_2$ follows immediately.

Application of (4.175) allows us to rewrite, for example, the expression for B in (4.174) in the following way

$$\begin{aligned}B &= \frac{1}{2}([w_m, \Delta^{-2}[w_m, w_m]], w_m)\\[4pt]&= \frac{1}{2}([w_m, w_m], \Delta^{-2}[w_m, w_m])\\[4pt]&= \frac{1}{2}(\Delta^{-1}[w_m, w_m], \Delta^{-1}[w_m, w_m]) = \frac{1}{2}\|\Delta^{-1}[w_m, w_m]\|^2.\end{aligned}$$

(4.179)

Here Δ^{-1} is the inverse of the Laplacian with zero boundary conditions

(Appendix E). The specific calculations necessary to obtain the numerical values of the coefficients B, C, D, α, β, are given as examples for B and D at the end of this Section.

It is shown in [94] that (4.163) can be derived as the Fréchet derivative of the energy expression

$$V(w) = \frac{1}{2}(\Delta w, \Delta w) + \frac{1}{8}(\Delta^{-2}[w, w], [w, w]) - \frac{1}{2}p(w_{,x}, w_{,x}), \qquad (4.180)$$

where the inner product is given by (4.170). The Fréchet derivative $V'(w)$ of V is the linear mapping operating on v

$$V'(w)v = (\Delta w, \Delta v) + \frac{1}{4}(\Delta^{-2}[w, v], [w, w]) +$$

$$+ \frac{1}{4}(\Delta^{-2}[w, w], [w, v]) - p(w_{,x}, v_{,x}).$$

From $V'(w)v = 0$ follows, using (4.175) and integration by parts, that

$$(\Delta^2 w, v) + \frac{1}{2}([\Delta^{-2}[w, w], w], v) + p(w_{,xx}, v) = 0 ,$$

which yields (4.163) in a similar rasoning as on p. 36 for the rod, since v is an arbitrary function. Rewriting (4.180) in engineering notation, we see that the energy can be given by

$$V = \frac{1}{2} \int_0^1 \int_0^\ell \{(\Delta w)^2 - p w_{,x}^2 + (\Delta f)^2\} dx dy, \qquad (4.181)$$

where we have used $(4.175)_1$ and (4.162).

Finally, we calculate the numerical values of the coefficients B, C, D for the first double eigenvalue at $\ell = \ell_c = \sqrt{2}$, that is, $m = 1$ and 2. The calculation follows the exposition given in the appendix of [28]. The orthonormal system of eigenfunctions, where we still keep m and ℓ_c variable,

$$w_m = \frac{2}{\sqrt{\ell_c}} \sin \frac{m\pi x}{\ell_c} \sin \pi y \qquad (4.182)$$

is substituted into (J.22). As before, we write w_m and omit the second index in w_{m1}. This results in

$$[w_m, w_m] = \frac{8m^2\pi^4}{\ell_c^3} \left(\sin^2 \frac{m\pi x}{\ell_c} \sin^2 \pi y - \cos^2 \frac{m\pi x}{\ell_c} \cos^2 \pi y \right)$$

$$= -\frac{4m^2\pi^4}{\ell_c^3} \left(\cos \frac{2m\pi x}{\ell_c} + \cos 2\pi y \right). \qquad (4.183)$$

We expand (4.183) into a Fourier series in the functions (4.182) and obtain

$$\cos \frac{2m\pi x}{\ell_c} = \sum_{p,q\,\text{odd}} a_{pq}^m \sin \frac{p\pi x}{\ell_c} \sin q\pi y \qquad (4.184)$$

$$\cos 2\pi y = \sum_{p,q\,\text{odd}} b_{pq} \sin \frac{p\pi x}{\ell_c} \sin q\pi y \qquad (4.185)$$

with the coefficients (in a_{pc}^m m has the meaning of an index)

$$a_{pq}^m = \frac{16p}{\pi^2 q(p^2 - 4m^2)} \qquad \text{and} \qquad b_{pq} = \frac{16q}{\pi^2 p(q^2 - 4)} \,.$$

We can calculate $\Delta^{-1}[w_m, w_m]$ in two different ways. Either we insert
(4.183) with (4.184) and (4.185) into (E.8) and divide by λ_{pq} according
to (E.26). Or we set

$$f(\xi, \eta) = [w_m, w_m](\xi, \eta)$$

with $[w_m, w_m](\xi, \eta)$ given by (4.183) into

$$\Delta^{-1}[w_m, w_m] = \iint_\Omega G(x, y, \xi, \eta) f(\xi, \eta) d\xi d\eta$$

where G is given by (E.27). Taking the norm we find

$$\frac{1}{2}\|\Delta^{-1}[w_m, w_m]\|^2$$

$$= \frac{8m^4}{\ell_c^6} \sum_{p,q\,\text{odd}} \frac{\ell_c}{4} \pi^4 (a_{pq}^m + b_{pq})^2 \left(\frac{p^2}{\ell_c^2} + q^2\right)^{-2}. \qquad (4.186)$$

Setting in (4.186) $m = 1$ or 2 and $\ell_c = \sqrt{2}$ yields according to (4.174)
and (4.179) B or D, respectively. The calculation of C can be done
similarly and is given in [28]. One conclusion that may be drawn from
this calculation of the coefficients in the bifurcation equations is that, in
general, the summation of infinite series is necessary.

The bifurcation equations (4.173) can be simplified further by the
following scalings

$$r = \sqrt{\frac{\alpha}{B}} q_1 \qquad \text{and} \qquad s = \sqrt{\frac{\beta}{D}} q_2 \qquad (4.187)$$

to

$$\begin{aligned} q_1^3 + \mu_1 q_1 q_2^2 - \tilde{\lambda} q_1 &= 0 \\ \mu_2 q_1^2 q_2 + q_2^3 - \tilde{\lambda} q_2 &= 0 \end{aligned} \qquad (4.188)$$

where

$$\mu_1 = \frac{C\beta}{\alpha D} \quad \text{and} \quad \mu_2 = \frac{C\alpha}{B\beta}. \qquad (4.189)$$

The parameters μ_1 and μ_2 are called *modal parameters*. They depend on the boundary conditions, the form of the plate and the type of the in-plane loading and allow to classify qualitatively different bifurcation behaviors of the plate ([122]). This will be explained in Section 6.6.1 when the solutions of (4.188) are discussed.

4.3.5   The pattern formation problem: buckling of complete spherical shells

A very interesting and also important problem in many fields of the sciences is the phenomenon of pattern appearance. That is, the formation of structured patterns emanating from an unstructured configuration at a critical parameter value for a quasistatic variation of one or several parameters. Pattern formation is obviously related to symmetry breaking bifurcations, which are treated more extensively in Chapter 5. In this section we are going to study three examples of buckling of thin elastic spherical shells under external pressure. We only want to indicate some of the difficulties which are met in treating this phenomenon. We have chosen shell buckling because this is a practically important subject and has some theoretically interesting properties leading to severe complications in the application of the bifurcation analysis developed in Chapter 3. The great complexity of shell buckling problems follows from their high degeneracy, expressed by the high multiplicity of the critical eigenvalue. Furthermore, it will turn out to be very important which type of equations are used to describe the behavior of the shell.

(a)   Shallow shell theory

In a first attempt to deal with the problem of pattern formation, the simplest nonlinear shell model, namely, the shallow shell equations (J.24) are taken as the governing equations. At first glance, this model does not seem to be a good choice to treat buckling of a complete sphere, because shallow shell theory is applicable only for a sector of the sphere. This follows from the fact that for these shallow shell equations a cartesian coordinate system x, y (Fig. J.1) is used. However, in [63] it is shown that the application of this representation to the complete sphere is consistent, if the characteristic buckling wave lengths are small compared to the shell radius. Then, it is possible to choose a shallow section of the shell surface in which the buckling pattern is duplicated many times. By shallow, it is understood that the slopes of the surface, measured from the section base, are small, and thus, the shallow shell approximations are valid. With shell radius a and shell stiffness K according to (J.12), the equations (J.24) are

$$K\Delta^2 W + \frac{1}{a}\Delta F - [F, W] = -p$$

$$\frac{1}{Eh}\Delta^2 F - \frac{1}{a}\Delta W + \frac{1}{2}[W, W] = 0 \tag{4.190}$$

where the operator $[\ ,\ ]$ is given by (J.22). In equation (4.190) E is Young's modulus, h is the constant shell thickness and p is the uniform external pressure.

As first step, the fundamental solution (W_0, F_0) is calculated. Basically, this is the membrane stress state in the concentrically contracted, perfectly spherical shell, which can be calculated from a simple equilibrium consideration of a semi-spherical shell to (see any introductory text into structural mechanics, for example [167])

$$N_{x0} = N_{y0} = -\frac{1}{2}pa. \tag{4.191}$$

The corresponding inward radial displacement W_0 follows to ([167])

$$W_0 = -(1-\nu)\frac{pa^2}{2Eh}. \tag{4.192}$$

From (4.191) and (J.18) the stress function F_0 is calculated as

$$F_0 = -\frac{1}{4}(x^2 + y^2)pa. \tag{4.193}$$

We perform the linearization of (4.190) about W_0 and F_0 in an engineering manner and set

$$W = W_0 + w = -(1-\nu)\frac{pa^2}{2Eh} + w \tag{4.194}$$

and

$$F = F_0 + f = -\frac{1}{4}(x^2 + y^2)pa + f. \tag{4.195}$$

The variables w and f are zero prior to buckling and are considered to be small after buckling so that the resulting equations can be linearized with respect to w and f. Inserting (4.194) and (4.195) into (4.190) yields

$$K\Delta^2 w + \frac{1}{a}\Delta f + \frac{1}{2}pa\Delta w = 0$$

$$\frac{1}{Eh}\Delta^2 f - \frac{1}{a}\Delta w = 0. \tag{4.196}$$

Solutions to this homogeneous eigenvalue problem can be given as the product of cosine functions, such as

$$w = C \cos \left(k_x \frac{x}{a} \right) \cos \left(k_y \frac{y}{a} \right)$$

$$f = D \cos \left(k_x \frac{x}{a} \right) \cos \left(k_y \frac{y}{a} \right)$$

(4.197)

where C and D are arbitrary constants and k_x and k_y wave numbers to be determined. The eigenvalue for p associated with this choice is obtained by substituting (4.197) into (4.196). This yields a linear homogeneous system of algebraic equations for C and D, which is

$$C \left\{ \frac{K}{a^4} (k_x^2 + k_y^2) - \frac{p}{2a} \right\} - D \frac{1}{a^3} = 0$$

$$C \frac{1}{a^3} + D \frac{1}{Eha^4} (k_x^2 + k_y^2) = 0.$$

(4.198)

Non-zero solutions for C, D are obtained if the determinant of the coefficient matrix vanishes. This yields the eigenvalue

$$p = \frac{2Eh}{a} \{ (k_x^2 + k_y^2)^{-1} + \delta^2 (k_x^2 + k_y^2) \}$$

(4.199)

where

$$\delta^2 = \left(\frac{h}{a} \right)^2 \frac{1}{12(1 - \nu^2)}.$$

(4.200)

The dimensionless quantity δ describes the "geometry" of the shell. To determine the classical buckling pressure p_c for the complete sphere from (4.199), we define a wave number k by

$$k^2 = k_x^2 + k_y^2.$$

(4.201)

That value of k^2 for which p, given by (4.199), takes its minimum value, must be determined. If we consider, for a moment, k^2 to be a continuous quantity we find from the condition $\partial p / \partial (k^2) = 0$

$$k^2 = \frac{2a}{h} \sqrt{3(1 - \nu^2)} = \frac{1}{\delta} ,$$

(4.202)

where use of (4.200) has been made. Inserting (4.202) into (4.199), we obtain

$$p_c = 2E \left(\frac{h}{a} \right)^2 \frac{1}{\sqrt{3(1 - \nu^2)}} = \frac{4Eh\delta}{a} = 4E\delta^2 \sqrt{12(1 - \nu^2)}.$$

(4.203)

It will be shown below that the critical pressure p_c given by (4.203) which is obtained on the basis of shallow shell theory is identical to the expression obtained from equations valid on the full sphere.

However, a remarkable fact concerning the buckling pattern must be noted. There is no specification for the buckling pattern, because from (4.202) only the value k can be obtained, but not k_x and k_y. This is because equation (4.201) can still be satisfied by an infinite number of pairs (k_x, k_y) lying on the critical circle. This property can also be expressed by writing $k_x = k \cos \varphi$ and $k_y = k \sin \varphi$ where φ remains undetermined. Thus, an infinite dimensional critical eigenvalue is given. Hence, a straightforward nonlinear analysis by a reduction of the shell equations to algebraic bifurcation equations, as it was done for the plate using the method of Ljapunov-Schmidt is not possible here.

(b)   Nonlinear shell equations on the complete sphere

In this section, we follow [72]. It will be shown that the unpleasant situation of an eigenvalue with infinite multiplicity, found above, is due to the fact that the shallow shell equations with an cartesian coordinate system were used. Replacing them by a set of equations valid for the complete sphere, the physically much more realistic result is obtained that the multiplicity of the critical eigenvalue is finite and depends on the ratio h/a or δ, respectively.

It is possible to write down the shell equations for the complete sphere as a set of two nonlinear partial differential equations for a deformation variable W (curvature) and a stress function F ([72]). They are as follows

$$\frac{1}{Eh} \left(\tilde{\Delta}^2 F + \frac{2}{a^2} \tilde{\Delta} F \right) - \frac{1}{a} \left(\tilde{\Delta} W + \frac{2}{a^2} W \right) +$$

$$+ \frac{1}{2} \left(\tilde{\varepsilon}^{\alpha\lambda} \tilde{\varepsilon}^{\beta\mu} W \Big|_{\lambda\mu} W_{,\alpha} \right) \Big|_\beta +$$

$$+ \frac{1}{2a^2} \left(\tilde{g}^{\alpha\beta} W_{,\alpha} W_{,b} + 2W \tilde{\Delta} W + \frac{2}{a^2} W^2 \right) = 0$$

$$\text{(4.204)}$$

$$K \left(\tilde{\Delta}^2 W + \frac{2}{a^2} \tilde{\Delta} W \right) + \frac{1}{a} \left(\tilde{\Delta} F + \frac{2}{a^2} F \right) -$$

$$- \left(\tilde{\varepsilon}^{\alpha\lambda} \tilde{\varepsilon}^{\beta\mu} F \Big|_{\lambda\mu} W_{,\alpha} \right) \Big|_\beta -$$

$$- \frac{1}{a^2} \left(\tilde{g}^{\alpha\beta} W_{,\alpha} F_{,\beta} + (F \tilde{\Delta} W + W \tilde{\Delta} F) + \frac{2}{a^2} FW \right) = -p.$$

The middle surface of the undeformed spherical shell is described by $r = a$, $0 \leq \vartheta \leq \pi$ and $0 \leq \varphi \leq 2\pi$. The ratio $1/a^2$ is the Gaussian curvature of the undeformed sphere and $\tilde{\Delta}$ denotes the Laplace-Beltrami operator relative to the metric tensor $\tilde{g}_{\alpha\beta}$, that is,

$$\tilde{\Delta}w = \tilde{g}^{\alpha\beta}w\big|_{\alpha\beta} .$$ (4.205)

Here $\tilde{g}_{\alpha\beta}$ and $\tilde{g}^{\alpha\beta}$ denote the covariant and contravariant metric tensors given by (F.12) and (F.14). Furthermore, $\tilde{\varepsilon}_{\alpha\beta}$ and $\tilde{\varepsilon}^{\alpha\beta}$ are the alternating tensors. For example, $\tilde{\varepsilon}_{\alpha\beta}$ is given by

$$\tilde{\varepsilon}_{\alpha\beta} = \begin{pmatrix} 0 & \sqrt{g} \\ -\sqrt{g} & 0 \end{pmatrix} = \begin{pmatrix} 0 & a^2\sin\vartheta \\ -a^2\sin\vartheta & 0 \end{pmatrix}$$ (4.206)

where $g = \det\tilde{g}_{\alpha\beta} = a^4\sin^2\vartheta$. The symbol "|" followed by additional subscripts denotes covariant differentiation relative to the metric tensor $\tilde{g}_{\alpha\beta}$ and the usual summation convention is used with Greek indices ranging over the values 1 and 2. The raising and lowering of indices is performed relative to the metric tensors $\tilde{g}_{\alpha\beta}$ and $\tilde{g}^{\alpha\beta}$.

In Appendix F the derivation of the above relations is given. It is convenient to introduce new variables $\tilde{f}$ and $\tilde{w}$ by

$$F = K\tilde{f}$$

and

$$W = \frac{h\tilde{w}}{2\sqrt{3(1-\nu^2)}}$$

in (4.204) and to multiply the resulting equations by a^4. This yields

$$\Delta^2\tilde{f} + 2\Delta\tilde{f} - \frac{1}{\delta}(\Delta\tilde{w} + 2\tilde{w}) + \frac{1}{2}\{\tilde{w},\tilde{w}\} = 0$$

$$\Delta^2\tilde{w} + 2\Delta\tilde{w} + \frac{1}{\delta}(\Delta\tilde{f} + 2\tilde{f}) - \{\tilde{f},\tilde{w}\} = -2\frac{\overline{\lambda}}{\delta}$$ (4.207)

where δ is given by (4.200). The loading parameter $\overline{\lambda}$ is given by

$$\overline{\lambda} = \frac{6pa^3(1-\nu^2)}{Eh^3} .$$ (4.208)

In (4.207), Δ denotes the Laplace-Beltrami operator on the unit sphere. Furthermore, the bracket operator $\{\,,\,\}$ is defined by

$$\{u,v\} = \left(\varepsilon^{\sigma\kappa}\varepsilon^{\beta\mu}u\big|_{\kappa\mu}\,v_{,\sigma}\right)\Big|_{\beta} + g^{\sigma\mu}u_{,\sigma}v_{,\mu} + u\Delta v + v\Delta u + 2uv$$ (4.209)

where $g^{\alpha\beta}$ and $\varepsilon^{\alpha\beta}$ are the metric and alternating tensors with respect to the unit sphere.

Using the Ricci identity ([72]), (4.209) can also be written as

$$\{u, v\} = \Delta u \Delta v - g^{\sigma\kappa} g^{\beta\mu} u\Big|_{\sigma\beta} v\Big|_{\kappa\mu} + u\Delta v + v\Delta u + 2uv. \qquad (4.210)$$

Spherical harmonics (Appendix F) form a natural basis for this problem. Note that the spherical harmonics of degree zero and degree one span the kernel of the linear operator $\Delta^2 + 2\Delta$, which appears in (4.207). Thus, it is natural to set

$$\tilde{f} = f_0 + f_1 + f \qquad \text{and} \qquad \tilde{w} = w_0 + w_1 + w, \qquad (4.211)$$

where f_0 and w_0 are spherical harmonics of degree zero (that is, constants), f_1 and w_1 are spherical harmonics of degree one, and f and w are functions to be determined. To start with the constant terms, w_0, f_0 are substituted into (4.207) and with (4.210) the following two equations

$$-(\Delta w_0 + 2w_0) + \frac{\delta}{2}\{w_0, w_0\} = -2w_0 + \delta w_0^2 = 0 \qquad (4.212)$$

$$(\Delta f_0 + 2f_0) - \delta\{f_0, w_0\} = 2f_0 - 2\delta f_0 w_0 = -2\overline{\lambda}$$

are obtained. System (4.212) has two solutions: $(f_0, w_0) = (-\overline{\lambda}, 0)$ and $(f_0, w_0) = (\overline{\lambda}, 2/\delta)$. The physically relevant solution is the first one, because this one reduces to zero, when the load $\overline{\lambda}$ is zero. With this choice of f_0, w_0 we substitute (4.211) into the system (4.207) to obtain

$$\Delta^2 f + 2\Delta f - \frac{1}{\delta}(\Delta w + 2w) + \frac{1}{2}\{w, w\} = 0 \qquad (4.213)$$

$$\Delta^2 w + 2\Delta w + \frac{1}{\delta}(\Delta f + 2f) + \overline{\lambda}(\Delta w + 2w) - \{f, w\} = 0.$$

It is interesting to note that the terms in (4.207) containing f_1 and w_1 do not appear in (4.213) and that the contribution of f_1 and w_1 to the stresses and curvature changes are also zero. Thus, in the model being used here the spherical harmonics of first degree play no role either in solving (4.207) or in determining the stresses and changes of curvature. This follows from the fact that any spherical harmonic $Y_{\ell m}(\vartheta, \varphi) = C_m P_\ell^m(\vartheta) e^{im\varphi}$ satisfies (Appendix F)

$$\Delta Y_{\ell m} = -\ell(\ell+1)Y_{\ell m} \qquad \ell = 0, 1, 2, \ldots \qquad (4.214)$$

At the end of Appendix F it is shown that (4.210) vanishes if (4.214) is inserted taking $\ell = 1$ and making use of (4.205).

Finally, the linear eigenvalue problem follows from (4.213) to

$$\Delta^2 f + 2\Delta f - \frac{1}{\delta}(\Delta w + 2w) = 0 \qquad (4.215)$$

$$\Delta^2 w + 2\Delta w + \frac{1}{\delta}(\Delta f + 2f) + \overline{\lambda}(\Delta w + 2w) = 0.$$

We substitute $f = C_m Y_{\ell m}$ and $w = D_m Y_{\ell m}$ into (4.215) and use

$$\mu_\ell = -\ell(\ell + 1) \tag{4.216}$$

to find

$$\begin{aligned}
\delta(\mu_\ell^2 + 2\mu_\ell)C_m - (\mu_\ell + 2)D_m &= 0 \\
(\mu_\ell + 2)C_m + \delta[\mu_\ell^2 + 2\mu_\ell + \overline{\lambda}(\mu_\ell + 2)]D_m &= 0.
\end{aligned} \tag{4.217}$$

For nontrivial solutions C_m and D_m, the determinant of the coefficient matrix in (4.217) must be zero. This yields the required value of $\overline{\lambda}$ to

$$\overline{\lambda} = -\frac{(1/\delta)^2 + \mu_\ell^2}{\mu_\ell} = -(\mu_\ell + \frac{1}{\delta^2 \mu_\ell}). \tag{4.218}$$

The dimension of the corresponding eigenspace is $(2\ell + 1)$ (see (F.47)).

If, for the moment, μ_ℓ is treated as a continuous variable, then the minimum value of $\overline{\lambda}$ as a function of μ_ℓ follows from

$$\frac{\partial \overline{\lambda}}{\partial \mu_\ell} = 0 = 1 - \frac{1}{(\delta \mu_\ell)^2} .$$

According to (4.216), the root with the negative sign

$$\mu_\ell = -\frac{1}{\delta} \tag{4.219}$$

must be taken. Substituting (4.216) and (4.200) into (4.219) yields

$$\ell(\ell + 1) = \frac{1}{\delta} = \frac{a}{h} 2\sqrt{3(1 - \nu^2)}. \tag{4.220}$$

Since δ according to (4.200) is a given constant, (4.220) gives the relationship between the ratio h/a and the finite multiplicity of the critical eigenvalue. This is one basic difference compared to shallow shell theory, where an infinite-dimensional eigenspace was found. Finally, using (4.219), it follows from (4.218) that

$$\overline{\lambda}_{\min} = \frac{2}{\delta} = 4\sqrt{3(1 - \nu^2)}\frac{a}{h} . \tag{4.221}$$

This value becomes exactly p_c given by (4.203) if (4.221) is substituted into (4.208).

According to (F.47) the multiplicity of (4.221) is $2\ell + 1$, where ℓ is obtained from (4.220). Let us consider as a typical example a metallic spherical shell with $a/h \simeq 1000$ and $\nu = 0.3$. From (4.220) we find $\ell \simeq 57$ and the multiplicity of the eigenvalue is $2\ell + 1 = 115$.

The buckling pattern is given by

$$w(\vartheta, \varphi) = \sum_{m=0}^{\ell} (C_m \cos m\varphi + D_m \sin m\varphi) P_\ell^m(\cos \vartheta). \qquad (4.222)$$

It is intuitively clear that such a high degeneracy, as it occurs here for a spherical shell, will create great difficulties in a nonlinear bifurcation analysis. Therefore, it is necessary to reduce further the dimension of the space of eigenfunctions. This can be done making use of results of buckling experiments. From the results of buckling experiments ([15]) it seems to be justified to assume that buckling patterns have certain symmetries. By postulating these symmetry properties for the pattern, one can reduce the number of buckling modes considerably by the use of group representation theory ([72], [119], see also Chapter 5). In general, however, as is shown in [72], there still remains a set of about 10 bifurcation equations. We will show in Chapter 6, where the bifurcation equations are analysed, that this is still a number of bifurcation equations much too large to be analysed systematically. For this reason, in the next section, a much simpler case of shell buckling will be treated, namely, the buckling of a spherical shell subject to axisymmetric deformations. Before dealing with this case, we make three comments:

Comment 1: There exists an interesting analogy between the elastic buckling problem of the spherical shell and the stability problem of a viscous fluid layer heated from below, usually called the *Bénard problem* ([24], [20], [68]). If one considers a plane fluid layer then a mathematically similar situation arises, as is found for the shallow shell equations, and again the smallest eigenvalue analogously to (4.201) has infinite multiplicity. On the other hand, for a spherical fluid shell heated from inside, the degeneracy is finite and depends on the thickness of the fluid layer. The Bénard problem compared to the shell buckling problem has the nicer property from a theoretical point of view that also thick-walled fluid shells are physically meaningful. Therefore, after application of group theoretic arguments the set of bifurcation equations can become quite small, such that an investigation as is described in Chapter 6 may be possible [46].

Comment 2: Post-buckling of thin-walled shells is characterized by the occurence of solutions with boundary layers (see p. 138). Hence, a classical bifurcation analysis is only a first step in the understanding of the shell behavior as is explained in more detail in the following section.

Comment 3: The problem of the occurence of eigenvalues with finite or infinite multiplicity can be, in some sense, quantified by introducing the notion of *aspect ratio* ([1]). This term refers to the ratio of a length scale

of the dimension of the system to a length scale which is characteristic
for the phenomenon. This can be best explained by the example of the
convection problem of the spherical fluid layer just mentioned. Here the
aspect ratio is proportional to the ratio a/h, where a is the outer radius
of the shell type layer and h the thickness of the layer (Fig. 4.22). The

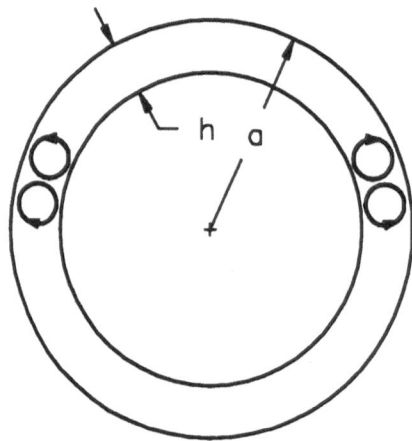

Figure 4.22. The aspect ratio a/h relates the length a characteristic for the
size of the system to the length h car-
acteristic for the dynamic phenomenon

behavior of the problem (convective motion in form of rolls or cells) can
be measured by a scale proportional to h, whereas the dimension of the
system is measured by a. Hence, if the layer becomes very thin, the
aspect ratio becomes very large, finally approaching for $h/a = 0$ the
plane layer for which an eigenvalue with infinite multiplicity occurs. On
the other hand, for example, if $h/a = 1$ (that is, for a point heat source
in a fluid sphere) the multiplicity is finite, small and equal to three. If,
for example, $h/a = 0.7$ the multiplicity is 5, and so on ([25]).

(c)   Shell equations subject to axisymmetrical deformation

In this case, the buckling pattern depends only on ϑ (Fig. 4.23), and
therefore, the governing differential equations are ordinary differential
equations. In [116] and in Appendix K is shown that similar to (4.190)
and (4.204) again two equations in two variables, namely, a stressfunction
$\psi(\vartheta)$ and the tangent rotation $\beta(\vartheta)$ (Fig. 4.23) can be obtained. These
equations are given by (K.38). We are going to study the axisymmetrical
buckling behavior under a uniformly distributed external pressure p.

First, we calculate the trivial solution, representing the contracted
perfectly spherical shell

$$
\begin{aligned}
\beta_0(\vartheta) &= 0 \\
\psi_0^\star(\vartheta) &= -\lambda \sin 2\vartheta
\end{aligned}
\tag{4.223}
$$

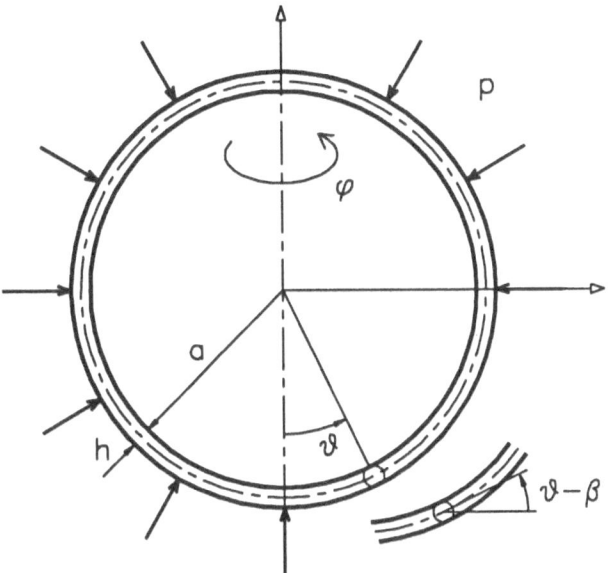

Figure 4.23. Rotationally symmetric spherical shell under uniform pressure p

where $\lambda = p/p_c$ and p_c is given by (4.203).

Second, new variables measuring the deviation from β_0 and $\psi_0^\star$ given by (4.223) are introduced. Since $\beta_0 = 0$, β can be used further on and we define ψ by

$$\psi = \psi^\star - \psi_0^\star. \tag{4.224}$$

Using the variables (β, ψ), the trivial solution is $(0,0)$ and is a concentrically deformed, perfect sphere. For the local bifurcation analysis, the equations can be simplyfied by expanding the nonlinear terms in power series. This results in the following system of differential equations

$$\beta'' + \beta' \cot \vartheta + \beta \left(\frac{2\lambda}{\delta} - \underline{\nu} - \cot^2 \vartheta \right) + \frac{\psi}{\delta}$$

$$= \frac{\beta^2 \cot \vartheta (3 - \nu)}{2} + \frac{\beta \psi \cot \vartheta}{\delta} -$$

$$- \beta^3 \left[\frac{2(\cot^2 \vartheta - 1)}{3} + \frac{\sin \vartheta (1 + \nu)}{6} - \frac{\lambda}{3\delta} \right] + \frac{\beta^2 \psi}{2\delta}$$

$$(4.225)$$

$$\psi'' + \psi' \cot \vartheta + \psi(\underline{\nu} - \cot^2 \vartheta) - \frac{\beta}{\delta} + \underline{\beta 2 \lambda (1 - \nu)}$$

$$= \psi[\underline{\beta \cot \vartheta (2 + \nu) + \nu \beta'}] - \frac{\beta^2 \cot \vartheta}{2\delta} + \underline{2\lambda[\nu\beta\beta' + \beta^2 \cot \vartheta]} +$$

$$+ \psi \left[\underline{\frac{\beta^2 \nu}{2} - \beta\beta'\nu \cot \vartheta + \beta^2 (1 - \cot \vartheta)} \right] - \frac{\beta^3}{6\delta} + \underline{\frac{2\lambda\beta^3 (4 - \nu)}{6}}.$$

To guarantee elastic buckling, the parameter δ given by (4.200) must be required to be very small (δ is of the order 10^{-3}). For the initial postbuckling problem which we only want to consider the magnitudes of β and ψ are bounded, that is, $\beta, \psi = O(1)$. Thus the underlined terms in (4.225) (see also [85]) can be neglected compared to the $1/\delta$ terms, giving

$$\beta'' + \beta' \cot \vartheta + \beta \left(\frac{2\lambda}{\delta} - \cot^2 \vartheta \right) + \frac{1}{\delta} \psi$$

$$= \frac{\cot \vartheta}{\delta} \beta\psi + \frac{\lambda}{3\delta} \beta^3 + \frac{1}{\delta} \beta^2 \psi$$

$$(4.226)$$

$$\psi'' + \psi' \cot \vartheta - \psi \cot^2 \vartheta - \frac{1}{\delta} \beta$$

$$= -\frac{\cot \vartheta}{2\delta} \beta^2 - \frac{1}{6\delta} \beta^3$$

with the corresponding boundary conditions

$$\begin{aligned} \beta(0) &= \beta(\pi) = 0 \\ \psi(0) &= \psi(\pi) = 0. \end{aligned} \qquad (4.227)$$

We note that in (4.226) still a singularity at $\vartheta = 0$, resulting from $\cot \vartheta$ is present. This singularity is a mathematical consequence of the polar coordinate system, and hence, of no relevance for the physical problem. In (4.226) dead pressure loading is assumed. This assumption is necessary if we want to solve the corresponding eigenvalue problem literally.

This is because only in the case of dead loading the integrals in (K.38) and (K.39) can be calculated explicitly. However, in [49] it is shown that there is no significant influence on the initial post-buckling behavior, no matter whether dead loading or follower force type of loading is assumed. The solution $(\beta, \psi) = (0, 0)$, represents a geometrically perfect concentrically contracted sphere for all values of λ and corresponds to (3.49). Equations (4.226) and (4.227) correspond to (3.48). To answer the question, whether or not there exists a smooth solution (3.51) through each value of the zero solution, the linear eigenvalue problem corresponding to (4.226) must be solved. This linear problem is given by the left-hand sides of (4.226)

$$\beta'' + \beta' \cot \vartheta + \beta \left(\frac{2\lambda}{\delta} - \cot^2 \vartheta \right) + \frac{1}{\delta} \psi \ = \ 0$$

$$\psi'' + \psi' \cot \vartheta - \psi \cot^2 \vartheta - \frac{1}{\delta} \beta \ = \ 0$$

(4.228)

and the boundary conditions (4.227). Following [77], [72], [85], a solution of (4.228) is sought in spherical harmonics, by setting

$$\beta(\vartheta) \ = \ C_n P_n^1(\cos \vartheta)$$

$$\psi(\vartheta) \ = \ D_n P_n^1(\cos \vartheta),$$

(4.229)

where according to (F.46)

$$P_n^m(z) = (1 - z^2)^{m/2} \frac{d^m}{dz^m} P_n(z)$$

and $P_n(z)$ is the Legendre polynomial of n-th order (Appendix F). The P_n^m form a complete set of eigenfunctions. Using the following relations (see also (F.33))

$$z = \cos \vartheta, \qquad \cot \vartheta = \frac{z}{\sqrt{1 - z^2}},$$

$$\frac{d^2}{d\vartheta^2} = (1 - z^2) \frac{d^2}{dz^2} - z \frac{d}{dz}, \qquad \frac{d}{d\vartheta} = -\sqrt{1 - z^2} \frac{d}{dz}$$

and inserting (4.229) into (4.228), we find

$$C_n (1 - z^2) P_n^{1''} - 2 C_n z P_n^{1'} + C_n P_n^1 \left(\frac{2\lambda}{\delta} - \frac{z^2}{1 - z^2} \right) + \frac{D_n}{\delta} P_n^1 \ = \ 0$$

(4.230)

$$D_n (1 - z^2) P_n^{1''} - 2 D_n z P_n^{1'} - D_n \frac{z^2}{1 - z^2} P_n^1 - \frac{C_n}{\delta} P_n^1 \ = \ 0.$$

In Appendix F is shown that the P_n^m are solutions of the differential equation (F.36)

$$(1 - z^2)P_n^{m''} - 2z P_n^{m'} + \left(n(n + 1) - \frac{m^2}{1 - z^2} \right) P_n^m = 0. \qquad (4.231)$$

Setting $m = 1$ in (4.231), multiplying (4.231) with C_n and D_n and subtracting these equations from each equation of (4.230) results in, respectively,

$$\left\{ \frac{2\lambda}{\delta} - n(n + 1) + 1 \right\} C_n + \frac{1}{\delta} D_n = 0$$

$$\frac{1}{\delta} C_n + \{ -n(n + 1) + 1 \} D_n = 0 . \qquad (4.232)$$

Nontrivial solutions C_n, D_n are obtained from (4.232) if the determinant of the coefficient matrix vanishes resulting in

$$\lambda_n = \frac{1}{2} \left(\delta \mu_n + \frac{1}{\delta \mu_n} \right) \qquad (4.233)$$

where

$$\mu_n = n(n + 1) - 1. \qquad (4.234)$$

Thus, λ_n is a function of δ and μ_n. Practically, δ (given by (4.200)) is kept at a fixed prescribed value for a given shell and the task is to find the smallest value of λ_n as a function of μ_n. If, for a moment, μ_n is treated as a continuous variable, we find the value μ_n for which the minimum value of λ_n is reached by taking the derivative with respect to μ_n in (4.233) and setting it equal to zero:

$$\frac{d\lambda_n}{d\mu_n} = \frac{1}{2} \left(\delta - \frac{1}{\delta \mu_n^2} \right) = 0 .$$

This yields

$$\mu_n = \frac{1}{\delta} . \qquad (4.235)$$

From (4.234) and (4.235) one can see that for small values of δ, which are the only practically important ones, n will be rather large. In Fig. 4.24, the eigenvalues λ_n as a function of δ are drawn for different values of n. Generically simple eigenvalues are obtained. However, there exist special values of $\delta = \delta_c$ for which double eigenvalues are obtained. Mathematically, this is the most complicated nongeneric case to occur. Similar to the calculations for the plate in Section 4.3.3, the critical values δ_c of δ can be calculated from the relation

$$\lambda_{n+1} = \lambda_n. \qquad (4.236)$$

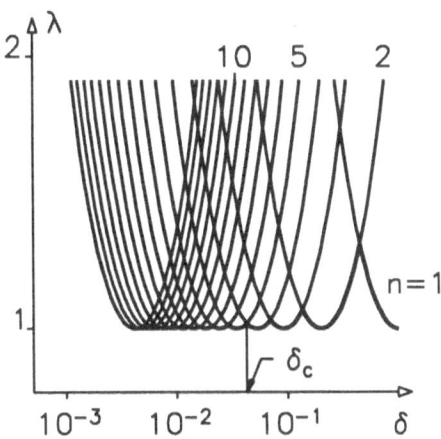

Figure 4.24. The envelope of λ_n yields the stability boundary in the λ, δ parameter space. For δ small closely spaced eigenvalues occur

Inserting (4.233) into (4.236) yields

$$\delta_c^2 = \frac{1}{\mu_n \mu_{n+1}}.$$

In Fig. 4.25 the buckling modes are drawn for $n = 15$ and 16 and their

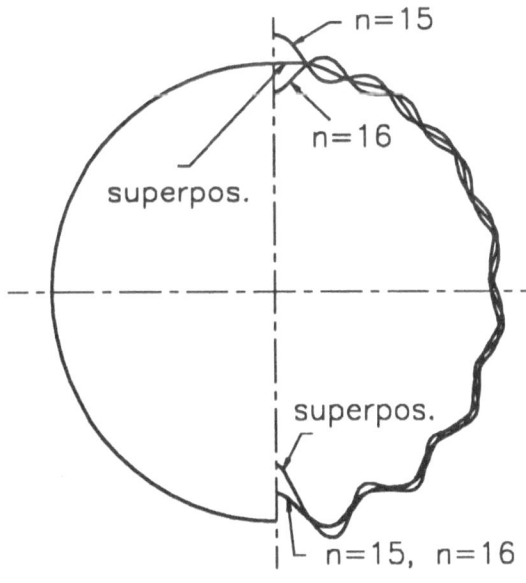

Figure 4.25. Adjacent buckling modes $n = 15$ and 16 and their superposition in case of a double eigenvalue. The superposition of the two modes results in a strong localised deflection at $\vartheta = 0$ and in no deflection at $\vartheta = \pi$

superposition in case of a double eigenvalue. This superposition results in a strong deflection at $\vartheta = 0$ and no deflection at $\vartheta = \pi$.

From Fig. 4.24, we see that for small δ closely spaced eigenvalues are found. That is, a large number of modes is packed together, which can lead to buckling deformations with sharp edges giving boundary layer type solutions. Such boundary layer type solutions can only be analysed using *singular perturbation* methods ([85], [49], [77]). Fig. 4.24 very much resembles Fig. 4.21 giving the eigenvalue curves for the rectangular plate. However, there is a subtle difference between these two problems. In the case of the plate, values of $1/\ell = 0(1)$ (that is, a short (quadratic) plate) are physically meaningful, whereas for the shell only small values of δ make sense. The consequence of the latter restriction for shells is that a local bifurcation analysis has a very restricted *range of parameter variation* and is therefore not of great practical importance. However, for both the shell equations for small values of δ and for the plate equations for $1/\ell$ small, a singularly perturbed boundary value problem is obtained. For the plate, it means that an *amplitude modulation* of the deflection pattern occurs at the ends of the long slender plate ([113]). For the spherical shell, for example, the so-called single dimple solution is obtained. This solution consists of two *reduced solutions* of the nonlinear shell equations. By *reduced solutions* we understand solutions of (K.38) and (K.39) for $\delta = 0$. They are the contracted shell and an inverted cap as depicted in Fig. 4.26. At the circle where these solutions meet they

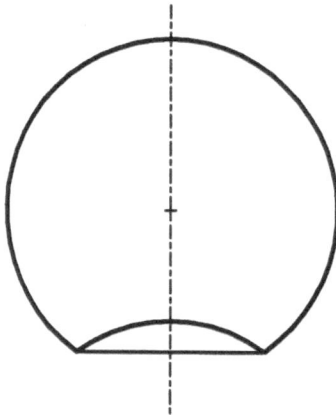

Figure 4.26. Reduced solutions ($\delta = 0$) of the singular perturbation problem of (K.38) and (K.39). They are a contracted sphere and an inverted cap

must be matched by a boundary layer ([85], [49]).

From what we said above, the local bifurcation analysis yields only the initial post-buckling behavior of the shell structure. This is because the bifurcation is subcritical and the branch, bifurcating off the trivial solution at λ_c corresponds to an unstable solution which leads to large deformations (see Fig. 6.20 and Fig. 6.21).

Ljapunov-Schmidt reduction

Despite the limited practical importance of a local bifurcation analysis for this problem, we perform the Ljapunov-Schmidt reduction of (4.226) to a set of algebraic bifurcation equations in the most complicated case possible, that is, for a double eigenvalue. We do this mainly for two reasons. One is to give another explicit example. The other is that for this problem a replacement of the Ljapunov-Schmidt reduction by the Galerkin method would lead to erroneous results (see p. 239).

We proceed in an engineering style of approach and represent the two unknowns β and ψ by infinite series

$$
\begin{aligned}
\beta &= \sum_i \beta_i P_i^1(\cos\vartheta) \\
\psi &= \sum_i \psi_i P_i^1(\cos\vartheta)
\end{aligned}
\tag{4.237}
$$

in terms of the eigenfunctions. For the amplitudes β_i and ψ_i that correspond to critical modes, bifurcation equations have to be established. After inserting (4.237) into (4.226), this system can be rewritten in the following way

$$
L_j(\beta, \psi, \lambda) = N_j(\beta, \psi, \lambda) , \qquad j = 1, 2,
\tag{4.238}
$$

where L_j and N_j designate the linear and nonlinear parts, respectively. If the equations (4.238) are projected onto the eigenfunctions by forming the inner product with $P_i^1(\cos\vartheta)$, we find for $j = 1, 2$

$$
\begin{aligned}
&\int_0^\pi L_j(\beta, \psi, \lambda) P_i^1(\cos\vartheta) \sin\vartheta d\vartheta \\
&= \int_0^\pi N_j(\beta, \psi, \lambda) P_i^1(\cos\vartheta) \sin\vartheta d\vartheta.
\end{aligned}
\tag{4.239}
$$

From (4.239), equations for β_i and ψ_i ($i = 1, 2, 3, \ldots$)

$$
\left(\frac{2\lambda}{\delta} - \mu_i\right)\beta_i + \frac{1}{\delta}\psi_i = \sum_{k,\ell} r_{ik\ell}\beta_k\beta_\ell -
$$

$$
- \sum_{k,\ell,m} r_{ik\ell m}\left(\frac{1}{2}\beta_k\beta_\ell\psi_m + \frac{\lambda}{3}\beta_k\beta_\ell\beta_m\right)
\tag{4.240}
$$

$$
-\frac{1}{\delta}\beta_i - \mu_i\psi_i = -\frac{1}{2}\sum_{k,\ell} r_{ik\ell}\beta_k\beta_\ell - \frac{1}{6}\sum_{k,\ell,m} r_{ik\ell m}\beta_k\beta_\ell\beta_m
$$

follow. The coefficients $r_{ik\ell}$ and $r_{ik\ell m}$ are given by ([125])

$$
\begin{aligned}
r_{ik\ell} &= \tfrac{1}{\delta}\tfrac{2i+1}{2i(i+1)} \int_0^\pi P_i^1(\cos\vartheta) P_k^1(\cos\vartheta) P_\ell^1(\cos\vartheta)\sin\vartheta\, d\vartheta \\
r_{ik\ell m} &= \tfrac{1}{\delta}\tfrac{2i+1}{2i(i+1)} \int_0^\pi P_i^1(\cos\vartheta) P_k^1(\cos\vartheta) P_\ell^1(\cos\vartheta) P_m^1(\cos\vartheta)\sin\vartheta\, d\vartheta .
\end{aligned}
$$
$$\tag{4.241}$$

From $(4.240)_2$, the amplitude ψ_i can be calculated in terms of β_i and inserted into $(4.240)_1$. Introducing the notation

$$
\omega_i = 2\frac{\lambda}{\delta} - \mu_i - \frac{1}{\delta^2\mu_i}
$$
$$\tag{4.242}$$

we find for $i = 1, 2, 3, \ldots$

$$
\omega_i \beta_i = -\frac{1}{\delta} \sum_{k,\ell} r_{ik\ell} \left(\frac{1}{2\mu_i} + \frac{1}{\mu_\ell} \right) \beta_k \beta_\ell +
$$

$$
+ \sum_{k,\ell,m} \beta_k \beta_\ell \beta_m \left\{ -\frac{r_{ik\ell m}}{6\delta\mu_i} + \frac{1}{2}\sum_j \frac{r_{ikj}r_{j\ell m}}{\mu_j} - \frac{r_{ik\ell m}}{2\delta\mu_m} + r_{ik\ell m}\frac{\lambda}{3} \right\} .
$$
$$\tag{4.243}$$

The elimination of the ψ_i basically corresponds to the elimination of the stress function from the two equations (4.226) similar to the approach used in the treatment of the plate in Section 4.3.3.

From (4.242), it can be seen that for those values of λ, given by (4.233), ω_i, and hence, the left-hand side of (4.243) vanishes. This is the bifurcation case. Depending on whether λ is a simple or a double eigenvalue, either one or two equations with vanishing linear part are obtained. These are the bifurcation equations. The corresponding variables are the critical (or active) variables and are designated by β_c for a simple eigenvalue and by β_c and β_{c+1} for a double eigenvalue. Hence, in the case of a double eigenvalue the bifurcation equations are

$$
0 = F_i(\beta_1, \beta_2, \ldots, \beta_c, \beta_{c+1}, \ldots), \qquad i = c, c+1 .
$$
$$\tag{4.244}$$

All other β_i ($i \neq c, c+1$) are the noncritical (or passive) variables which have to be eliminated from the two bifurcation equations. This can be done by expressing them as functions of the critical variables from those equations in (4.243) for which the left-hand sides do not vanish. These equations are

$$
\omega_i \beta_i = F_i(\beta_1, \beta_2, \ldots, \beta_c, \beta_{c+1}, \ldots), \qquad \begin{aligned} i &\neq c, c+1 \\ i &= 1, 2, 3, \ldots \end{aligned}
$$
$$\tag{4.245}$$

By applying the implicit function theorem, locally, in the neighborhood of the bifurcation point, we find solutions of (4.245) as

$$
\beta_i = h_i(\beta_c, \beta_{c+1}), \qquad i \neq c, c+1 .
$$
$$\tag{4.246}$$

Substituting (4.246) into the bifurcation equations (4.244), two equations are obtained which contain only active variables and the lowest order nonlinear terms in the bifurcation equations

$$F_i(\beta_c, \beta_{c+1}) = 0, \qquad i = c, c+1. \tag{4.247}$$

For the practical calculation of (4.246), the question of determinacy of the bifurcation equations is again of great importance. From (3.57) is known that

$$\beta_i = O(|\beta_c|^2 + |\beta_{c+1}|^2), \qquad i \neq c, c+1. \tag{4.248}$$

Thus, if the bifurcation equations are *two*-determinate, all the β_i ($i \neq c, c+1$) given by (4.246) can be set to zero. If, however, second order terms in (4.247) do not determine the problem, (4.246) must be calculated. As long as the analysis can be restricted to terms up to and including third order in (4.247), for reasons of determinacy, only quadratic terms must be retained from (4.246). Hence, we obtain from (4.243)

$$\beta_i = \frac{1}{\omega_i} \sum_{k,\ell=c}^{c+1} -\frac{1}{\delta} r_{ik\ell} \left(\frac{1}{2\mu_i} + \frac{1}{\mu_\ell} \right) \beta_k \beta_\ell + O(|\beta_c|^3 + |\beta_{c+1}|^3). \tag{4.249}$$

Finally, we can write down the bifurcation equations explicitly for the simple and the double eigenvalue.

(a) Simple eigenvalue

In this case, only one critical variable β_c is given. From (4.249) we obtain

$$\beta_i = \frac{1}{\omega_i} \left[-\frac{1}{\delta} r_{icc} \left(\frac{1}{2\mu_i} + \frac{1}{\mu_c} \right) \beta_c^2 \right] + O(|\beta_c|^3). \tag{4.250}$$

Inserting (4.250) into (4.243) with $i = c$, we find

$$\begin{aligned}
0 = & -\frac{1}{\delta} r_{ccc} \frac{3}{2\mu_c} \beta_c^2 + \left[\sum_{j \neq c} \frac{1}{\delta^2 \omega_j} r_{ccj} r_{jcc} \left(\frac{1}{2\mu_j} + \frac{1}{\mu_c} \right) \left(\frac{2}{\mu_c} + \frac{1}{\mu_j} \right) \right. \\
& \left. + \sum_j \frac{1}{2} \frac{1}{\mu_j} r_{ccj} r_{jcc} + r_{cccc} \left(\frac{\lambda}{3} - \frac{2}{3\delta\mu_c} \right) \right] \beta_c^3 + O(|\beta_c|^4).
\end{aligned} \tag{4.251}$$

In (4.251) the first term in the bracket results from the elimination of the non-critical variables. The summation in (4.251) is finite because $r_{ccj} = 0$ for $j > 2c$ ([124]).

(b) Double eigenvalue

Now the critical variables are β_c and β_{c+1}. If it can be verified that second order terms in the bifurcation equations determine the behavior, which is in fact the case (see p. 244), the bifurcation equations follow from (4.243) to (to simplify notation we set $d := c + 1$)

$$
\begin{aligned}
0 = & -\frac{1}{\delta}r_{ccc}\left(\frac{1}{2\mu_c} + \frac{1}{\mu_c}\right)\beta_c^2 - \frac{1}{\delta}r_{ccd}\left(\frac{1}{2\mu_c} + \frac{1}{\mu_d}\right)\beta_c\beta_d - \\
& -\frac{1}{\delta}r_{cdc}\left(\frac{1}{2\mu_c} + \frac{1}{\mu_c}\right)\beta_d\beta_c - \frac{1}{\delta}r_{cdd}\left(\frac{1}{2\mu_c} + \frac{1}{\mu_d}\right)\beta_d^2 \\
0 = & -\frac{1}{\delta}r_{dcc}\left(\frac{1}{2\mu_d} + \frac{1}{\mu_c}\right)\beta_c^2 - \frac{1}{\delta}r_{dcd}\left(\frac{1}{2\mu_d} + \frac{1}{\mu_d}\right)\beta_c\beta_d - \\
& -\frac{1}{\delta}r_{ddc}\left(\frac{1}{2\mu_d} + \frac{1}{\mu_c}\right) - \frac{1}{\delta}r_{ddd}\left(\frac{1}{2\mu_d} + \frac{1}{\mu_d}\right)\beta_d^2 \ .
\end{aligned}
$$

$$(4.252)$$

Equations (4.250), (4.251) and (4.252) will be discussed further in Section 6.6.1.

Chapter 5

Bifurcations under symmetries

5.1 Introduction

Symmetry properties belong to the most fundamental features of systems in all branches of science and engineering. For example, any engineer analysing the stresses and deformations of a structure makes implicitly repeated use of obvious and, sometimes, not so obvious symmetry properties to simplify the analysis.

We already saw in Section 4.3.4 when dealing with the buckling problem of the rectangular plate that the use of the simple reflectional symmetry resulted in a considerable reduction of the calculations necessary to obtain the bifurcation equations. In general, such symmetry considerations allow us to delete certain terms from the bifurcation equations before we actually start calculating their coefficients. This is very important from a practical engineering point of view because when we treat specific problems it can often be very difficult to decide whether or not to retain certain coefficients with small absolute values obtained from numerical calculations performing the Ljapunov-Schmidt or center manifold reduction. It will be shown below that in specific examples a reduction of the number of terms by symmetry arguments and in addition by normal form theory of orders of magnitude is possible.

In this section, we describe a systematic treatment of the derivation of the bifurcation equations for problems with symmetries. It will turn out that in systems with certain symmetry properties, the multiplicity of the critical eigenvalue is increased (for example doubled). On the other hand, the bifurcation equations must obey the same symmetry properties as the original problem, and therefore, can be strongly simplified from the outset.

143

Equivariance under a symmetry group

In order to substantiate the statements above, let us study some examples.

Statical problem

We consider the planar double pendulum with dead loading (Fig. 2.5a). This is a conservative finite-dimensional system, and therefore, a stability analysis of the straight equilibrium position ($\psi_1 = \psi_2 = 0$) can be performed by analysing the potential energy of the system. If a perfect system is assumed, that is, the springs are tensionless for $\psi_1 = \psi_2 = 0$ and P is acting centrally, then the potential energy $V(\psi_1, \psi_2, P)$ given by (2.17) must satisfy

$$V(\psi_1, \psi_2, P) = V(-\psi_1, -\psi_2, P) . \tag{5.1}$$

Equation (5.1) expresses the physically obvious fact that for the energy V buckling to the left must be the same as buckling to the right. From (5.1) the equilibrium equations follow to

$$\begin{aligned} G_1(\psi_1, \psi_2, P) &= \frac{\partial V}{\partial \psi_1} = 0 \\ G_2(\psi_1, \psi_2, P) &= \frac{\partial V}{\partial \psi_2} = 0. \end{aligned} \tag{5.2}$$

To guarantee the property expressed by (5.1), the potential energy V must be an even function in the variables ψ_1 and ψ_2. Thus the functions G_i in (5.2) are odd functions in their variables with the properties

$$\begin{aligned} G_1(\psi_1, \psi_2, P) &= -G_1(-\psi_1, -\psi_2, P) \\ G_2(\psi_1, \psi_2, P) &= -G_2(-\psi_1, -\psi_2, P). \end{aligned} \tag{5.3}$$

Equations (5.3) can be written in the form

$$\mathbf{T}_g \mathbf{G}(\boldsymbol{\psi}, P) = \mathbf{G}(\mathbf{T}_g \boldsymbol{\psi}, P) \tag{5.4}$$

where

$$\mathbf{T}_g = \begin{pmatrix} -1 & 0 \\ 0 & -1 \end{pmatrix}, \quad \mathbf{G} = \begin{pmatrix} G_1 \\ G_2 \end{pmatrix}, \quad \boldsymbol{\psi} = \begin{pmatrix} \psi_1 \\ \psi_2 \end{pmatrix}. \tag{5.5}$$

Equation (5.4) is the required relationship stating that the problem (5.2) is *equivariant* under the symmetry group Γ for which $\mathbf{T}_g$ and $\mathbf{E}$ are a matrix representation of the elements (Appendix G.3). With the notation of (5.5), (5.1) can be written as

$$V(\mathbf{T}_g \boldsymbol{\psi}, P) = V(\boldsymbol{\psi}, P) . \tag{5.6}$$

This equation states that the potential V is *invariant* under the symmetry group Γ.

Let us consider another simple example which shows that the system of bifurcation equations possesses the same symmetry properties as the original system. It is the buckling problem of the rod of Fig. 2.7. Again a symmetry relationship must hold for the energy (2.20) similar to (5.1) or (5.6) if $\psi(s)$ is replaced by $-\psi(s)$. Inserting into (2.20) yields

$$V(\psi, P) = V(-\psi, P) \ .$$

The corresponding equilibrium variational equation is given by (4.81). It is obvious that for this equation with $T_g = -1$ the relationship (5.4)

$$G(-\psi(s), p) = -G(\psi(s), p) \tag{5.7}$$

is satisfied. After reduction to a finite-dimensional problem, the corresponding potential $\overline{V}$ in the critical variable q according to (4.91) satisfies

$$\overline{V}(q, a) = \overline{V}(-q, a). \tag{5.8}$$

From (5.8) follows that

$$\frac{\partial \overline{V}}{\partial q} = \overline{G}(q, a) = 0 \tag{5.9}$$

with $\overline{G}$ given by (4.100) with the property

$$\overline{G}(-q, a) = -\overline{G}(q, a). \tag{5.10}$$

We consider now the concept of *equivariance* or *covariance*, as it is also sometimes called, more generally by stating the following theorem ([72]):

Theorem 5.1 (Equivariance) *It is assumed that* $\mathbf{G}(\boldsymbol{u}, \lambda) : \mathbf{H} \to \mathbf{H}$ *can be derived as the gradient of a potential* $V(\boldsymbol{u}, \lambda) : \mathbf{H} \to \mathbb{R}^1$, *where* $\mathbf{H}$ *is a Hilbert space. Furthermore* $g \to \mathbf{T}_g$ *is a unitary representation of a group* Γ *on the Hilbert space* $\mathbf{H}$ *(Appendix G.3), such that* $V(\boldsymbol{u}, \lambda)$ *is invariant under* Γ *in the sense that*

$$V(\mathbf{T}_g \boldsymbol{u}, \lambda) = V(\boldsymbol{u}, \lambda) \tag{5.11}$$

for all $g \in \Gamma$ *and all* $\boldsymbol{u} \in \mathbf{H}$. *Then it can be shown that* $\mathbf{G}$ *is equivariant (covariant) under* Γ *in the sense that*

$$\mathbf{T}_g \mathbf{G}(\boldsymbol{u}, \lambda) = \mathbf{G}(\mathbf{T}_g \boldsymbol{u}, \lambda) \ , \tag{5.12}$$

for all $g \in \Gamma$ *and all* $\boldsymbol{u} \in \mathbf{H}$.

The proof is as follows ([72]): Let $g \to \mathbf{T}_g$ be a unitary representation (Appendix G.3) of a group Γ on a Hilbert space $\mathbf{H}$ with inner product $(\cdot, \cdot)$ and $\mathbf{G}$ be the gradient of V. For any $g \in \Gamma$ and any $u \in \mathbf{H}$, the following expression can be formed

$$
\begin{aligned}
(\mathbf{T}_g \mathbf{G}(u, \lambda) &- \mathbf{G}(\mathbf{T}_g u, \lambda), \mathbf{T}_g h) \\
&= (\mathbf{T}_g \mathbf{G}(u, \lambda), \mathbf{T}_g h) - (\mathbf{G}(\mathbf{T}_g u, \lambda), \mathbf{T}_g h) \\
&= (\mathbf{G}(u, \lambda), h) - (\mathbf{G}(\mathbf{T}_g u, \lambda), \mathbf{T}_g h) \\
&= [V(u + h, \lambda) - V(u, \lambda)] - [V(\mathbf{T}_g(u + h), \lambda) - \\
&\quad - V(\mathbf{T}_g u, \lambda)] + o(|h|).
\end{aligned}
\tag{5.13}
$$

From (5.11) follows that the last line of (5.13) is $o(|h|)$ because the two expressions in the brackets are identical. Thus (5.13) yields

$$
(\mathbf{T}_g \mathbf{G}(u, \lambda) - \mathbf{G}(\mathbf{T}_g u, \lambda), \mathbf{T}_g h) = 0
$$

for all $h \in \mathbf{H}$. This implies that $\mathbf{T}_g \mathbf{G}(u, \lambda) = \mathbf{G}(\mathbf{T}_g u, \lambda)$.

Furthermore, an expression similar to (5.12), possibly being adapted to the finite-dimensional case, must hold for the bifurcation equation.

Dynamical problem

If a dynamical system is given by

$$
\dot{x} = f(x, \lambda)
$$

and a transformation of coordinates $x = \mathbf{T}_g y$ is performed, we obtain

$$
\dot{y} = \mathbf{T}_g^{-1} f(\mathbf{T}_g y, \lambda) .
$$

Now, we demand that the right-hand side which depends on y should have the same form as it had in x, that is,

$$
\mathbf{T}_g^{-1} f(\mathbf{T}_g y, \lambda) = f(y, \lambda)
$$

or

$$
f(\mathbf{T}_g y, \lambda) = \mathbf{T}_g f(y, \lambda) .
$$

This means that the function f must be *equivariant* under the transformation $\mathbf{T}_g$.

Another consequence of (5.12) is that if u is a solution of $\mathbf{G}(u, \lambda) = 0$ or y a solution of $f(y, \lambda) = 0$ then $\mathbf{T}_g u$ or $\mathbf{T}_g y$ are also a solution. This means that we get *sheets* or *orbits* of solutions $\{\mathbf{T}_g u$ or $\mathbf{T}_g y; \ g \in \Gamma\}$. However, in general $\mathbf{T}_g u \neq u$ and $\mathbf{T}_g y \neq y$. Simple examples are the spherical double pendulum (Section 5.1) and the fluid conveying tube (Section 5.3). For these examples we show in Section 6.6 that for certain parameter configurations a statically buckled planar state exists. The orbit of this solution is then obtained by rotating the plane, in which the pendulum or the tube is buckled, around the vertical axis.

Multiplicity of the critical eigenvalue

We assume a bifurcation problem to be given by

$$\mathbf{G}(\mathbf{u}, \lambda) = 0 . \tag{5.14}$$

Furthermore, we assume that (5.14) is equivariant under some symmetry group Γ with the representation $\mathbf{T}_g$, that is, (5.12) holds.

Suppose that $(\mathbf{u}_0, \lambda)$ is an equilibrium solution of (5.14). Then, because of (5.12), also $\mathbf{T}_g \mathbf{u}_0$ is a solution of (5.14). We assume now that

$$\mathbf{T}_g \mathbf{u}_0 = \mathbf{u}_0 \tag{5.15}$$

for all elements g of the group Γ. This is a very natural assumption for many physical problems because the solution $\mathbf{u}_0$, which is mostly a trivial state, is, therefore, in general *invariant* under the group action. The set $\mathbf{u}_0$ satisfying (5.15) is called the *Γ-fixed-point set*. Let us mention two examples to illustrate this point:

Example 1: The solutions $\psi_1 = \psi_2 = 0$ for the double pendulum (Section 4.1) and $\psi(s) \equiv 0$ of the rod buckling problem (Section 4.3.1) clearly possess reflectional symmetry;

Example 2: For the buckling problem of a spherical shell is shown in Section 4.3.5 that the fundamental solution is a uniformly contracted perfect sphere which, of course, is invariant under all rotations about the center of the sphere.

Using the Ljapunov-Schmidt procedure, the bifurcation problem (5.14) can be reduced to a finite-dimensional system of bifurcation equations

$$\mathbf{F}(\mathbf{q}, \lambda) = \mathbf{0} \tag{5.16}$$

in the neighborhood of a bifurcation point $(\mathbf{u}_0, \lambda_c)$.

Then the following *theorem* holds ([120]):

Theorem 5.2 (Equivariance) *Let $\mathbf{u}_0$ be invariant under the symmetry group Γ, let $\mathbf{G}_u$ be a Fredholm mapping (Appendix A.2) and let $\mathbf{G}$ be a differentiable mapping. Then, the kernel $\mathbf{N} = \ker \mathbf{G}_u(\mathbf{u}_0, \lambda_c)$ is invariant under $\mathbf{T}_g$ for all elements $g \in \Gamma$ and the bifurcation equations $\mathbf{F}(\mathbf{q}, \lambda) = 0$ are equivariant with respect to $\mathbf{S}_g$, the restriction of $\mathbf{T}_g$ on the finite-dimensional kernel $\mathbf{N}$.*

The proof of the theorem is given by stepping through the details of the Ljapunov-Schmidt procedure. The symmetry condition for the linear mapping follows from taking the Fréchet derivative of (5.12) at $\mathbf{u}_0$ to

$$\mathbf{T}_g \mathbf{G}_u(\mathbf{u}_0(\lambda), \lambda) = \mathbf{G}_u(\mathbf{T}_g \mathbf{u}_0, \lambda) \mathbf{T}_g . \tag{5.17}$$

From (5.17) and (5.15) we obtain the following relationship for the linear operator $\mathbf{A} := \mathbf{G}_u(\mathbf{u}_0, \lambda)$

$$\mathbf{T}_g \mathbf{A} = \mathbf{A} \mathbf{T}_g \tag{5.18}$$

which expresses the fact that $\mathbf{T}_g$ and $\mathbf{A}$ *commute*.

From (5.18) an important consequence concerning the eigenvectors and eigenfunctions can be derived. If, for example, $v \in \mathbf{N}(\mathbf{A})$ then v is an eigenfunction and $\mathbf{A}v = 0$. However, from (5.18) follows

$$\mathbf{T}_g \mathbf{A} v = \mathbf{A} \mathbf{T}_g v = 0 \, , \tag{5.19}$$

which shows that $\mathbf{T}_g v$ is also an eigenfunction of the operator $\mathbf{A}$, and hence, also an element of $\mathbf{N}(\mathbf{A})$. However, in general, v and $\mathbf{T}_g v$ are linearly independent.

As an example, let us consider the tube of Section 5.3. We take for v_1 the statically buckled planar tube configuration in the (x, z)-plane. If we rotate this plane by the angle $\pi/2$ we obtain a solution $v_2 = \mathbf{T}_{\pi/2} v_1$ in the (y, z)-plane which in $\mathbb{R}^3$ is linearly independent of v_1. These are the two linearly independent eigenfunctions. Any other planar state can be obtained from them by a linear combination. If, for example, we rotate v_1 by π then we obtain $v_3 = \mathbf{T}_\pi v_1$ in the (x, z)-plane which, of course, is linearly dependent on v_1 because $v_3 = (-1)v_1$.

The independence of v and $\mathbf{T}_g v$ has the important consequence that every eigenvalue has multiplicity *two* because for one eigenvalue two different eigenfunctions are obtained.

Furthermore follows for the projection operators, designated by $\mathbf{P}$ and $\mathbf{E} - \mathbf{P}$ in Chapter 3 that the relations $\mathbf{P}\mathbf{T}_g = \mathbf{T}_g\mathbf{P}$ and $(\mathbf{E} - \mathbf{P})\mathbf{T}_g = \mathbf{T}_g(\mathbf{E} - \mathbf{P})$ hold. Thus, $\mathbf{S}_g$ in Theorem 5.2 is given by $\mathbf{S}_g = \mathbf{P}\mathbf{T}_g$. Hence, for the bifurcation equations follows

$$\mathbf{S}_g \mathbf{F}(q, \lambda) = \mathbf{F}(\mathbf{S}_g q, \lambda). \tag{5.20}$$

From (5.20) conditions concerning the form of the bifurcation equations can be derived before the application of the Ljapunov-Schmidt or center manifold reduction. This has the important consequence that many terms in the bifurcation equations can be eliminated from the outset. Hence, a considerable reduction of the work necessary for the computation of coefficients of terms in the bifurcation equations is achieved. However, it must be pointed out that the calculation of the coefficients of those terms which fulfill the symmetry requirements and thus remain in the bifurcation equations must be done by means of either the center manifold theory or the Ljapunov-Schmidt method. That is, symmetry arguments determine only those terms that can be present in the bifurcation equations but they do not provide the numerical values of their coefficients.

Symmetry breaking bifurcation

If the equations of a problem are equivariant under the symmetry group Γ, then generally, the trivial solution u_0 is invariant under Γ (see (5.15)) but the bifurcated solutions are invariant only under a smaller symmetry group. That is, the symmetry group of the bifurcated solutions is a *subgroup* Γ' of Γ (Appendix G.6). In such a case *spontaneous symmetry breaking* occurs. The symmetry group of the equations remains unchanged, but the solutions which bifurcate spontaneously break symmetry or expressed in other words: the solutions are less symmetric than the equations. It is important to point out that this happens in the absence of an external symmetry breaking perturbation. Mathematically, one can express this fact in the following way. Whereas u_0 fulfills (5.15), the bifurcated solution according to (3.53) can be written in the form $u = u_c + u_s$, where $u_c \in N(A)$ and $u_s \perp N(A)$. Only the u_c part is of interest for the symmetry breaking phenomenon. $T_g u_c \in N(A)$ (see Theorem 5.2), but in general (see the comment on p. 148)

$$T_g u_c \neq u_c .$$

Incidentally, this relation does not contradict the assertion in Theorem 5.2 that the kernel $N(A)$ is invariant under T_g, because both eigenfunctions belong to the kernel. Hence, u_c is not invariant under Γ but possibly under a smaller group Γ' which is a subgroup of Γ (Appendix G.6).

Let us give some examples:

Example 1: The best known example of symmetry breaking is the Hopf bifurcation (Section 6.2) where temporal symmetry is broken. The autonomous system $\dot{x} = f(x, \lambda)$ $(x, f \in \mathbb{R}^n, \lambda \in \mathbb{R}^\ell)$ is invariant under the group of time translations $t \rightarrow t + \tau$ with arbitrary τ. For example, the time independent equilibrium solution is invariant under arbitrary time translations. The bifurcated time periodic solutions, however, are invariant only under the discrete group generated by $t \rightarrow t + T$, T being the minimal period of the oscillations.

Example 2: A nice physical example of spontaneous symmetry breaking is the appearance of hexagonal convection cells in the *Bénard problem* (see p. 131). In this case the *Euclidean symmetry* (Appendix G.8) (group of continuous translations and rotations) is broken to a discrete crystallographic subgroup.

Example 3: The *rotation group* (Appendix G.7) occurs frequently in physical applications. The breaking of rotational symmetry in $\mathbb{R}^3$ occurs for example in the buckling problem of spherical shells ([72]).

Example 4: Another interesting example of symmetry breaking is explained in [47]. It is the bifurcation of the shape of an incompressible
elastic solid having in its unloaded equilibrium position the form of a unit
cube. It is subjected to uniform tension p normal to each face (Fig. 5.1a).
It is further assumed that after loss of stability of the trivial normalized

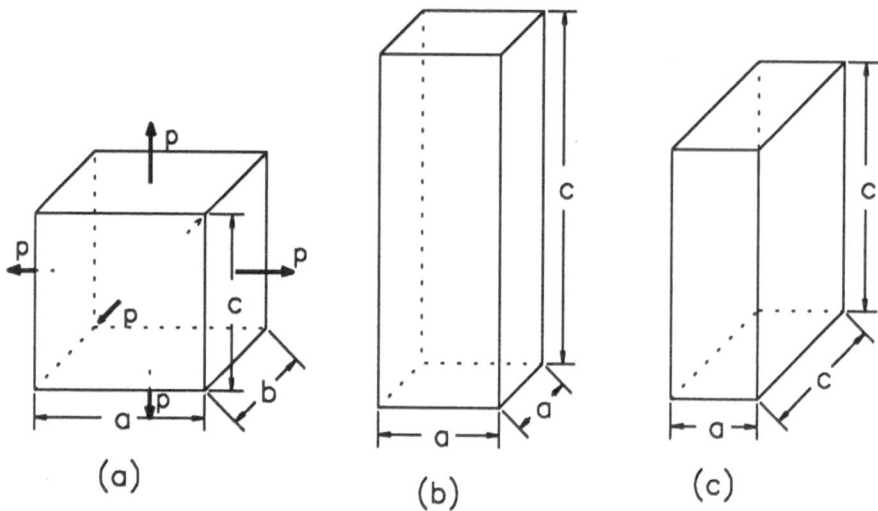

(a)   (b)   (c)

Figure 5.1. (a) Bifurcation of an incompressible elastic cube $a = b = c$ under
uniform tension p into (b) a rodlike $a = b$ and (c) platelike $b = c$ parallelepiped

equilibrium shape

(i) $a = b = c = 1$ ($abc = 1$ due to incompressibility),

the shape of the body remains a rectangular parallelepiped. We ask
now the question: is it possible to say something about the form of the
shape of the solid after bifurcation? The answer is: yes. The system
is equivariant under the *permutation group* S_3 (Appendix G.6) of three
symbols. It is shown in Section 6.6.4 that the solutions which appear
after a symmetry breaking bifurcation still have inherited some of the
symmetry of the original problem. Hence, the three possible solutions
after bifurcation

(ii) $a = b < c$

(iii) $a < b = c$

(iv) $a < b < c$

are (ii) a rodlike, (iii) a platelike and (iv) a parallelepiped with all sides of
unequal length (Fig. 5.1b–d). Solution (iv) has no nontrivial symmetries

and hence six different configurations are possible. Solutions (ii) and (iii) each have one nontrivial permutational symmetry (interchanging a and b for rods and b and c for plates), and hence, three solutions occur at a time (check the table on p. 335).

Therefore, the conclusion to be drawn is that in a physical experiment either (ii) or (iii) will be found. In order to determine which one actually occurs, requires the solution of the physical problem ([47]).

We study now some simple mechanical problems in detail.

5.2 Finite dimensional dynamical system: spherical double pendulum with elastic end support and follower force loading

The mechanical model, shown in Fig. 5.2, consists of two homogeneous rigid rods of lengths l_1 and l_2 and masses m_1 and m_2. The joints are considered to be viscoelastic. That is, a joint creates a moment proportional to both the relative angle and the relative angular velocity of the two rods. In addition to these internal damping moments external damping forces, which are assumed to be proportional to the absolute speed of the

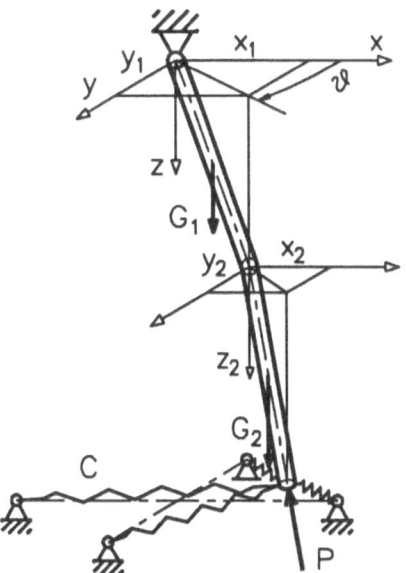

Figure 5.2. Spherical double pendulum with follower force loading, viscoelastic hinges and elastic end support

rods, are also included in the model. The loading is given by a tangential follower force P always acting in the direction of the second rod. This system can be assumed to be a discrete mechanical model for systems undergoing a flutter instability ([10]). For example, for a tube carrying fluid which as continuous model is considered in Section 5.3. The inclusion of the end support in the model is important for two reasons. First, such end supports can occur in physical applications, and second, by varying the stiffness of the end support an additional parameter is available to create more degenerate cases of loss of stability. Finally, it is assumed that all the moments and forces created by the joints and the end support are rotationally symmetric. For the end support, this requirement is fulfilled if infinitely many springs are arranged on the circle. If only a finite number of springs is given, as in Fig. 5.2, then the springs have to be very long (theoretically infinitely long) in order to create restoring forces always pointing through the z-axis.

For the derivation of the equations of motion, appropriate variables must be selected. A natural choice would be to use spherical polar coordinates. However, these are not suitable because from Fig. 5.2 it can be seen that the angle ϑ changes discontinuously by an amount of π if the upper rod moves through the z-axis. Following ([10]), we use as coordinates the distances x_1, x_2, y_1 and y_2 of the endpoints of the two rods as shown in Fig. 5.2. The equations of motion are derived in ([10]) and can be written in the form

$$\mathbf{M}\ddot{x} + \mathbf{D}\dot{x} + \mathbf{N}x = \begin{cases} K_1(\xi,\eta) \\ K_2(\xi,\eta) \end{cases}$$

$$\mathbf{M}\ddot{y} + \mathbf{D}\dot{y} + \mathbf{N}y = \begin{cases} K_3(\xi,\eta) \\ K_4(\xi,\eta) \end{cases} \tag{5.21}$$

where $x = (x_1, x_2)^T$, $y = (y_1, y_2)^T$, $\xi = (x_1, \dot{x}_1, x_2, \dot{x}_2)^T$ and $\eta = (y_1, \dot{y}_1, y_2, \dot{y}_2)^T$. The matrices $\mathbf{M}, \mathbf{D}$ are symmetric and $\mathbf{N}$ is not symmetric. The physical explanation for $\mathbf{N}$ being not symmetric is that the tangential load P is non-conservative. The nonlinear terms $K_i(\xi,\eta)$ in (5.21) follow from a series expansions up to and including third order terms of the nonlinear terms in the original equations of motion. These terms satisfy the relations

$$K_3(\xi,\eta) = K_1(\eta,\xi) \qquad \text{and} \qquad K_4(\xi,\eta) = K_2(\eta,\xi) \tag{5.22}$$

which express the reflectional symmetry of the model ([10]). We explain it below. Writing (5.21) as a first order system, we obtain

$$\begin{pmatrix} \mathbf{A} & 0 \\ 0 & \mathbf{A} \end{pmatrix} \begin{pmatrix} \dot{\xi} \\ \dot{\eta} \end{pmatrix} = \begin{pmatrix} \mathbf{B} & 0 \\ 0 & \mathbf{B} \end{pmatrix} \begin{pmatrix} \xi \\ \eta \end{pmatrix} + \tag{5.23}$$

$$+ (0, K_1(\xi,\eta), 0, K_2(\xi,\eta), 0, K_3(\xi,\eta), 0, K_4(\xi,\eta))^T .$$

The matrices $\mathbf{A}$ and $\mathbf{B}$ follow from a straightforward calculation from the matrices $\mathbf{M}, \mathbf{D}, \mathbf{N}$ and the additional equations defining the new variables (for example, $\dot{\xi}_1 = \xi_2$).

The symmetry properties of (5.21) are of great importance in the further treatment of this problem. An indication that there must be something special with these equations follows already from the identical linear parts in $\boldsymbol{x}$ and $\boldsymbol{y}$ in (5.21) or the block diagonal form with identical matrices in (5.23). Let us introduce the rotation matrix

$$
\mathbf{R}_\vartheta = \begin{pmatrix}
\cos\vartheta & & 0 & & \sin\vartheta & & 0 \\
 & \ddots & & & & \ddots & \\
0 & & \cos\vartheta & & 0 & & \sin\vartheta \\
-\sin\vartheta & & 0 & & \cos\vartheta & & 0 \\
 & \ddots & & & & \ddots & \\
0 & & -\sin\vartheta & & 0 & & \cos\vartheta
\end{pmatrix} .
\tag{5.24}
$$

Performing the matrix multiplication of $\mathbf{R}_\vartheta$ with the matrices in (5.23) we find that (5.18) holds because (5.23) is block diagonal in its linear parts. It is obvious from physical arguments that a rotation of the coordinate system by an angle ϑ must not have any influence on the form of (5.23) or (5.21). Therefore, these equations possess the rotational symmetry $\mathbf{SO}(2)$ (Appendix G.7).

Equations (5.23) have yet another symmetry. This is the reflectional symmetry $\mathbf{Z}_2$ with respect to any plane through the z-axis. In Appendix G.2 it is explained that a reflection about any plane through the z-axis can be composed by a reflection about a special plane (for example $\vartheta = \pi/4$) and a rotation. (See also the example of Fig. G.2, where, however, only a discrete number of reflections and rotations is possible.) From (5.22) follows that (5.23) possess the reflectional symmetry about the plane $\vartheta = \pi/4$ which can be represented by the matrix

$$
\mathbf{R}_s = \begin{pmatrix}
 & & 1 & & 0 \\
 & 0 & & \ddots & \\
 & & 0 & & 1 \\
1 & & 0 & & \\
 & \ddots & & & 0 \\
0 & & 1 & &
\end{pmatrix} .
\tag{5.25}
$$

Applications of (5.25) on (5.23) results in an exchange of the variables x, y and the two blocks in the equations. A similar behavior appears for the continuous model of the tube in Section 5.3. Hence, the equations of motion (5.23) possess the symmetry $\mathbf{SO}(2) \times \mathbf{Z}_2$ which is called $\mathbf{O}(2)$-symmetry (Appendix G.7).

Since the linear part of (5.23) decomposes into two identical parts, the loss of stability of the trivial equilibrium position can be calculated by computing the eigenvalues of only one half of the linear part of (5.23) for a quasistatic increase of the loading P. That is, we must solve

$$(\mathbf{A}\lambda - \mathbf{B})\boldsymbol{\xi}_0 = \mathbf{0} . \tag{5.26}$$

Each eigenvalue obtained from (5.26) must be doubled for (5.23), and hence, has multiplicity *two* for the full problem. In order to get a clear picture of all the cases which are possible, (5.23) is rewritten in the form

$$\dot{\boldsymbol{\zeta}} = \mathbf{C}(\lambda_c)\boldsymbol{\zeta} + \boldsymbol{f}(\boldsymbol{\zeta}, \lambda_c) \tag{5.27}$$

where $\boldsymbol{\zeta} = (\boldsymbol{\xi}, \boldsymbol{\eta})^T$. Transforming (5.27) into Jordan form by means of the linear change of variables $\boldsymbol{\zeta} = \mathbf{B}\boldsymbol{v}$ (Appendix B) results in

$$\dot{\boldsymbol{v}} = \mathbf{J}\boldsymbol{v} + \mathbf{B}^{-1}\boldsymbol{f}(\mathbf{B}\boldsymbol{v}, \lambda_c), \tag{5.28}$$

where

$$\mathbf{J} = \mathbf{B}^{-1}\mathbf{C}\mathbf{B} = \begin{pmatrix} \mathbf{J}_c & 0 \\ 0 & \mathbf{J}_s \end{pmatrix} . \tag{5.29}$$

All eigenvalues of $\mathbf{J}_s$ have a negative real part whereas the real parts of all eigenvalues of $\mathbf{J}_c$ are zero.

The stability boundary (Appendix H) in the (P, c) parameter plane for a fixed ratio d of internal to external damping but for increasing absolute values of internal damping k is the same as that given in Fig. 4.4 for the plane double pendulum of Section 4.1.2. However, because of the structure of (5.23) the essential difference to the problem treated in Section 4.1.2 is that now each eigenvalue has multiplicity two.

Hence, for c large, a zero eigenvalue with multiplicity two and for c small, an imaginary pair of eigenvalues with multiplicity two is found. That is, the matrix $\mathbf{J}_c$ in (5.29) has one of the following forms

$$\mathbf{J}_c = \begin{pmatrix} 0 & 0 \\ 0 & 0 \end{pmatrix} \quad \text{or} \quad \mathbf{J}_c = \begin{pmatrix} 0 & -\omega & 0 & 0 \\ \omega & 0 & 0 & 0 \\ 0 & 0 & 0 & -\omega \\ 0 & 0 & \omega & 0 \end{pmatrix} \tag{5.30}$$

which are the generic cases. Obviously, there exists a special value of the stiffness $c = c_c$ where a coupling between divergence and flutter occurs. However, the absolute value of damping k determines the eigenvalue structure of the matrix $\mathbf{J}_c$ in this case. If k is small a zero eigenvalue and a purely imaginary pair with multiplicity two is obtained given by the matrix

$$
\mathbf{J}_c = \begin{pmatrix}
0 & 0 & 0 & & & \\
0 & 0 & -\omega & & 0 & \\
0 & \omega & 0 & & & \\
& & & 0 & 0 & 0 \\
& 0 & & 0 & 0 & -\omega \\
& & & 0 & \omega & 0
\end{pmatrix}. \tag{5.31}
$$

For large k a double zero root in a Jordan block with multiplicity two is found which results in the matrix

$$
\mathbf{J}_c = \begin{pmatrix}
0 & 1 & 0 & 0 \\
0 & 0 & 0 & 0 \\
0 & 0 & 0 & 1 \\
0 & 0 & 0 & 0
\end{pmatrix}. \tag{5.32}
$$

These two cases occur generically with respect to the variation of damping k. At the transition of (5.31) to (5.32) there exists one special value of damping k_c for which a non-generic situation exists. This situation is easy to understand from Fig. 4.4 because in this figure the divergence and the flutter curve have a point of tangency. The corresponding $\mathbf{J}_c$ is given by

$$
\mathbf{J}_c = \begin{pmatrix}
0 & 1 & 0 & & & \\
0 & 0 & 1 & & 0 & \\
0 & 0 & 0 & & & \\
& & & 0 & 1 & 0 \\
& 0 & & 0 & 0 & 1 \\
& & & 0 & 0 & 0
\end{pmatrix}. \tag{5.33}
$$

For the reduction to bifurcation equations on the center manifold, we treat the cases given by (5.30). Especially, we explain now how the symmetry properties of the problem allow to simplify the bifurcation equations. For more complicated cases, see [138] and [137].

We point out that the content of the following two Sections 5.1.1 and 5.1.2 is quite general and completely independent of the example of the spherical pendulum. These two sections apply to any finite or infinite dimensional problem where a loss of stability of an equilibrium position of an $\mathbf{O}(2)$-symmetric system according to the eigenvalue structure of (5.30) occurs. An infinite dimensional application is given in Section 5.3.

5.2.1 Two zero roots

From center manifold theory (Section 3.1) follows that the corresponding system of bifurcation equations is two-dimensional. These bifurcation equations must be $\mathbf{O}(2)$-symmetric just as the original system. Use of

the symmetry properties can be best made by introducing a complex
variable

$$w = v_1 + iv_2 . \tag{5.34}$$

With (5.34), the unfolded (with one parameter a) bifurcation equation
up to third order terms formally reads

$$G(w, a) = \dot{w} - aw - \sum_{\substack{k,l=0 \\ k+l=3}}^{3} a_{3kl} w^k \overline{w}^l = 0 , \tag{5.35}$$

where $\overline{w}$ is the complex conjugate of w. Whether third order terms will
be sufficient to describe the local bifurcation behavior must be checked by
the concept of determinacy (Section 6.3) after the reduction process has
been performed. Only terms of odd order are included in (5.35) which
reflects a general property of symmetric oscillatory systems. To find out
which of the four nonlinear terms in the sum in (5.35) are equivariant
under an arbitrary rotation ϑ, is checked according to (5.12). The action
of the rotation operator R_ϑ on G, w and $\overline{w}$ in (5.34) and (5.35) is given
by

$$R_\vartheta G = e^{i\vartheta} G, \qquad R_\vartheta w = e^{i\vartheta} w, \qquad R_\vartheta \overline{w} = e^{-i\vartheta} \overline{w}.$$

Inserting these expressions into (5.35) yields the right-hand side of (5.12)

$$G(R_\vartheta w, a) = e^{i\vartheta} \dot{w} - a e^{i\vartheta} w - \sum_{\substack{k,l=0 \\ k+l=3}}^{3} a_{3kl} e^{i\vartheta(k-l)} w^k \overline{w}^l . \tag{5.36}$$

The left-hand side of (5.12) follows, as indicated above, from (5.35) by
multiplying it by $e^{i\vartheta}$. Inserting into (5.12), that is, comparing (5.36) with
$e^{i\vartheta} G$ results in

$$e^{i\vartheta} \dot{w} - a e^{i\vartheta} w - \sum_{\substack{k,\ell=0 \\ k+\ell=3}}^{3} a_{3k\ell} e^{i\vartheta(k-\ell)} w^k \overline{w}^\ell = e^{i\vartheta} \dot{w} - e^{i\vartheta} aw - \sum_{\substack{k,\ell=0 \\ k+\ell=3}}^{3} a_{3k\ell} e^{i\vartheta} w^k \overline{w}^\ell$$

from which the conditions

$$k - l = 1 \qquad \text{and} \qquad k + l = 3 \tag{5.37}$$

follow, which determine the admissible nonlinear terms. Equations (5.37)
have the unique solution $k = 2$, $l = 1$. Hence, from the four terms
in (5.35) only one term $w^2 \overline{w}$ is equivariant under R_ϑ. The bifurcation
equation reads

$$G(w, a) = \dot{w} - aw - a_{321} w^2 \overline{w} = 0 . \tag{5.38}$$

To find out whether the coefficient a_{321} and the parameter a in (5.38) are real or complex we make use of the fact that (5.38) must be also equivariant under reflectional symmetry. This is easy to show using complex variables because we only have to replace $w \to \overline{w}$ in (5.38) and make use of (5.12) to find

$$
\begin{aligned}
G(\overline{w}, a) &= \dot{\overline{w}} - a\overline{w} - a_{321}\overline{w}^2 w = \\
&= \overline{G}(w, a) = \dot{\overline{w}} - \overline{a}\overline{w} - \overline{a}_{321}\overline{w}^2 w \ .
\end{aligned}
\tag{5.39}
$$

From (5.39) the requirement results that $\overline{a} = a$ and $\overline{a}_{321} = a_{321}$. That is, a and a_{321} must be real. Hence, returning to (5.34) the bifurcation equations in real variables follow from (5.38) to $(A_3 := a_{321})$

$$
\begin{aligned}
\dot{v}_1 &= av_1 + A_3 v_1 (v_1^2 + v_2^2) + O(|v_1, v_2|^5) \\
\dot{v}_2 &= av_2 + A_3 v_2 (v_1^2 + v_2^2) + O(|v_1, v_2|^5) \ .
\end{aligned}
\tag{5.40}
$$

The important conclusion is that only one coefficient A_3 must be calculated in the center manifold reduction. The equations (5.40) are 3-determinate (Section 6.3) and a discussion of their solutions is given in Section 6.6.4.

5.2.2 Two purely imaginary pairs

The case of two purely imaginary pairs of eigenvalues at loss of stability of the trivial position is called *symmetric Hopf bifurcation* (Section 6.1). From (5.30)$_2$ follows that the bifurcation system on the center manifold is four-dimensional. In order to check for the $\mathbf{O}(2)$-symmetry of the terms in the bifurcation equations it is convenient to introduce again complex variables

$$
w_1 = v_1 + iv_2 \qquad \text{and} \qquad w_2 = v_3 + iv_4 \ .
\tag{5.41}
$$

This substitution allows us to write the one-parameter (a) bifurcation equations up to third order nonlinear terms in the form

$$
\dot{w}_1 = aw_1 - \omega w_2 + \sum_{\substack{j,k,l,m=0 \\ j+k+l+m=3}}^{3} a_{1jklm} w_1^j \overline{w}_1^k w_2^l \overline{w}_2^m
$$

$$
\tag{5.42}
$$

$$
\dot{w}_2 = \omega w_1 + aw_2 + \sum_{\substack{j,k,l,m=0 \\ j+k+l+m=3}}^{3} a_{2jklm} w_1^j \overline{w}_1^k w_2^l \overline{w}_2^m \ .
$$

There are 20 different cubic terms in each equation (5.42). To show which terms in (5.42) are equivariant under a rotation, we proceed as in Section 5.1.1. First, we specify the action of the rotation operator R_{ϑ} by

$$
w_1 \to e^{i\vartheta} w_1, \quad \overline{w}_1 \to e^{-i\vartheta} \overline{w}_1, \quad w_2 \to e^{i\vartheta} w_2 \quad \text{and} \quad \overline{w}_2 \to e^{-i\vartheta} \overline{w}_2
$$

and substitute these expressions into (5.42). This leads to two equations of the form

$$e^{i\vartheta}\dot{w}_1 = e^{i\vartheta}(aw_1 - \omega w_2) + \sum_{\substack{j,k,l,m=0 \\ j+k+l+m=3}}^{3} a_{1jklm}e^{i(j-k+l-m)\vartheta}w_1^j\overline{w}_1^k w_2^l\overline{w}_2^m . \quad (5.43)$$

To satisfy (5.12), we multiply (5.42)$_1$ by $e^{i\vartheta}$ and compare, analogously, as in Section 5.1.1 this equation with (5.43). This procedure results in the following two equations to determine those terms in (5.42)$_1$ which are equivariant under the symmetry conditions:

$$\begin{aligned} j + k + l + m &= 3 \\ j - k + l - m &= 1 . \end{aligned} \quad (5.44)$$

Equations (5.44) have the following six solutions. These, together with the corresponding coefficients and terms in (5.42)$_1$ are

$$\begin{array}{lll} j = 2, k = 1, l = m = 0 & a_{12100} & w_1^2\overline{w}_1 \\ j = 2, m = 1, k = l = 0 & a_{12001} & w_1^2\overline{w}_2 \\ l = 2, m = 1, j = k = 0 & a_{10021} & w_2^2\overline{w}_2 \\ l = 2, k = 1, j = m = 0 & a_{10120} & \overline{w}_1 w_2^2 \\ j = 1, k = 1, l = 1, m = 0 & a_{11110} & w_1\overline{w}_1 w_2 \\ j = 1, k = 0, l = 1, m = 1 & a_{11011} & w_1 w_2\overline{w}_2. \end{array} \quad (5.45)$$

Hence, only six of the original twenty terms in each of the two equations (5.42) satisfy the rotational symmetry requirement. The reflectional symmetry yields analogously to the case of the divergence bifurcation in Section 5.1.1 that the coefficients a_{1jklm} and the unfolding parameter a must be real. We remark that in the equivariance test above we should have worked with 4×4 rotation and reflection matrices and in addition also the conjugate complex equations to (5.42) should have been used. However the results are the same as obtained here.

In view of the further treatment of these equations by means of *normal form theory* (Section 6.6.4) we introduce new variables ([155])

$$z_1 = w_1 + iw_2 \quad \text{and} \quad z_2 = \overline{w}_1 + i\overline{w}_2 . \quad (5.46)$$

Inserting (5.46) into (5.42) under consideration of (5.45), we obtain

$$\begin{aligned} \dot{z}_1 &= (a + i\omega)z_1 + c_1 z_1|z_1|^2 + c_2 z_1^2 z_2 + \\ &\quad + c_3|z_1|^2\overline{z}_2 + c_4 z_1|z_2|^2 + c_5\overline{z}_1\overline{z}_2^2 + c_6\overline{z}_2|z_2|^2 + O(|z,\overline{z}|^5) \\ \dot{z}_2 &= (a + i\omega)z_2 + c_1 z_2|z_2|^2 + c_2 z_1 z_2^2 + \\ &\quad + c_3\overline{z}_1|z_2|^2 + c_4|z_1|^2 z_2 + c_5\overline{z}_1^2\overline{z}_2 + c_6|z_1|^2\overline{z}_1 + O(|z,\overline{z}|^5) . \end{aligned} \quad (5.47)$$

Here $|z|^2 = z\overline{z}$. We see that (5.47) is given in an especially simple form but again has complex coefficients. We remark at this point that the application of normal form theory (Section 6.6.4) to (5.47) will further reduce the number of nonlinear terms in each equation from six to two.

5.3 Infinite dimensional statical system: annular plate

This problem is studied in [45] as example for the application of symmetry concepts. These concepts allow to determine those terms in the bifurcation equations which under the symmetry conditions of the problem may appear. The calculation of the coefficients of these terms, however, requires the application of the Ljapunov-Schmidt method and is performed in [91], [90].

Plate equations and boundary conditions

The annular plate with the used notation is given in Fig. 5.3. The *von*

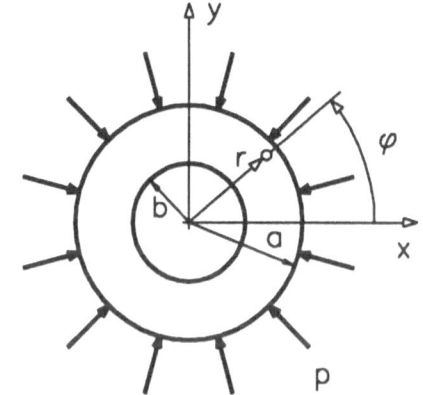

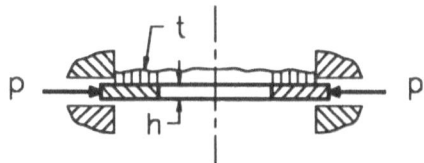

Figure 5.3. Annular plate under in-plane thrust p and transversal loading t. The outer radius a will be normalized to $a = 1$

Karman plate equations (J.23) are used as governing equations. They are in polar coordinates (transversal loading $t(r, \varphi) = 0$)

$$\Delta^2 w + p\frac{b^2}{1 - b^2}\left\{\left(\frac{1}{b^2} + \frac{1}{r^2}\right)\Delta w - \frac{2}{r^2}w_{,rr}\right\} - [w, f] = 0$$

$$\Delta^2 f + \frac{1}{2}[w, w] = 0 \tag{5.48}$$

with the boundary conditions in $r = a = 1$

$$w = 0, \qquad w_{,r} = 0 \tag{5.49}$$

$$N_r = \frac{1}{r}f_{,r} + \frac{1}{r^2}f_{,\varphi\varphi} = -p, \qquad N_{r\varphi} = \frac{1}{r^2}f_{,\varphi} - \frac{1}{r}f_{,r\varphi} = 0 \qquad (5.50)$$

and in $r = b$ $(0 \le b < 1)$

$$M_r = -K\left[w_{,rr} + \nu\left(\frac{1}{r}w_{,r} + \frac{1}{r^2}w_{,\varphi\varphi}\right)\right] = 0$$

$$V_r = Q_r + \frac{1}{r}M_{r\varphi,\varphi} \qquad\qquad\qquad\qquad (5.51)$$

$$= -K\left[(\Delta w)_{,r} + \frac{1-\nu}{r}\left(\frac{1}{r}w_{,r\varphi} - \frac{1}{r^2}w_{,\varphi}\right)_{,\varphi}\right] = 0$$

$$N_r = \frac{1}{r}f_{,r} + \frac{1}{r^2}f_{,\varphi\varphi} = 0, \qquad N_{r\varphi} = \frac{1}{r^2}f_{,\varphi} - \frac{1}{r}f_{,r\varphi} = 0. \qquad (5.52)$$

Furthermore, the operators Δ^2 and $[\cdot, \cdot]$ are given by

$$\Delta^2 w = w_{,rrrr} + \frac{2}{r}w_{,rrr} - \frac{1}{r^2}w_{,rr} + \frac{2}{r^2}w_{,rr\varphi\varphi} + \frac{1}{r^3}w_{,r} -$$

$$- \frac{2}{r^3}w_{,r\varphi\varphi} + \frac{1}{r^4}w_{,\varphi\varphi\varphi\varphi} + \frac{4}{r^4}w_{,\varphi\varphi}$$

$$(5.53)$$

$$[w,f] = w_{,rr}\left(\frac{1}{r}f_{,r} + \frac{1}{r^2}f_{,\varphi\varphi}\right) + f_{,rr}\left(\frac{1}{r} - w_{,r} + \frac{1}{r^2}w_{,\varphi\varphi}\right) -$$

$$- 2\left(\frac{1}{r}w_{,r\varphi} - \frac{1}{r^2}w_{,\varphi}\right)\left(\frac{1}{r}f_{,r\varphi} - \frac{1}{r^2}f_{,\varphi}\right).$$

Similar to the rectangular plate in Section 4.3.4, the stress function f can be eliminated from $(5.48)_1$. This is achieved by expressing f from $(5.48)_2$ as

$$f = -\frac{1}{2}\Delta^{-2}[w,w] . \qquad (5.54)$$

The calculation of Δ^{-2} in (5.54) depends on the boundary conditions (5.50) and (5.52) and is explained in Appendix E. Inserting (5.54) into $(5.48)_1$ yields the governing equation

$$G(w,p) = \Delta^2 w + p\frac{b^2}{1-b^2}\left\{\left(\frac{1}{b^2} + \frac{1}{r^2}\right)\Delta w - \frac{2}{r^2}w_{,rr}\right\} +$$

$$+ \frac{1}{2}[w, \Delta^{-2}[w,w]] = 0 , \qquad (5.55)$$

which with the boundary conditions (5.49) and (5.51) describes the buckling problem.

Symmetry properties

The buckling problem of the annular plate possesses the following two symmetries:

(1) Reflection about the middle plane of the plate. That is, replacing w by $-w$ implies according to (5.12)

$$G(-w,p) = -G(w,p) \ .$$

This is obviously the case for (5.55) and the boundary conditions. Physically this symmetry (called $\mathbf{Z}_2$-symmetry) means that buckling up is the same as buckling down.

(2) The rotational symmetry $\mathbf{O}(2)$ (Appendix G.7). This means that (5.55) is equivariant under a rotation ϑ that is, $T_\vartheta\colon \varphi \to \varphi + \vartheta$ and under a reflection, that is, $T_s\colon \varphi \to -\varphi$. The equivariance of (5.55) under these operations is obvious because first, the two operators (5.53) are invariant under these operations. This follows from the fact that φ is not explicitly contained in the coefficients of the two operators (Appendix G.4) and the derivatives with respect to φ in all expressions occur only in even order. Second, the boundary conditions N_r given by $(5.50)_1$ and $(5.52)_1$ and M_r and V_r given by (5.51) are also invariant. Third, $N_{r\varphi}$ given by $(5.50)_2$ and $(5.52)_2$ are equivariant. All these properties are physically quite clear, because the quantities having only the index r are independent of φ and those having indices $r\varphi$ change sign if φ changes the direction. Hence, for (5.55) the relationship (5.12) in the form $T_g G(w,p) = G(T_g w, p)$ holds.

Finally we explain the calculation of $T_g G(w,p)$ and $G(T_g w,p)$. If we apply the nonlinear operator G on a function $w(r,\varphi)$ we obtain, using the notation and the results of Appendix G.4, a new function $e(r,\varphi)$ given by $e(r,\varphi) = G(w(r,\varphi),p)$. Hence, $T_\vartheta e(r,\varphi) = e(r, T_\vartheta^{-1}\varphi) = e(r,\varphi - \vartheta)$ and $T_s e(r,\varphi) = e(r,-\varphi)$. On the other hand $G(T_\vartheta w(r,\varphi),p) = G(w(r,T_\vartheta^{-1}\varphi),p) = G(w(r,\varphi - \vartheta),p)$ and $G(T_s w(r,\varphi),p) = G(w(r,-\varphi),p)$.

The whole symmetry of the problem is $\mathbf{O}(2) \times \mathbf{Z}_2$.

The linear eigenvalue problem

The consequences of the symmetry properties are immediately seen when the linear eigenvalue problem is studied (Appendix E). From (5.55) follows

$$Lw = pMw \qquad\qquad (5.56)$$

where the operators L and M are given by

$$Lw = \Delta^2 w, \qquad Mw = -\frac{b^2}{1-b^2}\left\{\left(\frac{1}{b^2}+\frac{1}{r^2}\right)\Delta w - \frac{2}{r^2}w_{,rr}\right\}. \qquad (5.57)$$

The linear operator of the eigenvalue problem (5.56) is self-adjoint and positive definite (Appendix E). Hence, the eigenfunctions are orthogonal and the eigenvalues are real and positive. For the solution of (5.56) a separation of variables in the form

$$w_{nm}(r,\varphi) = g_{nm}(r)\left(\begin{array}{c}\cos n\varphi \\ \sin n\varphi\end{array}\right), \qquad \begin{array}{l}n = 0,1,2,\ldots, \\ m = 1,2,3,\ldots\end{array} \qquad (5.58)$$

can be used. It does not matter whether we choose $\cos n\varphi$ or $\sin n\varphi$. In any case we obtain for p and $g_{nm}(r)$ the eigenvalue problem

$$L_n g_{nm} = pM_n g_{nm} \qquad (5.59)$$

which depends only on the variable r. The operators L_n and M_n are given by

$$
\begin{aligned}
L_n g &= g_{,rrrr} + \frac{2}{r}g_{,rrr} - \frac{1+2n^2}{r^2}\left(g_{,rr} - \frac{1}{r}g_{,r}\right) + \frac{n^2(n^2-4)}{r^4}g \\
M_n g &= -\frac{b^2}{1-b^2}\left\{\left(\frac{1}{b^2}-\frac{1}{r^2}\right)g_{,rr} + \left(\frac{1}{b^2}+\frac{1}{r^2}\right)\left(\frac{1}{r}g_{,r} - \frac{n^2}{r^2}g\right)\right\}
\end{aligned}
\qquad (5.60)
$$

with the boundary conditions at $r = 1$

$$
\begin{aligned}
g_{nm} &= 0 \\
g_{nm,r} &= 0
\end{aligned}
\qquad (5.61)
$$

and at $r = b$

$$g_{nm,rr} + \nu\left(\frac{1}{r}g_{nm,r} + \frac{n^2}{r^2}g_{nm}\right) = 0 \qquad (5.62)$$

$$g_{nm,rrr} + \frac{1}{r}g_{nm,rr} - \frac{1+n^2(2-\nu)}{r^2}g_{nm,r} + \frac{n^2(3-\nu)}{r^3}g_{nm} = 0\,.$$

For $n = 0$ and $n = 1$ analytical solutions of (5.59)–(5.62) are given in form of Bessel functions in [95]. However, for $n \geq 2$ analytical solutions of the eigenvalue problem (5.59)–(5.62) are not known. In [91] the numerical procedure COLSYS is used to calculate the eigenfunctions for $n \geq 2$. These numerical results seem to be very accurate because there is excellent agreement with those which can be obtained analytically (that is, for $n = 0, 1$). The results are shown in Fig. 5.4, Fig. 5.5 and 5.6 for $\nu = 0.3$. The index m indicates that there is an infinite number of solutions of

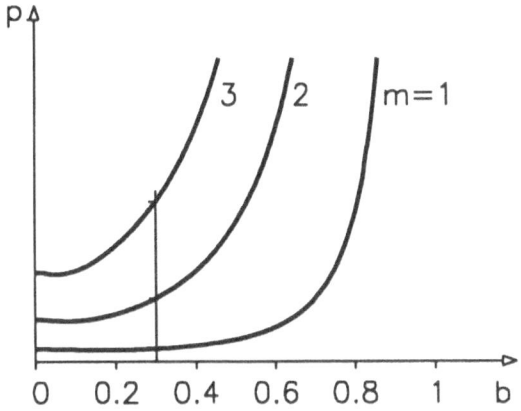

Figure 5.4. Eigenvalues p_{nm} of (5.59) depending on the radius of the hole b for $n = 0$ and different values of m; $m = 1$ corresponds to the smallest and, hence, physically relevant value

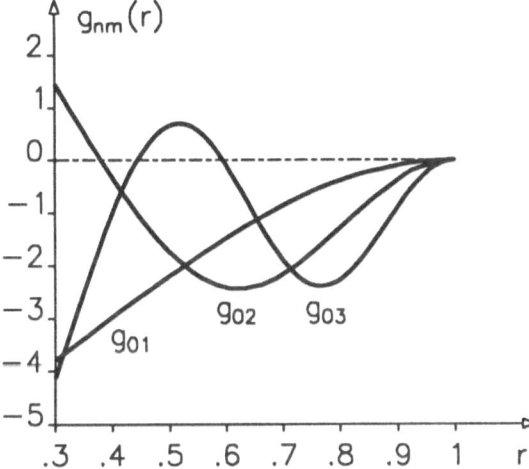

Figure 5.5. Mode shapes g_{nm} for $b = 0.3$, $n = 0$ and the values $m = 1, 2, 3$ of Fig. 5.4; $m = 1$ is the physically relevant value

(5.59)–(5.62) for each n. The numbering is ordered by $p_{nm} < p_{nm+1}$. From Fig. 5.4 follows immediately that only the case $m = 1$ needs to be studied, because this value corresponds to the lowest in-plane buckling load. In Fig. 5.5, the mode shapes corresponding to the three eigenvalues p_{nm} with $m = 1, 2, 3$ are shown for $n = 0$ and $b = 0.3$. The meaning of m for this problem is the same as that of n in (4.83) and (4.84) for the buckling problem of an axially compressed rod. The higher order modes are physically irrelevant and the physically relevant mode is g_{01} which corresponds to $m = 1$.

In Fig. 5.6 only curves for $m = 1$ are shown and it gives the influence of the size b of the hole, on the circumferential mode number n. For approximately $b < 0.51$, the plate buckles in the circular symmetric mode given by $n = 0$. It is interesting to note that this first eigenvalue is simple. This follows from the fact that the corresponding eigenfunction w is still

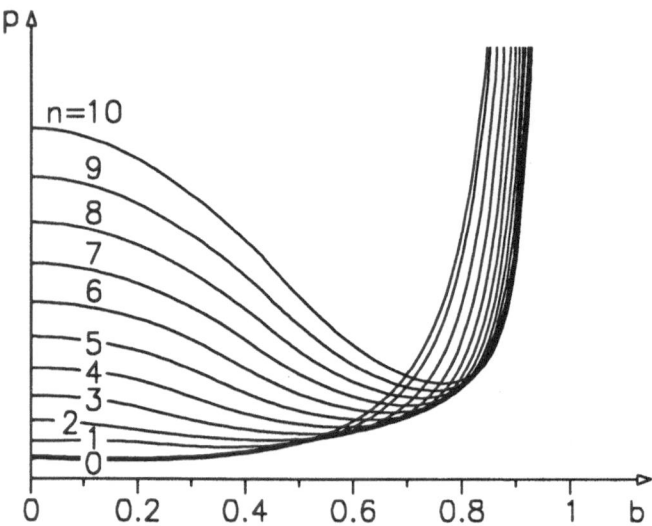

Figure 5.6. The stability boundary in the p, b parameter space is formed by the envelope of the p_{n1} curves for $n = 0, 1, \ldots, 10$

O(2)-symmetric. This means that if T_ϑ is a representation of **SO**(2) the relation $w = T_\vartheta w$ holds. Whereas for the double eigenvalue as it is given by (5.58) the relation $w \neq T_\vartheta w$ holds. This is easy to understand if we take $w(r, \varphi) = g_{n1}(r) \cos n\varphi$ from (5.58). Application of T_ϑ means to replace $\varphi \rightarrow \varphi + \vartheta$. Thus we obtain $w(r, \varphi) = g_{n1}(r) \cos n(\varphi + \vartheta)$ which in general is linearly independent from the original function (see also p. 148).

At approximately $b = 0.51$, the curves $n = 0$ and $n = 1$ have a point of intersection. For $n = 1$, there are according to (5.58) two solutions, one with $\cos \varphi$ and the other with $\sin \varphi$. Hence, this point of intersection corresponds to a triple eigenvalue. Increasing b further the next point of intersection is reached, where the curves $n = 1$ and $n = 2$ intersect. This and all other points of intersection for increasing values of b correspond to fourfold eigenvalues. The curves of the envelope together with the points of intersection form the stability boundary in the p, b parameter space. The curves correspond to double eigenvalues except for $0 \leq b \leq 0.51$ where the symmetric buckling mode corresponds to a simple eigenvalue. We see that the multiplicity of the critical eigenvalues is in general twice that for the rectangular plate. This fact is due to **O**(2)-symmetry of the problem.

Ljapunov-Schmidt reduction

At this stage of the analysis, the buckling problem of the annular plate appears to be much more complicated than the corresponding problem for the rectangular plate because the higher multiplicity of the critical eigenvalue has a strong influence on the dimension of the bifurcation equations. On the other hand due to the rotational symmetry it will be possible to simplify the bifurcation equations considerably.

The simplest set of bifurcation equations will be obtained if they are derived at a value of b which corresponds to simple ($n = 0$) or double eigenvalues. These are the generic cases. The non-generic cases correspond to multiple eigenvalues which are either triple or fourfold.

We perform the Ljapunov-Schmidt reduction at the triple eigenvalue. According to (3.53) we set

$$w(r, \varphi) = q_0 w_0 + q_{1c} w_{1c} + q_{1s} w_{1s} + w_s(q_0, q_{1c}, q_{1s}) , \qquad (5.63)$$

where q_0, q_{1c}, q_{1s} are the amplitudes of the buckling modes, and hence, the variables in the three bifurcation equations. $w_0(r) = g_{01}(r)$, $w_{1c}(r, \varphi) = g_{11}(r) \cos \varphi$ and $w_{1s}(r, \varphi) = g_{11}(r) \sin \varphi$ are the buckling modes (eigenfunctions) (5.58) corresponding to the triple eigenvalue. w_s includes all the non-critical variables corresponding to the modes which are orthogonal to the buckling modes. Again, according to (3.57) and similar to (4.165), we obtain

$$w_s(q_0, q_{1c}, q_{1s}) = O(|q_0|^2 + |q_{1c}|^2 + |q_{1s}|^2). \qquad (5.64)$$

Analogous to the analysis of the rectangular plate (Section 4.3.4) only third order terms must be included in the bifurcation equations because these terms completely determine the local bifurcation behavior.

Since this buckling problem is conservative, the three bifurcation equations must be derivable by taking the gradient of a discrete potential $V(q_0, q_{1c}, q_{1s})$. That is, a set of three scalar equations $\partial V / \partial q_0 = 0$, $\partial V / \partial q_{1c} = 0$ and $\partial V / \partial q_{1s} = 0$ will be obtained. As mentioned above we assume the bifurcation equations to be three-determinate (Section 6.3), and hence, the potential V will be formed by terms of at most fourth order. There are 10 terms of third and 15 terms of fourth order. Thus we obtain a total of 25 terms in the potential which could contribute to the nonlinear terms in the bifurcation equations. From the reflectional symmetry with respect to the middle plane of the plate, it is clear that only terms of fourth order are allowed to be included in V (see (5.1)). This requirement reduces the number of terms in V to 15. Due to rotational symmetry, the only way q_{1c} and q_{1s} can appear is in the form $q_{1c}^2 + q_{1s}^2$, as is shown in Appendix G.5. Hence, the terms allowed to occur in the potential V reduce from 15 to 3 and are

$$\begin{aligned} V &= \frac{1}{4} B q_0^4 + \frac{1}{2} C q_0^2 (q_{1c}^2 + q_{1s}^2) + \frac{1}{4} D (q_{1c}^2 + q_{1s}^2)^2 - \\ &\quad - \frac{1}{2} \alpha_0 \tilde{\lambda} q_0^2 - \frac{1}{2} \alpha_1 \tilde{\lambda} (q_{1c}^2 + q_{1s}^2), \end{aligned} \qquad (5.65)$$

where $\tilde{\lambda} = p - p_c$. The potential (5.65) has been supplemented by the quadratic terms accounting for the thrust p. This potential describes the so-called perfect or one parameter $(\tilde{\lambda})$ bifurcation problem. From (5.65) follow the bifurcation equations

$$
\begin{aligned}
Bq_0^3 + Cq_0(q_{1c}^2 + q_{1s}^2) - \alpha_0\tilde{\lambda}q_0 &= 0 \\
D(q_{1c}^2 + q_{1s}^2)q_{1c} + Cq_0^2q_{1c} - \alpha_1\tilde{\lambda}q_{1c} &= 0 \qquad (5.66) \\
D(q_{1c}^2 + q_{1s}^2)q_{1s} + Cq_0^2q_{1s} - \alpha_1\tilde{\lambda}q_{1s} &= 0 \; .
\end{aligned}
$$

Introducing a new variable q_1 defined by $q_1\cos\vartheta = q_{1c}$ and $q_1\sin\vartheta = q_{1s}$, the following two equations

$$
\begin{aligned}
Bq_0^3 + Cq_0q_1^2 - \alpha_0\tilde{\lambda}q_0 &= 0 \\
Cq_0^2q_1 + Dq_1^3 - \alpha_1\tilde{\lambda}q_1 &= 0
\end{aligned}
\qquad (5.67)
$$

are obtained. The physical meaning of q_1 and ϑ becomes clear by adding the two buckling modes

$$
\begin{aligned}
q_{1c}g_{11}\cos\varphi + q_{1s}g_{11}\sin\varphi &= q_1g_{11}\cos\vartheta\cos\varphi + q_1g_{11}\sin\vartheta\sin\varphi \\
&= q_1g_{11}\cos(\varphi - \vartheta) \; .
\end{aligned}
$$

Hence, q_1 is the amplitude of the nonsymmetric mode and ϑ describes its location in circumferential direction. The coefficients B, C, D, α_0 and α_1 are the only ones that appear in the one-parameter bifurcation equations. They must be determined by means of the Ljapunov-Schmidt method. These calculations are done in [91].

In conclusion, we can say that due to the triple eigenvalue, the system of bifurcation equations is three-dimensional. However, making use of the fact that the bifurcation equations must be also rotationally symmetric, the three-dimensional system is reduced to a two-dimensional one. Hence, the buckling behavior of the annular plate at a three-fold eigenvalue is formally governed by the same bifurcation equations as the rectangular plate at a double eigenvalue.

The reader may wonder why we dealt with the single special case of the three-fold eigenvalue instead of treating a four-fold eigenvalue which occurs infinitely often. However, the treatment of the four-fold eigenvalue is much more complicated. We only indicate this with the following comments. The potential (5.65) for the three-fold eigenvalue does not depend on the angle ϑ introduced above but only on the amplitude q_1. This follows from the fact that the mode with amplitude q_0 is rotationally symmetric, and hence, the mode with amplitude q_1 can be moved in circumferential direction without influencing the potential. However, for a four-fold eigenvalue two angles ϑ_1 and ϑ_2 are present and the corresponding potential is no longer independent of them. Due to the rotational symmetry it is still possible to reduce the dependence of the potential on ϑ_1 and ϑ_2 to a difference between these two angles called a *resonance angle* but the dependence of the potential on the relative location of the two modes cannot be eliminated completely. This is easy to understand physically because the potential will depend on the relative position of the two modes. This dependence on the resonance angle requires bifurcation equations of higher than third order to completely determine the problem.

5.4 Infinite dimensional dynamical system: three-dimensional fluid-conveying viscoelastic tube

In this section we give an application of the reduction process to an infinite-dimensional dynamical system with certain symmetry properties. It is the fluid-conveying tube of Fig. 5.7. Fluid-conveying tubes display a very interesting behavior after loss of stability of the trivial

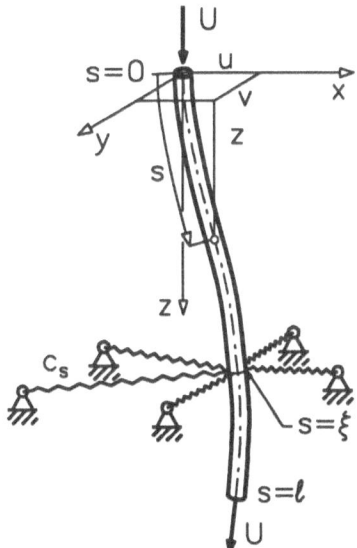

Figure 5.7. Model of the fluid conveying tube with rotationally symmetric intermediate elastic support at $s = \xi$

downhanging equilibrium position for which also experiments are available ([148]). The behavior of the tube without elastic support is studied in [11]. The presence of the rotationally symmetric elastic support can lead to multiple bifurcations. Analogously to the spherical double pendulum of Section 5.1 this problem is $O(2)$-symmetric, and hence, all critical eigenvalues appear with multiplicity two, doubling the dimension of the bifurcation equations on the center manifold. However, the use of symmetry properties and normal form theory allows to simplify the bifurcation equations significantly.

The equations of motion of the tube (L.25), the boundary and the conditions at the elastic support with the assumptions necessary for their derivation are given in Appendix L.

Symmetry properties

Equations (L.25) can be written in short-hand notation as

$$\mathbf{G}(\psi, \lambda) = \mathbf{0} . \tag{5.68}$$

Here, $\mathbf{G} = (G_1, G_2)^T$, $\psi = (u, v)^T$ and λ is the vector of the main parameters given by (5.78). First, we show the equivariance of (5.68) under a rotation about the z-axis and a reflection about any plane containing the z-axis. We know that we must show the reflectional symmetry only for one plane because the reflection about any other plane can be constructed by the reflection about the choosen plane and a rotation (see Appendices G.2, G.7 and Section 5.2). We select the plane of reflection such that an interchange of u and v occurs. This requires to set $\vartheta = \pi/2$ in (G.26). The location of this reflection plane follows from (G.27) and is at $\vartheta = \pi/4$. The representations of the rotation and reflection are given by (Appendix G.7)

$$\mathbf{T}_\vartheta = \begin{pmatrix} \cos\vartheta & \sin\vartheta \\ -\sin\vartheta & \cos\vartheta \end{pmatrix}, \qquad \mathbf{T}_s = \begin{pmatrix} 0 & 1 \\ 1 & 0 \end{pmatrix} . \tag{5.69}$$

From (5.12) follows the equivariance relation

$$\mathbf{T}\mathbf{G}(\psi, \lambda) = \mathbf{G}(\mathbf{T}\psi, \lambda) \tag{5.70}$$

where for $\mathbf{T}$ either $\mathbf{T}_\vartheta$ or $\mathbf{T}_s$ from (5.69) must be substituted. That (5.70) holds for (L.25) with (5.69) can be verified by substitution and calculation.

In case of $\mathbf{T}_s$, this is obvious because on the left-hand side of (5.70) the components of $\mathbf{G}$ are exchanged and on the right-hand side u and v are exchanged.

Taking $\mathbf{T}_\vartheta$ in (5.70) yields on the left-hand side

$$\cos\vartheta G_1 \quad + \quad \sin\vartheta G_2$$
$$-\sin\vartheta G_1 \quad + \quad \cos\vartheta G_2$$

whereas on the right-hand side $u \to u\cos\vartheta + v\sin\vartheta$ and $v \to -u\sin\vartheta + v\cos\vartheta$. For the linear terms, the equivariance is obvious. To show that it is also satisfied for the nonlinear terms, we pick as an example the term $(EJ_B\kappa^2 u')'$. From the left-hand side of (5.70) follows the first component

$$(EJ_B\kappa^2 u')'\cos\vartheta + (EJ_B\kappa^2 v')'\sin\vartheta = [EJ_B\kappa^2(u'\cos\vartheta + v'\sin\vartheta)]' .$$

Making use of (L.9), the first component on the right-hand side of (5.70) reads

$$[EJ_B\{(u'')^2 + (v'')^2 + (z'')^2\}u']' \rightarrow [EJ_B\{(u''\cos\vartheta + v''\sin\vartheta)^2 +$$
$$+ (-u''\sin\vartheta + u''\cos\vartheta)^2 + (z'')^2\}(u'\cos\vartheta + v'\sin\vartheta)]'$$
$$= [EJ_B\{(u'')^2 + (v'')^2 + (z'')^2\}(u'\cos\vartheta + v'\sin\vartheta)]'$$

which is exactly the same as the expression above. The same is true for the boundary conditions. Hence, the problem is both equivariant with respect to $\mathbf{T}_s$ and $\mathbf{T}_\vartheta$, and therefore $\mathbf{O}(2)$-symmetric.

Formulation of the dynamical system in dimensionless variables

It is convenient to introduce non-dimensional variables by the following change of variables

$$s \rightarrow \frac{s}{L}, \quad t \rightarrow \sqrt{\frac{EJ_B}{m_T + m_F}}\frac{t}{L^2}, \quad u \rightarrow \frac{u}{L}, \quad v \rightarrow \frac{v}{L}. \tag{5.71}$$

Result of this non-dimensionalization process are the following dimensionless parameters

$$\alpha^\star = \sqrt{\frac{J_B}{(m_T + m_F)E\,L^2}}\frac{\alpha}{L}, \qquad \beta = \frac{m_F}{m_T + m_F},$$

$$\gamma = \frac{m_T + m_F}{EJ_B}L^2 g, \qquad c = \frac{c_s L^3}{EJ_B}, \tag{5.72}$$

$$\mu^\star = \frac{\mu L^2}{\sqrt{EJ_B(m_T + m_F)}}, \qquad \varrho = \sqrt{\frac{m_F}{EJ_B}}UL.$$

Introducing new variables

$$w_1 = u, \quad w_2 = \dot{u}, \quad w_3 = v, \quad w_4 = \dot{v}, \tag{5.73}$$

the system (L.25) of two nonlinear partial differential equations of second order in time is transformed into a system of four first order partial differential equations of the form (3.61)

$$\dot{w} = \mathbf{A}(\lambda)w + \mathbf{g}(w, \lambda), \tag{5.74}$$

where $w(s,t) = (w_1, w_2, w_3, w_4)^T$ is a vector in Hilbert space. The linear operator $\mathbf{A}(\lambda)$ follows from a straightforward calculation from (L.25) to

$$\mathbf{A}(\lambda) = \begin{pmatrix} \mathbf{A}_1(\lambda) & 0 \\ 0 & \mathbf{A}_1(\lambda) \end{pmatrix} \tag{5.75}$$

where

$$\mathbf{A}_1(\lambda) = \begin{pmatrix} 0 & 1 \\ -C & -B \end{pmatrix} . \tag{5.76}$$

C and B are differential operators in the (dimensionless) position variable s and are defined by

$$\begin{aligned} Cw_1 &= w_1^{IV} + \varrho^2 w_1'' - \gamma[(1-s)w_1']' \\ Bw_2 &= \mu^* w_2 + 2\sqrt{\beta}\varrho w_2' + \alpha^* w_2^{IV} . \end{aligned} \tag{5.77}$$

Analogous to the spherical double pendulum of Section 5.1, where the linear problem decomposed into two identical problems, each having half of the dimension of the system, also $\mathbf{A}(\lambda)$ given by (5.75) decomposes into two identical operators $\mathbf{A}_1(\lambda)$ (5.76). Hence, only the eigenvalues of $\mathbf{A}_1(\lambda)$ must be calculated, but of course, each eigenvalue of $\mathbf{A}_1(\lambda)$ will be doubled.

The vector of main parameters λ is given by

$$\lambda = (\varrho, c, \beta, \alpha^*)^T \in \mathbb{R}^4 . \tag{5.78}$$

As its distinguished component we will select either ϱ or c. According to (5.72), ϱ is proportional to the flow rate U and c to the stiffness of the support c_s. Later it will turn out that the mass ratio β and the coefficient of the internal damping α^* are also important physical quantities for the behavior of the system.

Stability boundary in parameter space

The tube is assumed to be in a perfectly straight downhanging equilibrium position which is asymptotically stable if the flow rate U which is proportional to ϱ is small. Keeping all other parameters in (5.78) fixed, the flow rate is increased quasistatically (Section 3.1.1) until the stable equilibrium position loses its stability at the critical value $\varrho = \varrho_c$. The calculation of ϱ_c requires the solution of an inverse eigenvalue problem. This follows from the linear evolution equation

$$\dot{u} = \mathbf{A}_1(\lambda)u \tag{5.79}$$

where $\mathbf{A}_1$ is given by (5.76) and (5.77), λ by (5.78), and $u = (u, \dot{u})^T = (w_1, w_2)^T$.

The corresponding boundary and intermediate (that is, at the location of the elastic support) conditions are

$$\begin{aligned} s = 0 : \quad & u = 0, \ u' = 0 \\ s = \xi : \quad & u'''(\xi_+) + \alpha^* \dot{u}'''(\xi_+) - u'''(\xi_-) - \alpha^* \dot{u}'''(\xi_-) = -cu(\xi) \\ & u, u', u'' + \alpha^* \dot{u}'' \quad \text{are continuous} \\ s = 1 : \quad & u'' + \alpha^* \dot{u}'' = 0, \ u''' + \alpha^* \dot{u}''' = 0 . \end{aligned} \tag{5.80}$$

First, we look for a simple zero eigenvalue of $\mathbf{A}_1(\lambda)$. Performing a separation of variables, we substitute

$$u(s,t) = e^{\sigma t}x(s) ,$$

where $x(s) = (x_1(s), x_2(s))^T$, into (5.79) to obtain (with $\sigma = 0$)

$$(\mathbf{A}_1(\lambda) - 0\mathbf{E})x = \begin{pmatrix} 0 & 1 \\ -C & -B \end{pmatrix} \begin{pmatrix} x_1 \\ x_2 \end{pmatrix} = 0$$

or

$$\begin{aligned} x_2 &= 0 \\ -Cx_1 - Bx_2 &= 0 . \end{aligned} \tag{5.81}$$

From (5.81) follows

$$x_2 = 0 , \qquad Cx_1 = x_1^{IV} + \varrho^2 x_1'' + \gamma[(s-1)x_1']' = 0 \tag{5.82}$$

with the boundary and intermediate conditions

$$\begin{aligned} s = 0: \quad & x_1(0) = 0 , \quad x_1'(0) - 0 \\ s = \xi: \quad & x_1'''(\xi_+) - x_1'''(\xi_-) + cx_1(\xi) = 0 , \\ & x_1, x_1', x_1'' \text{ continuous} \\ s = 1: \quad & x_1''(1) = 0 , \quad x_1'''(1) = 0 . \end{aligned} \tag{5.83}$$

Because of the presence of the gravitational term $\gamma[(s-1)x_1']'$ in (5.82) the problem is non-autonomous and a closed form solution of (5.82) and (5.83) is not known. Therefore, in [138] the solution is given numerically using the boundary value problem solver BNDSCO ([106]). However, if we neglect the influence of gravity (as it is done in [11]) that is, if we set $\gamma = 0$ then (5.82) represents an autonomous fourth order system that can be easily solved. For its solution, we obtain

$$x_1(s) = \begin{cases} c_1 \cos \varrho s + c_2 \sin \varrho s + c_3 + c_4\varrho s & s \leq \xi \\ c_5 \cos \varrho s + c_6 \sin \varrho s + c_7 + c_8\varrho s & s > \xi . \end{cases} \tag{5.84}$$

Making use of the boundary conditions (5.83), we find the following equations that specify the constants $c_1, \ldots, c_8$:

$$x_1(0) = 0 : \qquad\qquad c_1 + c_3 = 0$$

$$x_1'(0) = 0 : \qquad\qquad c_2 + c_4 = 0$$

$$x_1''(1) = 0 : \qquad\qquad -c_5 \cos \varrho - c_6 \sin \varrho = 0$$

$$x_1'''(1) = 0 : \qquad\qquad c_5 \sin \varrho - c_6 \cos \varrho = 0$$

$$x_1''(\xi) \text{ is continuous:} \quad -\varrho^2 c_1 \cos \varrho\xi - \varrho^2 c_2 \sin \varrho\xi = 0 \qquad (5.85)$$

$$x_1'''(\xi_-) = c x_1(\xi) : \quad \varrho^3 c_1 \sin \varrho\xi - \varrho^3 c_2 \cos \varrho\xi =$$

$$c[c_1(\cos \varrho\xi - 1) + c_2(\sin \varrho\xi - \varrho\xi)]$$

$$\begin{aligned} x_1(\xi_-) &= x_1(\xi_+), \\ x_1'(\xi_-) &= x_1'(\xi_+) \end{aligned} \qquad \text{allow the calculation of } c_7 \text{ and } c_8 .$$

We immediately obtain that the constants $c_5 = c_6 = 0$. This result is already used in the conditions $(5.85)_5$ and $(5.85)_6$ at the location $s = \xi$ of the support. The critical parameter value is determined by the solution of the homogenous system of equations $(5.85)_5$ and $(5.85)_6$ for c_1 and c_2 which reads

$$\begin{pmatrix} -\varrho^2 \cos \varrho\xi & -\varrho^2 \sin \varrho\xi \\ \varrho^3 \sin \varrho\xi + c(1 - \cos \varrho\xi) & -\varrho^3 \cos \varrho\xi + c(\varrho\xi - \sin \varrho\xi) \end{pmatrix} \begin{pmatrix} c_1 \\ c_2 \end{pmatrix} = \begin{pmatrix} 0 \\ 0 \end{pmatrix} .$$

$$(5.86)$$

Nontrivial solutions for (c_1, c_2) of (5.86) are obtained if the determinant of the coefficient matrix is zero. We find

$$\varrho^2[\varrho^3 + c(\sin \varrho\xi - \varrho\xi \cos \varrho\xi)] = 0$$

or

$$c = \frac{\varrho^3}{\varrho\xi \cos \varrho\xi - \sin \varrho\xi} . \qquad (5.87)$$

Despite the neglection of gravity ($\gamma = 0$) (5.87) is a good approximation to the stability boundary corresponding to the zero eigenvalue in Fig. 5.8–Fig. 5.10. To see the influence of gravity in Fig. 5.8 the stability boundaries with ($\gamma = 21.4$) and without ($\gamma = 0$) the gravitational effect are drawn.

Second, we look for purely imaginary eigenvalues $\sigma = \pm i\omega$. The calculations are simplified if we introduce complex variables

$$u(s, t) = e^{i\omega t} z(s) ,$$

where

$$z = (z_1, z_2)^T \quad \text{and} \quad z_j(s) = x_j(s) + i y_j(s) . \qquad (5.88)$$

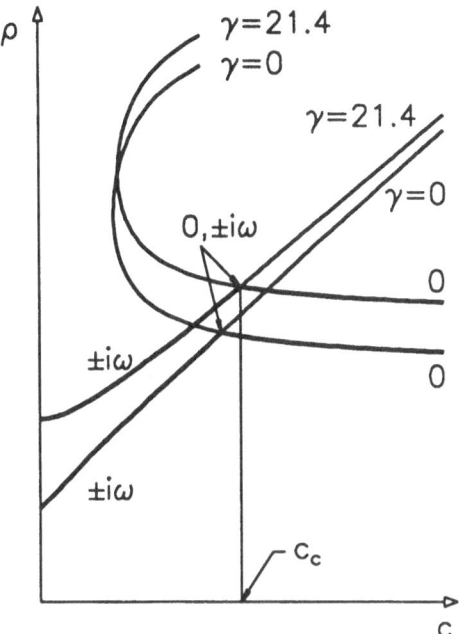

Figure 5.8. Stability boundaries in ϱ, c parameter space with ($\gamma = 21.4$) and without ($\gamma = 0$) consideration of gravity in (5.82) and for $\xi = 0.5$, that is, the elastic support is at mid span. All eigenvalues shown $(0, \pm i\omega)$ must be doubled

Inserting (5.88) into (5.79) and considering (5.76) yields

$$i\omega \begin{pmatrix} z_1 \\ z_2 \end{pmatrix} = \begin{pmatrix} 0 & 1 \\ -C & -B \end{pmatrix} \begin{pmatrix} z_1 \\ z_2 \end{pmatrix}$$

or

$$\begin{aligned} z_2 &= i\omega z_1 \\ -C z_1 - B z_2 &= i\omega z_2 \,. \end{aligned} \tag{5.89}$$

Substituting $z_j = x_j + i y_j$ into (5.89) yields

$$\begin{aligned} x_2 + i y_2 &= i\omega x_1 - \omega y_1 \\ -C x_1 - B x_2 &= -\omega y_2 \\ -C y_1 - B y_2 &= \omega x_2 \,. \end{aligned} \tag{5.90}$$

Elimination of y_1 and y_2 from the second and third equation of (5.90) by means of the first equation yields

$$\begin{aligned} (C - \omega^2) x_1 + B x_2 &= 0 \\ -\omega^2 B x_1 + (C - \omega^2) x_2 &= 0 \end{aligned} \tag{5.91}$$

with the operators B and C defined by (5.77). The boundary and intermediate conditions are

$$s = 0: \quad x_1 = 0,\; x_1' = 0,\; x_2 = 0,\; x_2' = 0$$

$$s = \xi: \quad x_1'''(\xi_+) + \alpha^* x_2'''(\xi_+) - x_1'''(\xi_-) - \alpha^* x_2'''(\xi_-) = -c x_1(\xi)$$

$$-x_2'''(\xi_+) + \alpha^* \omega^2 x_1'''(\xi_+) + x_2'''(\xi_-) - \alpha^* \omega^2 x_1'''(\xi_-) = c x_2(\xi)$$
$$x_1, x_2, x_1', x_2', x_1'' + \alpha^* x_2'', \alpha^* \omega^2 x_1'' - x_2'' \quad \text{are} \quad \text{continuous}$$

$$s = 1: \quad x_1'' + \alpha^* x_2'' = 0, \ \alpha^* \omega^2 x_1'' - x_2'' = 0$$
$$x_1''' + \alpha^* x_2''' = 0, \ \alpha^* \omega^2 x_1''' - x_2''' = 0 . \tag{5.92}$$

The solutions $x_1(s)$ and $x_2(s)$ of the eigenvalue problem (5.91), (5.92) yield the real part of the eigenfunctions. The imaginary parts follow from the first equation (5.90).

Setting $\omega = 0$ in (5.91) (see Fig. 4.5 plate 4) yields the eigenvalue problem for the double zero eigenvalue in a Jordan block

$$\begin{aligned} C x_1 + B x_2 &= 0 \\ C x_2 &= 0 \end{aligned} \tag{5.93}$$

with $x_2(s)$ as eigenvector (eigenfunction) and $x_1(s)$ as first principal vector (Appendix B).

The result of these calculations are the stability boundaries in ϱ, c space given in Fig. 5.8–Fig. 5.10 for the elastic support located at $\xi = 0.5, 0.75$ and 1.0, respectively, and for various values of damping. The data used are the same as in [148], where qualitatively similar results are obtained and shown to be in close agreement with experimental findings.

From the block diagonal form of (5.75) follows immediately that the multiplicity of each eigenvalue obtained above must be doubled. Hence, we have the following different cases of critical eigenvalues at loss of stability for a quasistatic increase of the distinguished parameter ϱ:

(1) For a stiff support (c large), the loss of stability occurs at a double zero eigenvalue.

(2) For a soft support (c small) there are two identical pairs of purely imaginary eigenvalues.

(3) It can be seen from Fig. 5.8 that for a special value $c = c_c$ two zero eigenvalues and two pairs of purely imaginary eigenvalues are present.

(4) From Fig. 5.9 and Fig. 5.10 follows that shifting the support changes the qualitative type of the coupling between divergence and flutter from case (3) to two two-dimensional Jordan blocks with a zero eigenvalue.

(5) Finally, from Fig. 5.9, a coupling between two flutter bifurcations with different frequencies ω_1 and ω_2 is seen to be possible.

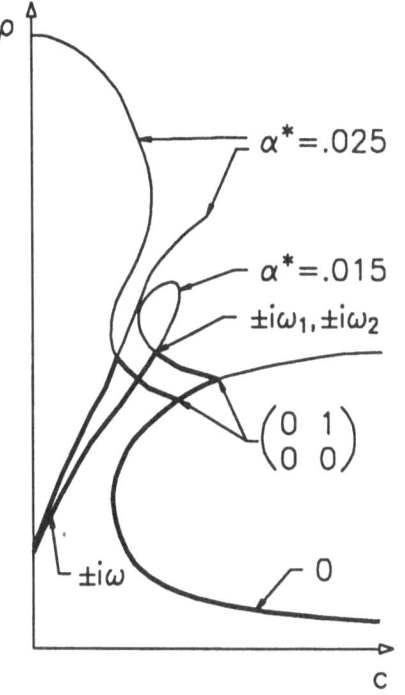
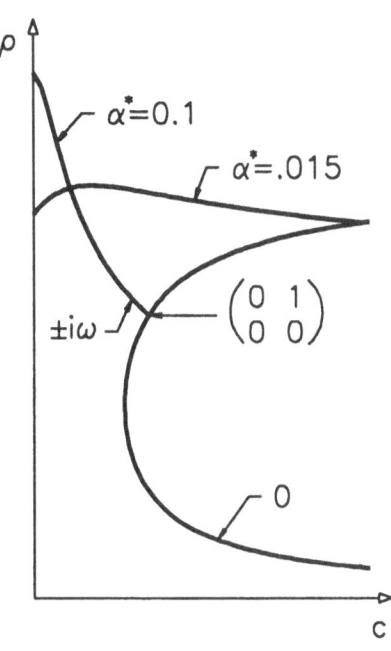

Figure 5.9.   Stability boundary
in ϱ, c parameter space for $\xi =$
0.75 and different values of internal
damping $\alpha^\star$

Figure 5.10. As in Fig. 5.9 but for
$\xi = 1.0$

The cases (1) and (2) are the generic types of loss of stability. Case (1)
is a divergence bifurcation and leads to a two-dimensional bifurcation
system and the matrix $\mathbf{A}_c$ in (3.65) is given by $(5.30)_1$. In case (2) $\mathbf{A}_c$
is given by $(5.30)_2$. This case is a flutter instability and leads to a four-
dimensional system of bifurcation equations. Cases (3) and (4) represent
couplings between flutter and divergence according to (5.31) and (5.32).
Only case (5) is new and has not been observed with the spherical double
pendulum in Section 5.1. This coupling between two different frequences
ω_1 and ω_2 leads to the 8×8 matrix $\mathbf{A}_c$

$$
\mathbf{A}_c =
\begin{pmatrix}
0 & -\omega_1 & & & & & & \\
\omega_1 & 0 & & & & & 0 & \\
& & 0 & -\omega_1 & & & & \\
& & \omega_1 & 0 & & & & \\
& & & & 0 & -\omega_2 & & \\
& & & & \omega_2 & 0 & & \\
& 0 & & & & & 0 & -\omega_2 \\
& & & & & & \omega_2 & 0
\end{pmatrix} .
\qquad (5.94)
$$

The consequence of (5.94) is an eight-dimensional system of bifurcation equations.

A nice and simple physical verification of the results of Fig. 5.10 can be performed with a shower tube. For this experiment we consider c as the distinguished parameter. That is, we keep ϱ at a fixed value and simulate the elastic constraint by touching the tube with a finger. For small values of ϱ the tube buckles statically and for large values of ϱ the tube begins to oscillate.

Center manifold reduction to bifurcation equations

We follow the exposition given in Section 3.2.2. Further, we restrict the investigation to the adjacent (local) nonlinear motion in the neighborhood of the downhanging equilibrium position. Mathematically this means that the bifurcation equations are derived only up to and including third order terms. Under this restriction, similarly as it occurred with the Ljapunov-Schmidt method applied to the plate buckling problem in Section 4.3.4, center manifold theory reduces to the Galerkin method with an ansatz in the critical modes only. This is because the contribution of $h(u_c)$ in (3.67) is at least of fourth order. Therefore, only the critical part u_c in (3.63) must be used which is spanned by the eigenfunctions calculated above. However, in this problem the linear operator $\mathbf{A}$ given by (5.75) is not self-adjoint. For the calculation of the projection operator $\mathbf{P}$, the adjoint operator $\mathbf{A}^\star$ must be calculated (Appendix C.1).

To perform the reduction process by means of the Galerkin method, we proceed as explained at the end of Appendix D. In the ansatz $\hat{w}(s,t)$ for $w(s,t)$, the eigenfunctions obtained from the linearized problem are used. Hence we have

$$\hat{w}(s,t) = \sum_{j=1}^{n_c} v_j(t)\boldsymbol{x}_j(s) \qquad (5.95)$$

with the components of $\boldsymbol{x}_j(s)$ calculated from the linear eigenvalue problems (5.81) and (5.90).

The bifurcation equations then follow from inserting (5.95) into (5.74) and by projecting this equation onto the eigenvectors. This results in a system of n_c equations of the form

$$\int_0^1 \boldsymbol{x}_k^{\star T}\dot{\hat{w}}\,ds = \int_0^1 \boldsymbol{x}_k^{\star T}\mathbf{A}\hat{w}\,ds + \int_0^1 \boldsymbol{x}_k^{\star T}\mathbf{g}(\hat{w},\lambda_c)ds \ , \qquad k=1,\dots,n_c$$

$$(5.96)$$

where $\boldsymbol{x}_k^\star$ is the adjoint eigenvector to $\boldsymbol{x}_k$.

In the case of the double zero eigenvalue $n_c = 2$ and

$$\hat{w}(s,t) = v_1(t) \begin{pmatrix} x_1(s) \\ 0 \\ 0 \\ 0 \end{pmatrix} + v_2(t) \begin{pmatrix} 0 \\ 0 \\ x_1(s) \\ 0 \end{pmatrix} .$$

For the purely imaginary pair $n_c = 4$ and

$$\hat{w}(s,t) = v_1(t) \begin{pmatrix} x_1(s) \\ x_2(s) \\ 0 \\ 0 \end{pmatrix} + v_2(t) \begin{pmatrix} y_1(s) \\ y_2(s) \\ 0 \\ 0 \end{pmatrix} + $$

$$+ v_3(t) \begin{pmatrix} 0 \\ 0 \\ x_1(s) \\ x_2(s) \end{pmatrix} + v_4(t) \begin{pmatrix} 0 \\ 0 \\ y_1(s) \\ y_2(s) \end{pmatrix} . \tag{5.97}$$

The resulting bifurcation equations are (5.35) and (5.42) or, after making use of the equivariance condition, (5.40) and (5.47) respectively. It is important to note when performing the numerical calculations in (5.96) that only those terms must be calculated which appear in (5.40) and (5.45). Hence, applying the symmetry properties to the bifurcation equations results in a strong reduction of the amount of numerical calculations needed to be performed.

Concluding remarks

At the end of this chapter we stress once more the importance of making use of the symmetry group concept, by explaining that problems could appear if a system like the fluid-carrying tube were to be analysed without taking its symmetry properties into account.

Of course, the investigator would realize from the form of the linear part of the equations of motion in the form (L.25) or (5.74) and (5.75) that only an eigenvalue problem with half of the dimension of the original problem must be solved. This, at first glance, pleasant effect for the solution of the inverse eigenvalue problem has the unpleasant consequence that each eigenvalue with a zero real part appearing at a loss of stability must be doubled. Hence, after reduction to bifurcation equations the dimension of the bifurcation equations is doubled. If at this point the investigator does not make use of the symmetry properties, he would run into troubles because for performing the reduction process the coefficients must be calculated numerically. However, many of them are forced to be zero because of symmetry. From the numerical calculations in some cases it will be difficult to decide whether a small coefficient is actually zero or is only of small absolute value. This has two unpleasant consequences. First, unnecessary numerical calculations of coefficients which turn out to be zero anyway must be performed. Second, some of them which should be zero are erroneously still retained in the bifurcation equations and could complicate the further analysis of the problem.

Chapter 6

Discussion of the bifurcation equations

The cases considered in Chapters 3 and 4 show that the bifurcation equations are either a set of n_c nonlinear ordinary differential equations of first order ((3.19) and (3.67)), a set of n_c difference equations (3.47), or a set of n_c nonlinear algebraic equations (3.58). The number n_c is given by the number of eigenvalues with zero real part for differential equations, or of modulus *one* for difference equations, or by the multiplicity of the critical eigenvalue for the linear operator G_u for static systems in Section 3.2.1, respectively. The number of nonlinear terms in the bifurcation equations can become quite large, as the examples in Chapters 4 and 5 show. Hence, the question arises whether a simplification of the nonlinear part in the equations is possible. By simplification, we mean a reduction of the number of terms without changing the qualitative behavior of the flow in the phase space defined by the bifurcation system. We shall see that such a simplification is possible and will be provided by *normal form theory*. In general, a strong reduction of the number of terms can be achieved. However, we remark that the reduction to a normal form is not unique. This non-uniqueness need not bother those who apply normal form theory because in the cases of low codimension, that is, low number of essential parameters, the normal form into which the problem should be transformed is known. This will be shown in the next sections when we discuss the classified cases.

If a list of classified cases is given, a geometric interpretation of the results leads to so-called *bifurcation diagrams*. Bifurcation diagrams give a stratification of the parameter space into domains of qualitatively similar behavior. Here, the fact that for most dynamical systems one- or two-dimensional bifurcation systems are obtained plays an important role because of Theorem 2.1 the property of structural stability is generic for one- and two-dimensional systems. Therefore, structurally unstable systems are exceptional systems and form boundary surfaces in the parameter space. Besides the dimension n_c, there are other important quantities, characterizing the bifurcation equations, namely *codimension* and *determinacy* which are explained in this chapter.

6.1 Transformation to normal form

6.1.1 Time-continuous dynamical systems

In this section, we follow [7] ch. 5 and [50] ch. 3. First, we treat the time continuous case. The basic idea of *normal form* theory is to reduce the number of nonlinear terms in the bifurcation equations by means of a nonlinear change of variables. As is shown below, the difficulty in performing this program comes from the fact that, typically, in the bifurcation equations the linear part is degenerate. By degenerate we mean that there are eigenvalues with zero real part.

Let us assume to be given a set of nonlinear differential equations for which the linearization about an equilibrium position has only eigenvalues with nonzero real part. We recall that a linear system that has only eigenvalues with nonzero real part is called *hyperbolic*. For the hyperbolic case, it is known from Hartman's theorem ([6] p. 144) that locally, in the neighborhood of the equilibrium position, the flow of the nonlinear system is topologically equivalent to its linear part (see Fig. 6.1 for the

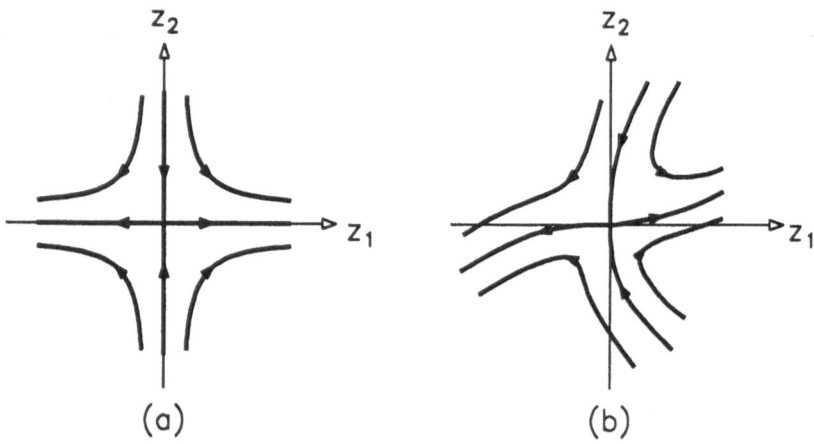

Figure 6.1. Near a hyperbolic equilibrium (a) the linear and (b) the nonlinear flow are topologically equivalent

planar case). This means that there exists a (nonlinear) transformation of variables which, locally, in the neighborhood of the equilibrium position, replaces the nonlinear system by its linear part. In other words it is possible to eliminate all nonlinear terms by this transformation of variables. In treating the *non-hyperbolic* or *bifurcation* case, it is shown below that this is no longer true.

We now turn to the *non-hyperbolic* case. We start with (3.19), which is written in the form

$$\dot{y} = \mathbf{J}y + \mathbf{g}(y). \tag{6.1}$$

The dimension of (6.1) is n_c and $\mathbf{g}(y)$ contains the nonlinear terms up to order k. Mathematically, $\mathbf{g}(y)$ is the *k-jet of a function* $\mathbf{G}(y)$. By the k-jet $j^k\mathbf{G}(y)$ we mean the Taylor series expansion at $\mathbf{0}$ up to terms of order k. That is, $\mathbf{g}(y) = j^k\mathbf{G}(y)$ is a polynomial function of degree $\leq k$. The order k, up to which terms must be retained in the Taylor series expansion, is determined by the concept of *determinacy*, which is treated in Section 6.3.

To reduce the number of terms, we proceed in the following manner. First, we introduce a nonlinear change of coordinates $y \rightarrow z$ by

$$y = z + \mathbf{h}(z) , \tag{6.2}$$

where $\mathbf{h}(z)$ is a vector-valued function whose components are homogeneous polynomials of order $r \geq 2$. Next, we substitute (6.2) into (6.1) to find

$$(\mathbf{E} + \mathbf{h}'(z))\dot{z} = \mathbf{J}z + \mathbf{J}\mathbf{h}(z) + \mathbf{g}(z + \mathbf{h}(z)) . \tag{6.3}$$

As usual $\mathbf{E}$ denotes the unit matrix. The function $\mathbf{h}(z)$ is chosen in such a way that in the transformed system

$$\dot{z} = \mathbf{J}z + \mathbf{f}(z) \tag{6.4}$$

the nonlinear terms $\mathbf{f}(z)$ are as simple as possible. Substituting (6.4) into (6.3) we find the lowest order terms

$$\mathbf{J}z + \mathbf{f}(z) + \mathbf{h}'(z)\mathbf{J}z = \mathbf{J}z + \mathbf{J}\mathbf{h}(z) + \mathbf{g}(z) + o(\mathbf{h}(z))$$

where $o(\mathbf{h}(z))$ designates terms of higher order than those included in $\mathbf{h}(z)$). From this expression follows

$$\mathbf{f}(z) = \mathbf{J}\mathbf{h}(z) + \mathbf{g}(z) - \mathbf{h}'(z)\mathbf{J}z . \tag{6.5}$$

The most desirable situation would be to eliminate all terms in $\mathbf{f}(z)$. That is,

$$\mathbf{f}(z) = \mathbf{J}\mathbf{h}(z) + \mathbf{g}(z) - \mathbf{h}'(z)\mathbf{J}z = 0 . \tag{6.6}$$

Suppose now that it was possible either to eliminate all terms of order up to and including $m - 1$ or that terms of order up to $m - 1$ that could not be annihilated did not determine the problem (Section 6.3). Therefore, the terms of order m in $\mathbf{g}(z)$ must be considered. We note that terms of order $m-1$ will not be affected by the following calculations. We assume $\mathbf{g}(\mathbf{z})$ to be given in the form

$$
\mathbf{g}(\mathbf{z}) \;=\; \sum_{\substack{i=1 \\ m_1+\ldots+m_{n_c}=m}}^{n_c} g_{i,m} \mathbf{z}^m \mathbf{e}_i \;;
$$

$$
\mathbf{z}^m \;=\; z_1^{m_1} z_2^{m_2}\ldots z_{n_c}^{m_{n_c}}; \qquad m_1+\ldots+m_{n_c}=m\;,
$$

(6.7)

where $\mathbf{e}_i = (0,\ldots,0,1,0,\ldots,0)^T$ is the i-th unit vector. The expression $g_{i,m}$ denotes the coefficient of the monomial $\mathbf{z}^m$ in the i-th equation.

   Let us explain the notation in (6.7) for the case of two equations in two variables z_1, z_2 with terms of third order

$$
\mathbf{g}(\mathbf{z}) = \sum_{\substack{i=1 \\ m_1+m_2=3}}^{2} g_{i,m}\mathbf{z}^m \mathbf{e}_i \;=\; (g_{1,30}z_1^3 + g_{1,21}z_1^2 z_2 + g_{1,12}z_1 z_2^2 + g_{1,03}z_2^3)\mathbf{e}_1 +
$$

$$
+ (g_{2,30}z_1^3 + g_{2,21}z_1^2 z_2 + g_{2,12}z_1 z_2^2 + g_{2,03}z_2^3)\mathbf{e}_2\,.
$$

With this form of $\mathbf{g}(\mathbf{z})$, $\mathbf{h}(\mathbf{z})$ is chosen to be

$$
\mathbf{h}(\mathbf{z}) = \sum_{\substack{i=1 \\ m_1+\ldots+m_{n_c}=m}}^{n_c} h_{i,m}\mathbf{z}^m \mathbf{e}_i\,. \tag{6.8}
$$

From (6.8) the term at location $(i,1)$ of the matrix $\mathbf{h}'$ in (6.6) can be determined to be

$$
h_{i,m}m_1 z_1^{m_1-1}z_2^{m_2}\ldots z_{n_c}^{m_{n_c}} = h_{i,m}m_1 z_1^{-1}\mathbf{z}^m\,.
$$

Hence, writing down only the i-th row of the third term in (6.6) yields

$$
\mathbf{h}'\mathbf{J}\mathbf{z} \;=\; \mathbf{z}^m h_{i,m} \begin{pmatrix} \vdots & \vdots & & \vdots \\ m_1 z_1^{-1} & m_2 z_2^{-1} & \cdots & m_{n_c} z_{n_c}^{-1} \\ \vdots & \vdots & & \vdots \end{pmatrix}\cdot
$$

$$
\cdot \begin{pmatrix} \mu_1 & & & & \\ & \ddots & & & \\ & & \mu_i & & \\ & & & \ddots & \\ & & & & \mu_{n_c} \end{pmatrix}\begin{pmatrix} z_1 \\ \vdots \\ z_{n_c} \end{pmatrix}
$$

$$
= \begin{pmatrix} \vdots \\ h_{i,m}\mathbf{z}^m(\mu_1 m_1 + \ldots + \mu_{n_c}m_{n_c}) \\ \vdots \end{pmatrix}\,,
$$

where for simplicity $\mathbf{J}$ is assumed to be a diagonal matrix. The i-th rows of the other two terms in (6.6) are

$$\mathbf{Jh}(z) = \begin{pmatrix} \vdots \\ h_{i,m}\mu_i z^m \\ \vdots \end{pmatrix} , \qquad \mathbf{g}(z) = \begin{pmatrix} \vdots \\ g_{i,m} z^m \\ \vdots \end{pmatrix} .$$

Combining the three terms yields for the i-th row in (6.6)

$$h_{i,m} z^m (\mu_1 m_1 + \ldots + \mu_{n_c} m_{n_c}) - h_{i,m}\mu_i z^m - g_{i,m} z^m = 0 . \tag{6.9}$$

Consequently, the elimination of the i-th component of the terms of order m in (6.5) can be obtained if (6.9) can be solved for $h_{i,m}$. This is possible only if in the solution

$$h_{i,m} = \frac{g_{i,m}}{(\mu_1 m_1 + \ldots + \mu_{n_c} m_{n_c}) - \mu_i} \tag{6.10}$$

the denominator is not equal to zero. In other words, this means that all nonlinear terms in (6.5) can be eliminated if the *resonance condition*

$$(\mu_1 m_1 + \ldots + \mu_{n_c} m_{n_c}) - \mu_i = 0 \tag{6.11}$$

is *not* satisfied, where the integers $m_j \geq 0$. If (6.11) is satisfied, the eigenvalues are said to be *resonant* and

$$|m| = \sum_{j=1}^{n_c} m_j \geq 2 \tag{6.12}$$

is the *order* of resonance.

Example: Hopf bifurcation ([56])

As mentioned in the Introduction (equations (1.16), (1.20)), and found in several examples in Chapters 4 and 5, it will be shown in Section 6.5.1 that bifurcation at a pair of purely imaginary eigenvalues occurs in a generic way in one parameter families. Hence, the bifurcation at a pair of imaginary eigenvalues is very important in applications. This bifurcation is called the *Hopf bifurcation*. The corresponding bifurcation equations are two-dimensional (for example (4.11), (4.17)) and can be written in the following form

$$\begin{aligned} \dot{y}_1 &= -\omega y_2 + g_{12}(y_1, y_2) + g_{13}(y_1, y_2) + O(|y|^4) \\ \dot{y}_2 &= \omega y_1 + g_{22}(y_1, y_2) + g_{23}(y_1, y_2) + O(|y|^4) \end{aligned} \tag{6.13}$$

where the g_{12}, g_{22} contain quadratic and the g_{13}, g_{23} cubic terms, that is,

$$
\begin{aligned}
g_{j2} &= a_{j20}y_1^2 + a_{j11}y_1y_2 + a_{j02}y_2^2 , \\
g_{j3} &= a_{j30}y_1^3 + a_{j21}y_1^2y_2 + a_{j12}y_1y_2^2 + a_{j03}y_2^3 ,
\end{aligned}
\qquad j = 1, 2 . \qquad (6.14)
$$

In other words, the nonlinear terms in (6.13) are the 3-jet of the Taylor expansions.

The questions to answer are: are there any resonances? And if the answer is yes: of which order are these resonances? The eigenvalues of (6.13) are $\mu_1 = i\omega$, $\mu_2 = -i\omega$. Thus, the relation $\mu_1 + \mu_2 = 0$, put in the form (6.11), reads

$$
\mu_1 = 2\mu_1 + \mu_2 . \qquad (6.15)
$$

With regard to more complicated cases treated in Sections 6.5.1, 6.6.2 and 6.6.4 we write (6.11) in the form (see also p. 249, p. 217 and p. 278)

$$
m_1(i\omega) + m_2(-i\omega) - \mu_i = 0 \qquad (6.16)
$$

where for $\mu_1 = i\omega$ resonance is given by $m_1 = 2$ and $m_2 = 1$.

Therefore, according to (6.12) the lowest order $|m|$ of the resonance is three. This means that it will not be possible to eliminate all third order terms in (6.13).

However, it should be possible to eliminate all second order terms. To show this, we make the following change of coordinates

$$
\begin{aligned}
y_1 &= z_1 + h_{12}(z_1, z_2) \\
y_2 &= z_2 + h_{22}(z_1, z_2)
\end{aligned}
\qquad (6.17)
$$

where h_{12} and h_{22} are homogeneous quadratic polynomials of the form

$$
h_{j2}(z_1, z_2) = \alpha_{j20}z_1^2 + \alpha_{j11}z_1z_2 + \alpha_{j02}z_2^2 , \qquad j = 1, 2 . \qquad (6.18)
$$

From (6.17), we obtain the left-hand side of (6.13)

$$
\dot{y}_j = \dot{z}_j + \frac{\partial h_{j2}}{\partial z_1}\dot{z}_1 + \frac{\partial h_{j2}}{\partial z_2}\dot{z}_2 , \qquad j = 1, 2 . \qquad (6.19)
$$

The right-hand side of (6.13) reads

$$
\begin{aligned}
\dot{y}_1 &= -\omega(z_2 + h_{22}) + g_{12}(z_1 + h_{12}, z_2 + h_{22}) \\
\dot{y}_2 &= \omega(z_1 + h_{12}) + g_{22}(z_1 + h_{12}, z_2 + h_{22}) .
\end{aligned}
\qquad (6.20)
$$

Combining (6.19) and (6.20), we find

$$
\begin{aligned}
\dot{z}_1 &= -\omega z_2 - \omega h_{22} + g_{12} - \\
&\quad - \left[-\frac{\partial h_{12}}{\partial z_1} \omega z_2 + \frac{\partial h_{12}}{\partial z_2} \omega z_1 \right] + O(|z|^3) \\
\dot{z}_2 &= \omega z_1 + \omega h_{12} + g_{22} - \\
&\quad - \left[-\frac{\partial h_{22}}{\partial z_1} \omega z_2 + \frac{\partial h_{22}}{\partial z_2} \omega z_1 \right] + O(|z|^3) .
\end{aligned}
\tag{6.21}
$$

Now all quadratic terms in (6.21) are required to vanish. That is,

$$
\begin{aligned}
h_{22} - \frac{\partial h_{12}}{\partial z_1} z_2 + \frac{\partial h_{12}}{\partial z_2} z_1 &= \frac{1}{\omega} g_{12} \\
- h_{12} - \frac{\partial h_{22}}{\partial z_1} z_2 + \frac{\partial h_{22}}{\partial z_2} z_1 &= \frac{1}{\omega} g_{22} .
\end{aligned}
\tag{6.22}
$$

Inserting (6.14) and (6.18) into (6.22), we find

$$
(\alpha_{220} + \alpha_{111}) z_1^2 + (\alpha_{211} - 2\alpha_{120} + 2\alpha_{102}) z_1 z_2 + (\alpha_{202} - \alpha_{111}) z_2^2
$$
$$
= \frac{1}{\omega} (a_{120} z_1^2 + a_{111} z_1 z_2 + a_{102} z_2^2)
$$

and a similar second equation. By comparing coefficients, we immediately obtain the following system of linear algebraic equations for the $\alpha_{ik\ell}$:

$$
\begin{pmatrix}
0 & 1 & 0 & 1 & 0 & 0 \\
-2 & 0 & 2 & 0 & 1 & 0 \\
0 & -1 & 0 & 0 & 0 & 1 \\
-1 & 0 & 0 & 0 & 1 & 0 \\
0 & -1 & 0 & -2 & 0 & 2 \\
0 & 0 & -1 & 0 & -1 & 0
\end{pmatrix}
\begin{pmatrix}
\alpha_{120} \\
\alpha_{111} \\
\alpha_{102} \\
\alpha_{220} \\
\alpha_{211} \\
\alpha_{202}
\end{pmatrix}
= \frac{1}{\omega}
\begin{pmatrix}
a_{120} \\
a_{111} \\
a_{102} \\
a_{220} \\
a_{211} \\
a_{202}
\end{pmatrix} .
$$

The determinant of the matrix of coefficients is not equal to zero (its value is 9), and hence, a unique solution for the $\alpha_{jk\ell}$ exists, and therefore, all quadratic terms can be removed. One could continue in this manner and also try to eliminate the third order terms. However, it is computationally more efficient to use complex variables

$$
w = y_1 + i y_2 , \qquad \overline{w} = y_1 - i y_2 .
\tag{6.23}
$$

Introducing complex variables allows us to rewrite the bifurcation equations in the simpler form

$$
\dot{w} = i\omega w + g_2(w, \overline{w}) + g_3^*(w, \overline{w}) + O(|w|^4)
\tag{6.24}
$$

with

$$
\begin{aligned}
g_2 &= b_{20} w^2 + b_{11} w\overline{w} + b_{02} \overline{w}^2 \\
g_3^* &= b_{30}^* w^3 + b_{21}^* w^2 \overline{w} + b_{12}^* w\overline{w}^2 + b_{03}^* \overline{w}^3 ,
\end{aligned}
\tag{6.25}
$$

where the coefficients b and b^* are complex.

From the calculations above, we notice that the elimination of the second order terms also changes the third order terms in the original equations. Using complex variables, this can also be taken care of quite easily.

To eliminate the second order terms, one starts with

$$\begin{aligned} w &= v + h_2(v, \bar{v}) \\ \bar{w} &= \bar{v} + \bar{h}_2(v, \bar{v}) \end{aligned}$$
(6.26)

where h_2 is given by

$$h_2(v, \bar{v}) = \beta_{20} v^2 + \beta_{11} v \bar{v} + \beta_{02} \bar{v}^2 .$$
(6.27)

The second equation in (6.26) and a conjugate complex equation to (6.24), which we do not write down, must be added to use (6.6) where $\mathbf{J}$ is the diagonal matrix given by

$$\mathbf{J} = \begin{pmatrix} i\omega & 0 \\ 0 & -i\omega \end{pmatrix} .$$
(6.28)

We obtain for the first component of the terms appearing in (6.6)

$$\begin{aligned} \mathbf{h}_2' \mathbf{J} \begin{pmatrix} v \\ \bar{v} \end{pmatrix}_1 &= (2\beta_{20} v + \beta_{11} \bar{v}) i\omega v - \\ &\quad - (\beta_{11} v + 2\beta_{02} \bar{v}) i\omega \bar{v} \end{aligned}$$
(6.29)

$$\mathbf{J} \begin{pmatrix} h_2 \\ \bar{h}_2 \end{pmatrix}_1 = i\omega h_2(v, \bar{v}) .$$

Inserting (6.29) into (6.6) under consideration of $(6.25)_1$ and (6.27), we find

$$f_2(v, \bar{v}) = (b_{20} - i\omega\beta_{20}) v^2 + (b_{11} + i\omega\beta_{11}) v\bar{v} + (b_{02} + 3i\omega\beta_{02}) \bar{v}^2 .$$
(6.30)

Obviously, all coefficients β_{ij} of the transformation can be calculated to annihilate $f(v)$, and hence, all second order terms in (6.24) can be removed. Furthermore, the influence of the transformation (6.27) on the third order terms can be easily derived by following the steps leading to (6.4) and (6.5). That is, we must insert

$$\dot{v} = i\omega v + f_3(v, \bar{v})$$
(6.31)

and its complex conjugate one which correspond to (6.4), into the equation corresponding to (6.3) which has the form

$$\dot{v} + \frac{\partial h_2}{\partial v}\dot{v} + \frac{\partial h_2}{\partial \overline{v}}\dot{\overline{v}} = i\omega(v + h_2(v,\overline{v})) +$$
$$+ g_2(v + h_2(v,\overline{v}), \overline{v} + \overline{h}_2(v,\overline{v})) + g_3(v,\overline{v}) . \tag{6.32}$$

After inserting (6.31) into (6.32) third order terms on the left-hand side of (6.32) are given only by $f_3(v,\overline{v})$ whereas on the right-hand side we find $g_3(v,\overline{v})$ and the terms of cubic order of $g_2(v+h_2(v,\overline{v}), \overline{v}+\overline{h}_2(v,\overline{v}))$. It will turn out below that the coefficient of $v^2\overline{v}$ is the only term of relevance for the normal form of third order. Hence, we present only the coefficient b_{21} of $v^2\overline{v}$ as

$$b_{21} = b_{21}^\star + 2b_{20}\beta_{11} + b_{11}(\overline{\beta}_{11} + \beta_{20}) + 2b_{02}\overline{\beta}_{02} . \tag{6.33}$$

The first term $b_{21}^\star$ gives the contribution from the third order terms in the original equations. The other three terms follow from inserting (6.26) into (6.25) and collecting the contribution to the coefficient of $v^2\overline{v}$. Inserting from (6.30), (6.33) takes the form

$$b_{21} = b_{21}^\star + \frac{i}{\omega}(b_{20}b_{11} - b_{11}\overline{b}_{11} - \frac{2}{3}b_{02}\overline{b}_{02}) . \tag{6.34}$$

After having eliminated the second order terms, we may now assume the transformed equations (6.13) and (6.24) to be given by

$$\dot{w} = i\omega w + g_3(w,\overline{w}) \tag{6.35}$$

where $g_3(w,\overline{w})$ is given by (6.25) but with coefficients b instead of $b^\star$. To eliminate the cubic terms in (6.35), we proceed as before. We begin by setting

$$w = v + h_3(v,\overline{v}) , \tag{6.36}$$

where h_3 is given by

$$h_3(v,\overline{v}) = \beta_{30}v^3 + \beta_{21}v^2\overline{v} + \beta_{12}v\overline{v}^2 + \beta_{03}\overline{v}^3 .$$

Inserting (6.36) into (6.35) results in

$$\dot{v} + \frac{\partial h_3}{\partial v}\dot{v} + \frac{\partial h_3}{\partial \overline{v}}\dot{\overline{v}} = i\omega(v + h_3(v,\overline{v})) + g_3(v,\overline{v}) .$$

Inserting (6.31) into this equation and collecting corresponding terms yields

$$f_3 = (-2i\omega\beta_{30} + b_{30})v^3 + b_{21}v^2\overline{v} + $$
$$+ (2i\omega\beta_{12} + b_{12})v\overline{v}^2 + (4i\omega\beta_{03} + b_{03})\overline{v}^3 . \tag{6.37}$$

We see that in the coefficient of the term $v^2\overline{v}$ in (6.37) β_{21} does not appear, and therefore, it is not possible to assign a value to β_{21} such that the coefficient of $v^2\bar{v}$ becomes zero. Only the terms $v^3, v\bar{v}^2, \bar{v}^3$ can be removed from f_3, if we set

$$\beta_{30} = \frac{1}{2i\omega}b_{30}\,, \qquad \beta_{12} = -\frac{1}{2i\omega}b_{12}\,, \qquad \beta_{03} = -\frac{1}{4i\omega}b_{03}\,.$$

Hence, the resulting normal form up to third order is

$$\dot{v} = i\omega v + b_{21}v^2\bar{v} = i\omega v + b_{21}|v|^2 v\,. \tag{6.38}$$

Returning in (6.38) to real variables z_1, z_2 by $v = z_1 + iz_2$, we find from

$$(\dot{z}_1 + i\dot{z}_2) = i\omega(z_1 + iz_2) + (b_{21}^R + ib_{21}^I)(z_1^2 + z_2^2)(z_1 + iz_2)$$

the equations in normal form in real variables

$$\begin{aligned}
\dot{z}_1 &= -\omega z_2 + (b_{21}^R z_1 - b_{21}^I z_2)(z_1^2 + z_2^2) \\
\dot{z}_2 &= \omega z_1 + (b_{21}^I z_1 + b_{21}^R z_2)(z_1^2 + z_2^2)\,.
\end{aligned} \tag{6.39}$$

Hence, the number of the eight real coefficients of the nonlinear third order terms in $(6.14)_2$ or of the four complex coefficients in $(6.25)_2$ is reduced to two or one, respectively, given in the normal form (6.39) or (6.38).

What remains to be done is to give the relation between the two coefficients b_{21}^R and b_{21}^I and the original coefficients, for example, those given in $(6.14)_2$ or (6.35). This can be done by solving (6.23) for y_i, that is,

$$y_1 = \frac{1}{2}(w + \bar{w}), \qquad y_2 = \frac{i}{2}(\bar{w} - w)\,, \tag{6.40}$$

and inserting into (6.13) considering (6.14).

We perform these calculations for the second order terms in (6.25). With (6.40) we obtain

$$\begin{aligned}
y_1^2 &= \frac{1}{4}(w^2 + 2w\bar{w} + \bar{w}^2) \\
y_1 y_2 &= \frac{i}{4}(\bar{w}^2 - w^2) \\
y_2^2 &= -\frac{1}{4}(w^2 - 2w\bar{w} + \bar{w}^2)\,.
\end{aligned}$$

Inserting these expressions into (6.13) with (6.14) and making use of $\dot{w} = \dot{y}_1 + i\dot{y}_2$ the coefficients in (6.24) follow immediately to

$$\begin{aligned}
b_{20} &= \frac{1}{4}(a_{120} + a_{211} - a_{102}) + \frac{i}{4}(a_{220} - a_{111} - a_{202}) \\
b_{11} &= \frac{1}{2}(a_{120} + a_{102}) + \frac{i}{2}(a_{220} + a_{202}) \\
b_{02} &= \frac{1}{4}(a_{120} - a_{211} - a_{102}) + \frac{i}{4}(a_{220} + a_{111} - a_{202})\,.
\end{aligned}$$

Proceeding analogously as before by inserting (6.40) into (6.14)$_2$, the coefficient b_{21}^* of $w^2\overline{w}$ can be calculated as

$$
\begin{aligned}
b_{21}^* ={} & \frac{3}{8}a_{130} + \frac{1}{8}a_{112} + \frac{1}{8}a_{221} + \frac{3}{8}a_{203} + \\
& + i\left(-\frac{1}{8}a_{121} - \frac{3}{8}a_{103} + \frac{3}{8}a_{230} + \frac{1}{8}a_{212}\right) .
\end{aligned}
\tag{6.41}
$$

The coefficient b_{21} follows from inserting b_{21}^* into (6.34). If no quadratic terms were present in (6.25) then $b_{21} = b_{21}^*$ and (6.41) gives already the required result.

A second possibility to arrive at the normal form (6.39) is given by application of the *averaging principle* ([69], [29], [118]) which we consider next.

Averaging principle

Consider a dynamical system to be given in the form

$$\dot{\boldsymbol{x}} = \mathbf{f}_0(\boldsymbol{x}) + \varepsilon \mathbf{f}_1(\boldsymbol{x}) \tag{6.42}$$

where ε is a small parameter and the system $\dot{\boldsymbol{x}} = \mathbf{f}_0(\boldsymbol{x})$ is assumed to be *integrable*. Important examples of integrable systems are:

(a) $\mathbf{f}_0(\boldsymbol{x})$ is linear with constant coefficients

(b) $\mathbf{f}_0(\boldsymbol{x})$ describes a conservative one degree of freedom system. Additionally, there exist some special cases with more than one degree of freedom ([5]).

Hence, (6.42) is a small perturbation of an integrable case. For the solution of (6.42) we define a function $\hat{\boldsymbol{x}}(t)$ by

$$\boldsymbol{x}(t) = \hat{\boldsymbol{x}}(t) + \varepsilon \boldsymbol{u}(\hat{\boldsymbol{x}}(t)) , \tag{6.43}$$

where $\boldsymbol{u}(\hat{\boldsymbol{x}})$ is a function to be determined. The transformation (6.43) is an *almost identical transformation* since for $\varepsilon = 0$ (6.43) becomes $\boldsymbol{x}(t) = \hat{\boldsymbol{x}}(t)$. The equation which is satisfied by $\hat{\boldsymbol{x}}$ follows from inserting (6.43) into (6.42) to obtain

$$\dot{\boldsymbol{x}} = \left(\mathbf{E} + \varepsilon\frac{\partial \boldsymbol{u}}{\partial \hat{\boldsymbol{x}}}(\hat{\boldsymbol{x}})\right)\dot{\hat{\boldsymbol{x}}} = \mathbf{f}_0(\hat{\boldsymbol{x}} + \varepsilon \boldsymbol{u}(\hat{\boldsymbol{x}})) + \varepsilon \mathbf{f}_1(\hat{\boldsymbol{x}} + \varepsilon \boldsymbol{u}(\hat{\boldsymbol{x}})) .$$

Making use of the Taylor expansions

$$
\begin{aligned}
\left(\mathbf{E} + \varepsilon\frac{\partial \boldsymbol{u}}{\partial \hat{\boldsymbol{x}}}\right)^{-1} &= \mathbf{E} - \varepsilon\frac{\partial \boldsymbol{u}}{\partial \hat{\boldsymbol{x}}} + O(\varepsilon^2) \\
\mathbf{f}_0(\hat{\boldsymbol{x}} + \varepsilon \boldsymbol{u}(\hat{\boldsymbol{x}})) &= \mathbf{f}_0(\hat{\boldsymbol{x}}) + \varepsilon\frac{\partial \mathbf{f}_0}{\partial \boldsymbol{x}}(\hat{\boldsymbol{x}})\boldsymbol{u}(\hat{\boldsymbol{x}}) + O(\varepsilon^2) \\
\varepsilon \mathbf{f}_1(\hat{\boldsymbol{x}} + \varepsilon \boldsymbol{u}(\hat{\boldsymbol{x}})) &= \varepsilon \mathbf{f}_1(\hat{\boldsymbol{x}}) + O(\varepsilon^2)
\end{aligned}
$$

we obtain

$$\dot{\hat{x}} = \mathbf{f}_0(\hat{x}) + \varepsilon \left[\frac{\partial \mathbf{f}_0}{\partial x}(\hat{x})\mathbf{u}(\hat{x}) - \frac{\partial \mathbf{u}}{\partial \hat{x}}(\hat{x})\mathbf{f}_0(\hat{x}) + \mathbf{f}_1(\hat{x}) \right] + O(\varepsilon^2) . \quad (6.44)$$

We do not want to give a detailed mathematical discussion of the averaging method. For that we refer the reader to [69], but we proceed rather heuristically. Since one wants to replace (6.42) by a system for $\hat{x}$ which is as simple as possible, one requires that the terms of order ε in (6.44) should be as simple as possible.

Obviously, there exists a strong formal similarity between averaging and normal form theory by comparing (6.5) with the expression of order ε in (6.44). This is clearly evident if $\mathbf{f}_0, \mathbf{f}_1$ and $\mathbf{u}$ are replaced by $\mathbf{J}\hat{x}, \mathbf{g}(\hat{x})$ and $\mathbf{h}(\hat{x})$, respectively.

We explain now the further treatment of (6.44) with an example. Again, we treat equations (6.13) but only with third order terms. These are

$$\begin{aligned}
\dot{y}_1 &= -\omega y_2 + a_{130}y_1^3 + a_{121}y_1^2 y_2 + a_{112}y_1 y_2^2 + a_{103}y_2^3 \\
\dot{y}_2 &= \omega y_1 + a_{230}y_1^3 + a_{221}y_1^2 y_2 + a_{212}y_1 y_2^2 + a_{203}y_2^3 .
\end{aligned} \quad (6.45)$$

First, we introduce polar coordinates $y_1 = \sqrt{\varepsilon r}\cos\varphi$, $y_2 = \sqrt{\varepsilon r}\sin\varphi$ into (6.45) to find

$$\begin{aligned}
\dot{r}\cos\varphi - r\dot{\varphi}\cos\varphi &= -\omega r\sin\varphi + \varepsilon r^3(a_{130}\cos^3\varphi + a_{121}\cos^2\varphi\sin\varphi + \\
&\quad + a_{112}\cos\varphi\sin^2\varphi + a_{103}\sin^3\varphi) \\
\dot{r}\sin\varphi + r\dot{\varphi}\sin\varphi &= \omega r\cos\varphi + \varepsilon r^3(a_{230}\cos^3\varphi + a_{221}\cos^2\varphi\sin\varphi + \\
&\quad + a_{212}\cos\varphi\sin^2\varphi + a_{203}\sin^3\varphi) .
\end{aligned} \quad (6.46)$$

Second, we multiply $(6.46)_1$ by $\cos\varphi$ and $(6.46)_2$ with $\sin\varphi$ and add the two equations. This results in the first equation of (6.47). Multiplying $(6.46)_1$ by $-\sin\varphi$ and $(6.46)_2$ by $\cos\varphi$ and adding the two equations results in the second one of the following two equations

$$\begin{aligned}
\dot{r} &= \varepsilon r^3[a_{130}\cos^4\varphi + (a_{121} + a_{230})\cos^3\varphi\sin\varphi + \\
&\quad + (a_{112} + a_{221})\cos^2\varphi\sin^2\varphi + (a_{103} + a_{212})\cos\varphi\sin^3\varphi + \\
&\quad + a_{203}\sin^4\varphi] = \varepsilon S(r,\varphi) \\
\dot{\varphi} &= \omega + \varepsilon r^2[a_{230}\cos^4\varphi + (a_{221} - a_{130})\cos^3\varphi\sin\varphi + \\
&\quad + (a_{212} - a_{121})\cos^2\varphi\sin^2\varphi + (a_{203} - a_{112})\cos\varphi\sin^3\varphi - \\
&\quad - a_{103}\sin^4\varphi] = \omega + \varepsilon T(r,\varphi) .
\end{aligned} \quad (6.47)$$

To apply (6.44) to (6.47), we introduce the following notations

$$x = \begin{pmatrix} r \\ \varphi \end{pmatrix}, \quad \hat{x} = \begin{pmatrix} \hat{r} \\ \hat{\varphi} \end{pmatrix}, \quad \mathbf{f}_0 = \begin{pmatrix} 0 \\ \omega \end{pmatrix}, \quad \mathbf{f}_1 = \begin{pmatrix} S \\ T \end{pmatrix}, \quad \mathbf{u} = \begin{pmatrix} s(\hat{r},\hat{\varphi}) \\ t(\hat{r},\hat{\varphi}) \end{pmatrix} .$$

With this notation, we find from (6.44)

$$
\begin{aligned}
\dot{\hat{r}} &= \varepsilon\left[-\frac{\partial s}{\partial\hat{\varphi}}\omega + S(\hat{r},\hat{\varphi})\right] + O(\varepsilon^2) \\
\dot{\hat{\varphi}} &= \omega + \varepsilon\left[-\frac{\partial t}{\partial\hat{\varphi}}\omega + T(\hat{r},\hat{\varphi})\right] + O(\varepsilon^2)
\end{aligned}
\tag{6.48}
$$

where the functions s and t have not yet been specified. The most desirable situation would now be to remove all terms of order ε. However, this cannot always be done. In general only the oscillating part of S and T can be removed ([69]). We explain this elimination for S. Of course, similar expressions also hold for T. To start, one splits S into an averaged and an oscillating part. These are

$$
S(\hat{r},\hat{\varphi}) = \hat{S}(\hat{r}) + \tilde{S}(\hat{r},\hat{\varphi})
\tag{6.49}
$$

where

$$
\hat{S}(\hat{r}) = \frac{1}{2\pi}\int_0^{2\pi} S(\hat{r},\hat{\varphi})d\hat{\varphi} .
\tag{6.50}
$$

From (6.49) follows the oscillating part to

$$
\tilde{S} = S - \hat{S} .
\tag{6.51}
$$

To eliminate the oscillating part of S, we obtain, after inserting (6.49) into the bracket of $(6.48)_1$, the following condition

$$
-\omega\frac{\partial s}{\partial\hat{\varphi}} + \tilde{S} = 0 .
\tag{6.52}
$$

Keeping $\hat{r}$ fixed, integration of (6.52) yields

$$
s(\hat{r},\hat{\varphi}) = \frac{1}{\omega}\int_0^{\hat{\varphi}} \tilde{S}(\hat{r},\hat{\psi})d\hat{\psi} .
\tag{6.53}
$$

As result of these transformations (6.48) is replaced by

$$
\begin{aligned}
\dot{\hat{r}} &= \varepsilon\hat{S}(\hat{r}) + O(\varepsilon^2) \\
\dot{\hat{\varphi}} &= \omega + \varepsilon\hat{T}(\hat{r}) + O(\varepsilon^2) .
\end{aligned}
\tag{6.54}
$$

Again, it is important to point out that the transformation (6.43) influences the terms of order ε^2. Averaging over terms of order ε^2 has to take this change into account.

However, for the example under investigation, we only need to calculate the two integrals (6.50) for $S(\hat{r},\hat{\varphi})$ and $T(\hat{r},\hat{\varphi})$ which are given by the right-hand sides of (6.47).

This results in two equations

$$\dot{r} = \varepsilon A \hat{r}^3$$
$$\dot{\varphi} = \omega + \varepsilon B \hat{r}^2 \,,$$

where the brackets in the following integrals for A and B are those from (6.47). Hence, we obtain

$$
\begin{aligned}
A &= \frac{1}{2\pi} \int_0^{2\pi} [a_{130} \cos^4 \varphi + \ldots + a_{203} \sin^4 \varphi] d\varphi \\
&= \frac{3}{8}(a_{130} + a_{203}) + \frac{1}{8}(a_{112} + a_{221})
\end{aligned}
$$

$$
\begin{aligned}
B &= \frac{1}{2\pi} \int_0^{2\pi} [a_{230} \cos^4 \varphi + \ldots - a_{103} \sin^4 \varphi] d\varphi \\
&= \frac{3}{8}(a_{230} - a_{103}) + \frac{1}{8}(a_{212} - a_{121}) \,.
\end{aligned}
$$

The expressions for A and B are identical to b_{21}^R and b_{21}^I in (6.41).

This example shows that not only a strong reduction of the complexity of the problem is possible by means of the averaging method, but also the close connection to normal form theory which is elaborated in [128].

A complete listing of the different normal forms for cases of codimension *one* and *two* together with the dependence of the coefficients in the normal form on the coefficients in the bifurcation equations which are obtained from the reduction process is given in Section 6.5.1. We give this listing in Section 6.5.1 because there we also include the unfolding parameters.

6.1.2   Time-discrete dynamical systems

We discuss the three cases of codimension *one* mentioned in Section 4.2.4.

We start with μ real. We can deal with the cases $\mu = +1$ (transcritical bifurcation) and $\mu = -1$ (flip bifurcation) simultaneously because after the center manifold reduction both are given by a one-dimensional bifurcation equation similar to (4.53). We write this equation following the notation of Section 6.1.1 in the form

$$y_{t+1} = \pm y_t + a_2 y_t^2 + a_3 y_t^3 + O(|y_t|^4) \,. \tag{6.55}$$

Similar to (6.2) the change of variables

$$y_t = z_t + h(z_t) \tag{6.56}$$

is introduced. First, we try to eliminate the second order term in (6.55). For this purpose, we assume

$$h(z_t) = \alpha_2 z_t^2 . \tag{6.57}$$

Considering (6.57) and inserting (6.56) into (6.55) we find

$$
\begin{aligned}
z_{t+1} + \alpha_2 z_{t+1}^2 &= \pm (z_t + \alpha_2 z_t^2) + \\
&\quad + a_2(z_t + \alpha_2 z_t^2)^2 + a_3 z_t^3 + O(|z_t|^4)
\end{aligned}
$$

or

$$
\begin{aligned}
z_{t+1} &= \pm (z_t + \alpha_2 z_t^2) - \\
&\quad - \alpha_2[\pm z_t \pm \alpha_2 z_t^2 + a_2 z_t^2 - \alpha_2 z_t^2]^2 + \\
&\quad + a_2(z_t + \alpha_2 z_t^2)^2 + a_3 z_t^3 + O(|z_t|^4)
\end{aligned}
$$

and finally

$$
\begin{aligned}
z_{t+1} &= \pm z_t + (\pm \alpha_2 - \alpha_2 + a_2) z_t^2 + \\
&\quad + [-2\alpha_2^2 \mp 2\alpha_2 a_2 \pm 2\alpha_2^2 + 2\alpha_2 a_2 + a_3] z_t^3 + \tag{6.58} \\
&\quad + O(|z_t|^4) .
\end{aligned}
$$

From (6.58) follows immediately that for $\mu = +1$, that is for the upper signs, the second order term cannot be removed. Hence, the normal form in this case is

$$z_{t+1} = z_t + A_2 z_t^2 + O(|z_t|^3) \tag{6.59}$$

where $A_2 = a_2$. However, if $\mu = -1$, that is for the lower signs in (6.58), the quadratic term can be removed for $\alpha_2 = a_2/2$ and the normal form reads

$$z_{t+1} = -z_t + A_3 z_t^3 + O(|z_t|^4) . \tag{6.60}$$

The coefficient A_3 is obtained from (6.58) by taking the lower signs and inserting the value $\alpha_2 = a_2/2$ to

$$A_3 = -4\alpha_2^2 + 4\alpha_2 a_2 + a_3 = a_2^2 + a_3 . \tag{6.61}$$

Second, we treat the Hopf bifurcation, that is, $\mu_{1,2} = \nu \pm i\eta$, $|\mu_{1,2}| = 1$. In this case after the center manifold reduction, the two-dimensional bifurcation system (4.59) is obtained which reads

$$w_{t+1} = \mu w_t + g(w_t, \overline{w}_t) \tag{6.62}$$

where

$$
\begin{aligned}
g(w_t, \overline{w}_t) &= a_{20} w_t^2 + a_{11} w_t \overline{w}_t + a_{02} \overline{w}_t^2 + \\
&\quad + a_{30} w_t^3 + a_{21} w_t^2 \overline{w}_t + a_{12} w_t \overline{w}_t^2 + a_{03} \overline{w}_t^3 + \ldots \tag{6.63}
\end{aligned}
$$

We give the result for (6.62) in form of a *theorem* ([35] p. 242):

Theorem 6.1 (Hopf bifurcation of mappings) *If $g(w_t, \overline{w}_t)$ is given by (6.63) and μ is not a k-th root of unity for $k = 1, \ldots, 5$ then there exists a smooth change of coordinates*

$$w_t = v_t + h(v_t, \overline{v}_t) \tag{6.64}$$

that transforms (6.62) into

$$v_{t+1} = \mu v_t + A_{21} v_t^2 \overline{v}_t + O(|v_t|^5 + |\overline{v}_t|^5) \ . \tag{6.65}$$

Similar to the time-continuous case a very simple normal form (6.65) is obtained. There are no quadratic and fourth order terms, and only one cubic term.

In order to prove the theorem we begin by eliminating the second order terms in (6.63). We assume (6.64) to be given as

$$w_t = v_t + \alpha_{20} v_t^2 + \alpha_{11} v_t \overline{v}_t + \alpha_{02} \overline{v}_t^2 \ . \tag{6.66}$$

Inserting (6.66) into (6.62), we find

$$
\begin{aligned}
v_{t+1} &+ \alpha_{20} v_{t+1}^2 + \alpha_{11} v_{t+1} \overline{v}_{t+1} + \alpha_{02} \overline{v}_{t+1}^2 \\
&= \mu(v_t + \alpha_{20} v_t^2 + \alpha_{11} v_t \overline{v}_t + \alpha_{02} \overline{v}_t^2) + \\
&\quad + a_{20} v_t^2 + a_{11} v_t \overline{v}_t + a_{02} \overline{v}_t^2 + \\
&\quad + [a_{21} + 2a_{20}\alpha_{11} + a_{11}(\alpha_{20} + \overline{\alpha}_{11}) + 2a_{02}\overline{\alpha}_{02}] v_t^2 \overline{v}_t + \ldots
\end{aligned}
\tag{6.67}
$$

Similarly as it was done for the time-continuous case the coefficient of the third order term $v_t^2 \overline{v}_t$ is also calculated. Inserting on the left-hand side of (6.67) for v_{t+1}^2 from the difference equation we obtain

$$
\begin{aligned}
v_{t+1} &= \mu v_t + \mu(\alpha_{20} v_t^2 + \alpha_{11} v_t \overline{v}_t + \alpha_{02} \overline{v}_t^2) + \\
&\quad + a_{20} v_t^2 + a_{11} v_t \overline{v}_t + a_{02} \overline{v}_t^2 - \\
&\quad - \alpha_{20}\mu^2 v_t^2 - \alpha_{11}\mu\overline{\mu} v_t \overline{v}_t - \alpha_{02}\overline{\mu}^2 \overline{v}_t^2 + O(|v_t|^3 + |\overline{v}_t|^3)
\end{aligned}
$$

and a similar expression for $\overline{v}_{t+1}$. Writing out only the quadratic terms this expression becomes

$$
\begin{aligned}
v_{t+1} &= \mu v_t + (-\mu^2 \alpha_{20} + \mu\alpha_{20} + a_{20}) v_t^2 + \\
&\quad + (-\mu\overline{\mu}\alpha_{11} + \mu\alpha_{11} + a_{11}) v_t \overline{v}_t + \\
&\quad + (-\overline{\mu}^2 \alpha_{02} + \mu\alpha_{02} + a_{02}) \overline{v}_t^2 \ .
\end{aligned}
\tag{6.68}
$$

The second order terms in (6.68) can be annihilated by setting

$$\alpha_{20} = -\frac{a_{20}}{\mu - \mu^2} \ , \qquad \alpha_{11} = -\frac{a_{11}}{\mu - \mu\overline{\mu}} \ , \qquad \alpha_{02} = -\frac{a_{02}}{\mu - \overline{\mu}^2} \ . \tag{6.69}$$

However, $\mu - \overline{\mu}^2 = (\mu^3 - 1)/\mu^2$ vanishes if $\mu^3 - 1 = 0$. Hence, in this case of third order resonance the quadratic term $\overline{v}^2$ cannot be removed. The coefficient A_{21} of the third order term $v_t^2 \overline{v}_t$, after the second order terms had been eliminated, follows from (6.67) to

$$A_{21} = a_{21} + 2\alpha_{11}a_{20} + (\overline{\alpha}_{11} + \alpha_{20})a_{11} + 2\overline{\alpha}_{02}a_{02} .
\tag{6.70}$$

Thus, we can write the difference equation with third order terms as

$$w_{t+1} = \mu w_t + A_{30}w_t^3 + A_{21}w_t^2\overline{w}_t + A_{12}w_t\overline{w}_t^2 + A_{03}\overline{w}_t^3 .
\tag{6.71}$$

Here, the precise expressions for the coefficients A_{30}, A_{12}, A_{03} must not be given since they can be removed from (6.71). The calculation is straightforward. We insert

$$w_t = v_t + \alpha_{30}v_t^3 + \alpha_{21}v_t^2\overline{v}_t + \alpha_{12}v_t\overline{v}_t^2 + \alpha_{03}\overline{v}_t^3$$

into (6.71) and find

$$
\begin{aligned}
v_{t+1} = \ & \mu v_t + (-\mu^3\alpha_{30} + \mu\alpha_{30} + A_{30})v_t^3 + \\
& + (-\mu^2\overline{\mu}\alpha_{21} + \mu\alpha_{21} + A_{21})v_t^2\overline{v}_t + \\
& + (-\mu\overline{\mu}^2\alpha_{12} + \mu\alpha_{12} + A_{12})v_t\overline{v}_t^2 + \\
& + (-\overline{\mu}^3\alpha_{03} + \mu\alpha_{03} + A_{03})\overline{v}_t^3 + O(|v_t|^4 + |\overline{v}_t|^4) .
\end{aligned}
\tag{6.72}
$$

To remove the third order terms, we have to set to zero the coefficients in (6.72). This results in

$$\alpha_{30} = -\frac{A_{30}}{\mu(1 - \mu^2)} , \qquad \alpha_{12} = -\frac{A_{12}}{\mu(1 - \overline{\mu}^2)} , \qquad \alpha_{03} = -\frac{A_{03}}{\mu - \overline{\mu}^3} .$$

However, we do not obtain a solution for α_{21} because the denominator $\mu(1 - \mu\overline{\mu})$ vanishes for any μ on the unit circle, which is exactly the situation under study. We also note that $\mu - \overline{\mu}^3 = (\mu^4 - 1)/\mu^3$. If $\mu^4 = 1$, the term $A_{03}\overline{w}^3$ cannot be removed as well. Continuing in this manner the remaining assertions of the theorem can also be proved.

Finally, we note that many of the transformation procedures of normal form and averaging theory can be performed automated using symbolic manipulation packages such as REDUCE, MACSYMA or SMP ([114]).

6.1.3 Statical systems

We have already mentioned in the Introduction that for statical systems described by a potential in one critical variable, transformation to normal form is trivial, because the lowest order nonvanishing nonlinear term is also the term appearing in the normal form, when the main (distinguished) parameters take their critical value. However, we shall see in

Section 6.5.3 that the (trivial) transformation into normal form, including the unfolding parameters, can lead to an unpleasant mixing of the distinguished and the imperfection parameters.

The next case which we want to consider is that of potentials in two critical variables. This problem is similar to that treated in Section 6.1.1. However, now we shall use only linear transformations of coordinates to remove as many terms as possible from a given potential. Let us assume that after having performed the Ljapunov-Schmidt reduction the degenerate potential (no quadratic terms) in two critical variables q_1, q_2 takes the form of a homogeneous cubic

$$j^3 V = a_{30} q_1^3 + a_{21} q_1^2 q_2 + a_{12} q_1 q_2^2 + a_{03} q_2^3 . \tag{6.73}$$

The question to be asked is: into which canonical (normal) form can (6.73) be transformed? The answer is given in [111] p. 27, where it is shown how the following four different cases

$$
\begin{array}{lll}
\text{(i)} & p_1^2 p_2 - p_2^3 & \\
\text{(ii)} & p_1^2 p_2 + p_2^3 & \\
\text{(iii)} & p_1^2 p_2 & \\
\text{(iv)} & p_1^3 &
\end{array}
\tag{6.74}
$$

are arranged in 4-dimensional space (a_{ij}) of cubic forms (6.73). The classification (6.74) can be motivated from the fact that the zero set of (6.73) consists of straight lines through the origin of the coordinate system. By coordinate transformations these lines are transformed into coordinate axes and into the line $s_1 + s_2 = 0$ (see below equation (6.78)).

We follow here the description given in [121] p. 33 and explain how to transform (6.73) to one of (6.74). The first step in treating (6.73) is to find out the structure of the solution set of (6.73) which determines how (6.73) can be transformed. First, we note that if $a_{03} = 0$ the axis $q_1 = 0$ is a solution and the remaining solutions follow from the treatment of the quadratic form

$$a_{30} q_1^2 + a_{21} q_1 q_2 + a_{12} q_2^2 = 0 .$$

Second, if $a_{03} \neq 0$ we may set $q_2 = 1$ and solve the cubic equation

$$a_{30} q_1^3 + a_{21} q_1^2 + a_{12} q_1 + a_{03} = 0 . \tag{6.75}$$

As stated above, if we know the solutions for $q_2 = 1$ we know the solutions for any q_2 because they are on rays through the origin. Depending on the structure of the solutions of (6.75) (either three real roots or one real root and a conjugate complex one) one obtains either one or three rays

of solutions through the origin as is discussed in detail in the following
four distinct cases:

(i) If (6.75) admits three real distinct roots then (6.73) can be rewritten in the form

$$j^3 V = (a_1 q_1 + b_1 q_2)(a_2 q_1 + b_2 q_2)(a_3 q_1 + b_3 q_2) , \qquad (6.76)$$

where no two of the ratios a_i/b_i are equal, one obtains

$$j^3 V = (a r_1 + b r_2) r_1 r_2 . \qquad (6.77)$$

Here $r_1 = a_2 q_1 + b_2 q_2$ and $r_2 = a_3 q_1 + b_3 q_2$ are used. The new coefficients
a and b follow to

$$a = (a_1 b_3 - b_1 a_3)/\Delta$$
$$b = (-a_1 b_2 + a_2 b_1)/\Delta$$

where $\Delta = a_2 b_3 - a_3 b_2$. By the scaling $a r_1 = \gamma s_1$, $b r_2 = \gamma s_2$, (6.77) follows
to

$$j^3 V = (s_1 + s_2) \frac{s_1}{a} \frac{s_2}{b} \gamma^3 .$$

If we choose $\gamma = \sqrt[3]{ab}$ we obtain

$$j^3 V = (s_1 + s_2) s_1 s_2 . \qquad (6.78)$$

Using $s_1 = p_1 - p_2$ and $s_2 = -p_1 - p_2$, (6.78) is transformed to

$$j^3 V = 2 p_2 (p_1^2 - p_2^2) \sim p_1^2 p_2 - p_2^3 .$$

This is the canonical form of the *elliptic umbilic*. The symbol $\sim$ means
"is right-equivalent to" (see Appendix M.1)

(ii) If (6.75) has one real root and a conjugate complex pair of roots
$j^3 V$ must be given by

$$j^3 V = (a_1 q_1 + b_1 q_2)(a_2 q_1 + b_2 q_2)(\bar{a}_2 q_1 + \bar{b}_2 q_2)$$

where $\bar{a}_2$ and $\bar{b}_2$ are the conjugate complex of a_2, b_2 then

$$(a_2 q_1 + b_2 q_2)(\bar{a}_2 q_1 + \bar{b}_2 q_2) = ((\Re a_2) q_1 + (\Re b_2) q_2)^2 + ((\Im a_2) q_1 + (\Im b_2) q_2)^2 .$$

Taking the expressions in the parentheses on the right-hand side as the
definition of the change of variables, we obtain

$$j^3 V \sim (a r_1 + b r_2)(r_1^2 + r_2^2) .$$

Using $p_2 = a r_1 + b r_2$ and $p_1 = b r_1 - a r_2$ results in

$$j^3 V \sim p_2 (p_1^2 + p_2^2) = p_1^2 p_2 + p_2^3$$

which is the form of the *hyperbolic umbilic* given in (6.74). However,
sometimes an alternative form is used which follows from

$$(p_1 + p_2)^3 + (p_2 - p_1)^3 = 2p_2^3 + 6p_1^2 p_2 \sim s_1^2 s_2 + s_2^3 .$$

Using the transformation $t_1 = p_1 + p_2$, $t_2 = -p_1 + p_2$, another canonical
form of the hyperbolic umbilic is

$$j^3 V \sim t_1^3 + t_2^3 . \tag{6.79}$$

(iii) If two real roots coincide, that is $a_1/b_1 = a_2/b_2$, then case (iii) in
(6.74) is obtained, because

$$j^3 V \sim (a_1 q_1 + b_1 q_2)^2 (a_3 q_1 + b_3 q_2) \sim s_1^2 s_2 . \tag{6.80}$$

(iv) If all three roots coincide, that is, all ratios are equal the fourth
case in (6.74) results.

The four canonical forms (6.74) obviously possess the mentioned zero
sets: (i) has three real roots (three lines or rays), (ii) one real root (one
line), (iii) two coincident real roots (two lines) and (iv) three coincident
real roots (one line).

We do not list the canonical forms of low codimension now, because
they are given in their unfolded form in Section 6.5.2.

6.2   Codimension

In Chapters 4 and 5, in almost all cases so-called one-parameter bifurcation equations are derived where a (possibly rescaled) distinguished
parameter appears. After having simplified the nonlinear terms in the
bifurcation equations by means of normal form theory, we want to solve
the bifurcation equations. First, we try to do this for the one-parameter
bifurcation equations. However, in applications in engineering more general results are desired, namely, to obtain all qualitatively different solutions which are possible for the class of equations to which the bifurcation
equations belong. Hence, the (one-parameter) bifurcation equations must
be embedded into a *versal* family (defined below) of equations. In the
Introduction it is shown that for the bifurcation equation (4.79), (4.100)
or (4.127) this will be the family of equations (1.17), where one additional parameter appears. Hence, the first question to be answered is:
what is the minimum number of parameters necessary to obtain a versal
unfolding? The number giving the answer to this question is called the
codimension of the problem. Codimension in connection with bifurcation equations is related to their dimension n_c and to the order of the
nonlinear terms appearing in the bifurcation equations.

Before we discuss bifurcation equations, let us make some general comments on codimension, following [121]. If for a moment, we think of geometric objects, the codimension of such an object can be defined as the difference between its dimension and the dimension of the space in which it is embedded. For example, the codimension of a curve C in 3-dimensional space is *two* (Fig. 6.2). The number *two* follows also from

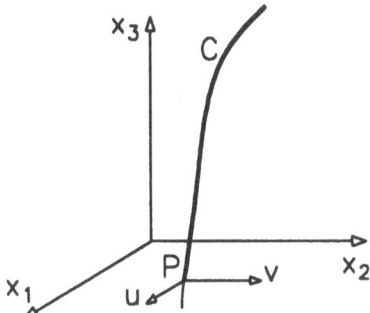

Figure 6.2. The curve C in $\mathbb{R}^3$ and the point P in $\mathbb{R}^2$ have the same codimension

the fact that two equations are needed to describe a curve in $\mathbb{R}^3$. One equation would give us a surface in $\mathbb{R}^3$ which is of codimension *one*. With the example of the curve in $\mathbb{R}^3$, one can show an important property of codimension, namely that it is preserved under a change of the dimension of the system. This is because if we neglect the third dimension x_3 in Fig. 6.2, $\mathbb{R}^3$ reduces to $\mathbb{R}^2$ and the curve reduces to a single point P, provided the curve has a point of intersection with the x_1, x_2 plane. In the plane this point has again codimension two, although the system is now of different dimension compared to the original one.

A very important fact for the following is that a continuous one-parameter family of objects of dimension p forms an object of dimension $p+1$. That is, a one-parameter family of points forms a curve, a one-parameter family of curves a surface, and so on. This allows us immediately to conclude a very important property. If a singular (degenerate) object (like a point in a plane) of codimension ℓ is given then at least a ℓ-dimensional parameter family will be needed to include each system lying in the neighborhood of the original system. For the example of the point P in the plane, this must be a two-parameter family. Of course, the two vectors u, v giving the displacement of the point in the plane (Fig. 6.2) must be linearly independent such that any point in the neighborhood of P can be reached. Such a family is called a *versal family*. More generally, we have the definition:

Definition 6.1 (Versal and universal family) *A family of objects is called a versal family if every possible member of this class of objects is included in the family. A versal family with the minimum number of parameters is called* universal.

This allows us to give the following definition:

Definition 6.2 (Codimension) *The number of parameters in a universal family is called the codimension of the degenerate problem.*

In the case of the example of the point in the plane, the universal family must include precisely two parameters, but not more.

From what we have said, one can easily conclude that if one studies a single object then a special degenerate situation is comparable to a particular point in the plane, and hence, making a random choice it is quite unlikely to be met. Further, such a case can be avoided by a small perturbation. However, if one studies a versal family then each member of the family will inevitably appear under a parameter variation. This will occur even if it is the most degenerate situation possible in this family. Therefore, it makes sense to study degenerate cases.

6.2.1   Static bifurcation

We start with an important example. From the buckling problems in Sections 4.3.1–4.3.3, we saw that frequently a one-parameter bifurcation equation is given by: $Aq^3 + a_1 q = 0$. Since all these problems are conservative, it is convenient to introduce the corresponding potential which is given by

$$V = \frac{1}{4}Aq^4 + \frac{1}{2}a_1 q^2 .\tag{6.81}$$

In (6.81) A is a fixed quantity obtained from the reduction process, wheras a_1 depends on a parameter. Since (6.81) depends only on one variable q, terms of higher than fourth order do not have any influence on the local stability behavior, because for $a_1 = 0$, that is, at the bifurcation point (critical parameter value) the lowest order nonlinear term is of fourth order, provided $A \neq 0$, and determines the behavior. However, in the case of several variables the lowest order nonlinear term does not always determine the behavior and then the so-called problem of *determinacy* arises which is treated in Section 6.3.

The first question to answer is: what is the codimension of (6.81) if $a_1 = 0$? Or in other words: how many parameters are needed to obtain all qualitatively different potentials V of fourth order? The notion "qualitatively different" will be explained below.

Before we answer this question, let us first study the simpler case of the potential of third order

$$V = \frac{1}{3}q^3 .\tag{6.82}$$

The graph of (6.82) is given in Fig. 6.3a. A perturbation of (6.82) is either of the topological type of

$$\frac{1}{3}q^3 + q \qquad \text{or} \qquad \frac{1}{3}q^3 - q$$

(Fig. 6.3b and Fig. 6.3c). These can be obtained from a one-parameter

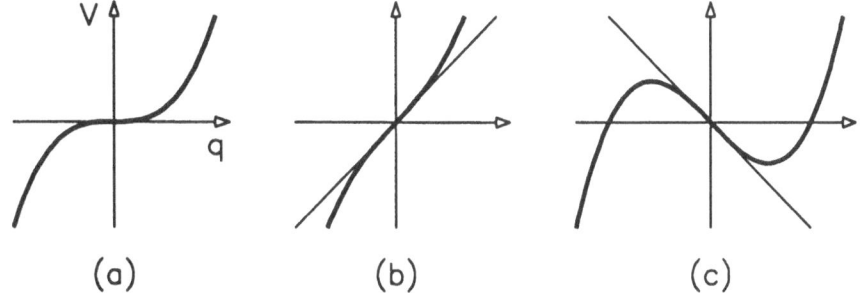

$$(a) \qquad\qquad (b) \qquad\qquad (c)$$

Figure 6.3. (a) Potential (6.82) and its two possible topological perturbations with either (b) no or (c) two singular points

family. Therefore, equation (6.82) has codimension one and the universal unfolding is given by

$$V = \frac{1}{3}q^3 + aq. \tag{6.83}$$

We remark that a constant term does not matter and a quadratic term could be removed by a coordinate shift (see Section 6.5.3).

From (6.83) follows the bifurcation equation

$$\frac{\partial V}{\partial q} = q^2 + a = 0 \tag{6.84}$$

with the solutions $q = \pm\sqrt{-a}$.

The corresponding *bifurcation graph* and *bifurcation diagram* are given in Fig. 6.4.

Let us now return to potential (6.81) and the question concerning its codimension. In the physical problem described by (4.79) or (4.100), the axial load appears quite naturally as one parameter. This parameter appears such that (6.81) is obtained which yields all symmetric deformations of potentials of fourth order (Fig. 6.5a–c). However, nonsymmetric deformations of (6.81) (Fig. 6.6) are obviously not included in (6.81), and therefore, at least a second parameter b_1 is yet needed for a versal (in fact it will be a universal) unfolding yielding

$$V = \frac{1}{4}Aq^4 + \frac{1}{2}a_1q^2 + b_1q. \tag{6.85}$$

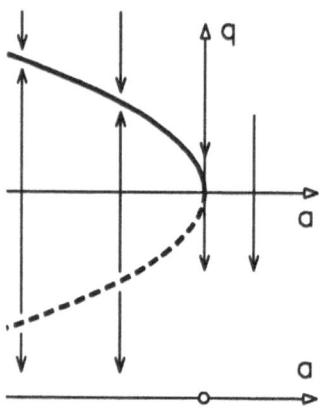

Figure 6.4. Bifurcation diagram (lower part)
and bifurcation graph (upper part) of (6.84)
giving the solutions and their stability

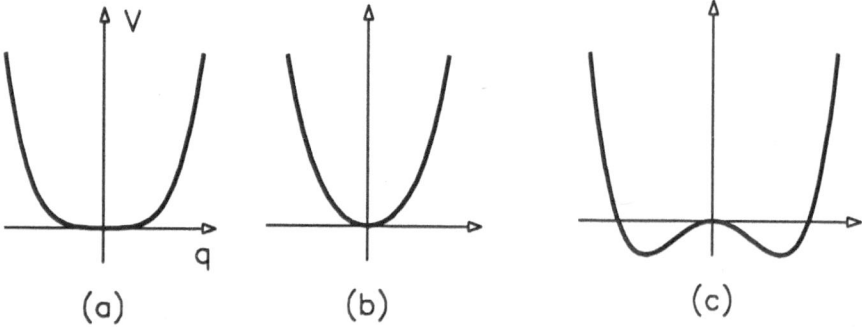

(a)   (b)   (c)

Figure 6.5. Symmetric deformations (6.81) of $V = \frac{1}{4}Aq^4$

Why just the terms q^2 and q appear in the universal unfolding (6.85) is
obvious in this simple example, but will be explained more generally on
p. 203. The bifurcation graphs corresponding to (6.81) and (6.85) are

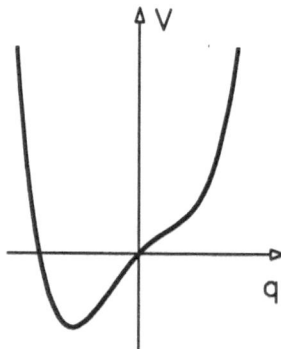

Figure 6.6. Nonsymmetric (generic) deformation (6.85) of (6.81)

given in Fig. 1.5a,b and follow from the bifurcation equation

$$Aq^3 + a_1 q + b_1 = 0 . \tag{6.86}$$

In Fig. 6.5, Fig. 6.6 and Fig. 1.5, the case of $A > 0$ is drawn. Dividing by A in (6.86), the bifurcation equation can be rewritten in the form

$$q^3 + aq + b = 0 . \tag{6.87}$$

When dividing by A, we must be careful if A is negative. In fact, the two potentials $V_1 = 1/4\, q^4 - 1/2\, aq^2$ and $V_2 = -1/4\, q^4 + 1/2\, aq^2$ both lead to an equilibrium equation $q^3 - aq = 0$. However, the stability of their solutions is different. This is easy to understand from graphs similar to Fig. 6.5.

One can represent (6.87) in a three-dimensional space (q, a, b) to obtain the surface in Fig. 6.7. This is the *cusp surface* from *elementary catastrophe theory* ([121]). We discuss it in Section 6.6.1.

Mathematical definition of codimension of potentials

For potentials $V(q_1, \ldots, q_n)$, it is rather easy to give a precise mathematical definition of codimension ([111], [141]):

Definition 6.3 (Codimension of potentials) *The non-negative integer $\ell = \operatorname{codim} V$ of a potential $V(q_1, \ldots, q_n)$ is defined as the number of terms which cannot independently be generated by the partial derivatives $\partial V/\partial q_i = V_{,i}$.*

A smooth function $\varphi(\mathbf{q})$ is said to be generated by the $V_{,i}$, if smooth functions $\psi_i(\mathbf{q})$ exist such that (constant terms can be ignored)

$$\varphi(\mathbf{q}) = \sum_{i=1}^{n} \psi_i(\mathbf{q}) V_{,i}(\mathbf{q}). \tag{6.88}$$

Let us explain the definition of codimension with some examples:

Example 1: $V_1 = (1/4)q_1^4$ according to (6.81). $V_{1,1} = q_1^3$. Obviously, the functions $\varphi(q_1) = q_1$ and q_1^2 cannot be generated by $V_{1,1} = q_1^3$. Therefore, $\operatorname{codim} V_1 = 2$ and the functions q and q^2 must be used for unfolding as it is done in (6.85).

Example 2: $V_2(q_1, q_2) = q_1^3 + q_2^3$ according to (6.79). $\operatorname{Codim} V_2 = 3$, because q_1, q_2 and $q_1 q_2$ cannot be generated from $V_{2,1} = 3q_1^2, V_{2,2} = 3q_2^2$.

Example 3: $V_3(q_1, q_2) = q_1^2 q_2$ according to (6.80). $\operatorname{Codim} V_3 = \infty$, since q_2^k cannot be generated from the $V_{3,1} = 2q_1 q_2$ and $V_{3,2} = q_1^2$ for any k.

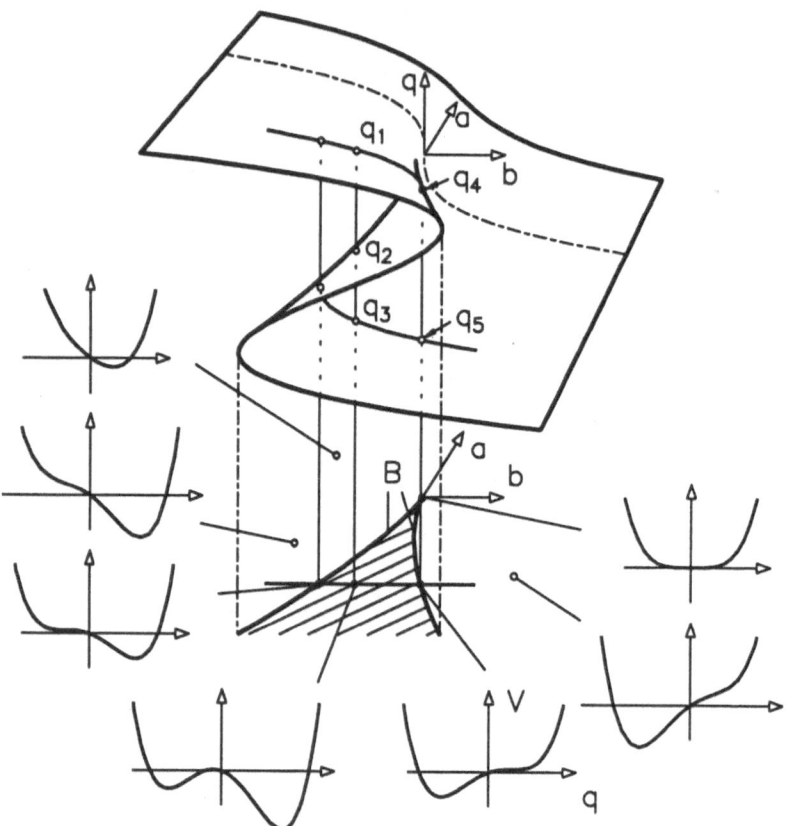

Figure 6.7. Cusp surface (6.87) in (q, a, b) space. The potentials V corresponding to those values in the a, b parameter plane which are indicated by the circles are depicted qualitatively. The cusp shaped curve in the a, b parameter plane is the bifurcation set

Example 4: $V_4(r, s) = (1/4)Br^4 + (1/2)Cr^2s^2 + (1/4)Ds^4$, which is the potential (4.172) describing the buckling problem of the rectangular plate at a double eigenvalue. $V_{4,r} = Br^3 + Crs^2$, $V_{4,s} = Cr^2s + Ds^3$. The terms $r, s, r^2, rs, s^2, rs^2, r^2s, r^2s^2$ cannot be generated. Therefore codim $V_4 = 8$.

The behavior of a function $V(q_1, \ldots, q_n)$ near a nondegenerate critical point $(V_{,i} = 0, \det V_{,ij} \neq 0)$ is determined by Morse's lemma which states that an appropriate transformation of coordinates $\mathbf{p} = \mathbf{p}(\mathbf{q})$ will lead to the Morsian normal form: $-p_1^2 - \ldots - p_\nu^2 + p_{\nu+1}^2 + \ldots + p_n^2 + \text{const.}$ Furthermore, it is known that near an isolated critical point, even if it is degenerate, an *analytic* function can be transformed into the leading part of its Taylor series by an appropriate analytic change of variables. We recall that a function is said to be *analytic* if its Taylor series converges to it at every point. In this sense, a function near an isolated critical point can always be reduced to polynomial normal form. However, only for degenerate critical points with codimension $\ell < 6$, a simple classification is possible ([4]). In general, for codimension $\ell \geq 6$ such a classification is not possible because the resulting normal forms contain parameters. For special values of these parameters an infinity of qualitatively different forms can be obtained.

A physical example of infinite codimension

An example for the occurence of a situation where a simple classification is not possible occurs in a special case of the buckling problem of a rectangular plate at a double eigenvalue which we discussed in Section 4.3.4. The potential describing this problem is the potential V_4, mentioned above, and given by (4.172). For general values of B, C, D it has codimension eight. We show now that for special values of B, C, D this is no longer the case.

Assuming that $B > 0$ and $D > 0$, we change coordinates in (4.172) by

$$r = q_1 B^{-1/4}, \quad s = q_2 D^{-1/4} \qquad \text{and introduce} \qquad \mu = \frac{C}{\sqrt{BD}}.$$

Then $V = V_\mu$ takes the form $(\lambda = 0)$

$$V_\mu = \frac{1}{4}q_1^4 + \frac{1}{2}\mu q_1^2 q_2^2 + \frac{1}{4}q_2^4 . \tag{6.89}$$

If the special value $\mu_0 = 1$ is chosen, (6.89) reduces to

$$V_{\mu_0} = \frac{1}{4}(q_1^2 + q_2^2)^2$$

which has infinite codimension. This is not too difficult to understand, if we consider

$$V_{\mu_0,1} = q_1(q_1^2 + q_2^2) \qquad \text{and} \qquad V_{\mu_0,2} = q_2(q_1^2 + q_2^2) .$$

From (6.88) follows

$$\varphi(q_1, q_2) = \psi_1(q_1, q_2)V_{\mu_0,1} + \psi_2(q_1, q_2)V_{\mu_0,2} = (q_1^2 + q_2^2)(q_1\psi_1 + q_2\psi_2) .$$

From this expression, it is obvious that one cannot generate a function $\varphi(q_1, q_2)$ which depends only on one variable q_1 or q_2. See also the comments on p. 212.

Listing of normal forms for degenerate potentials in one and two active variables

A complete list (6.90) of normal forms for a function $V(q_1, \ldots, q_n)$ in the neighborhood of a critical point corresponding to a simple or at most double eigenvalue appears as follows (it is assumed that $V(0) = 0$) ([3])

$$
\begin{array}{llll}
A_k: & V = \pm q_1^{k+1} \pm q_2^2 + Q & k \geq 1 & \text{codim } A_k = k-1 \\
D_k: & V = q_1^2 q_2 \pm q_2^{k-1} + Q & k \geq 4 & \text{codim } D_k = k-1 \\
E_6: & V = q_1^3 \pm q_2^4 + Q & & \text{codim } E_6 = 5 \qquad (6.90) \\
E_7: & V = q_1^3 + q_1 q_2^3 + Q & & \text{codim } E_7 = 6 \\
E_8: & V = q_1^3 + q_2^5 + Q & & \text{codim } E_8 = 7 .
\end{array}
$$

In the above list

$$
Q = -q_3^2 - \ldots - q_\nu^2 + q_{\nu+1}^2 + \ldots + q_n^2
$$

denotes a quadratic form depending on the passive variables $q_3, \ldots, q_n$. From (6.90) follows table (6.91) which gives all simple singularities of codimension $\ell < 6$:

ℓ	1	2	3	4	5
Singularity	A_2	A_3	A_4, D_4	A_5, D_5	A_6, D_6, E_6

$$(6.91)$$

The importance of the existence of only few different normal forms is that any problem of the same codimension and the same number of active variables is right-equivalent (Appendix M) to the corresponding normal form, and hence, can be transformed by a smooth change of variables into it. Therefore, only these few cases must be studied!

6.2.2 Dynamic bifurcation

We return to the example of Hopf bifurcation given by the normal form (6.39) and introduce polar coordinates by $z_1 = r \cos \varphi, z_2 = r \sin \varphi$ in (6.39) as on p. 190. This results in two equations for r and φ

$$
\begin{aligned}
\dot{r} &= b^R r^3 + h.o.t. \\
\dot{\varphi} &= \omega + b^I r^2 + h.o.t.
\end{aligned}
\qquad (6.92)
$$

In (6.92), the subscripts of b, used in (6.39), have been omitted.
    The significant lowest order terms are given by

$$
\dot{r} = b^R r^3 \qquad \text{and} \qquad \dot{\varphi} = \omega .
\qquad (6.93)
$$

The variable r represents the amplitude of the oscillation that appears at loss of stability.

Only a symmetry preserving unfolding is necessary in (6.93) contrary, for example, to the case of (6.87), because constant terms can be removed by a transformation of coordinates. We explain this in comparison with the divergence bifurcation at a zero root. For the Hopf bifurcation we have (6.38)

$$\dot{v} = i\omega v + c + O(|v|^3) , \qquad v \in \mathbb{C} .$$

The constant c can be removed by a linear change of variables $v = \tilde{v} - c/i\omega$. For the divergence bifurcation we have

$$\dot{z} = \lambda z + c + O(|z|^3) , \qquad z \in \mathbb{R} .$$

At bifurcation we have $\lambda = 0$, and therefore, c cannot be removed. Hence, the codimension of the one-dimensional bifurcation equation at a zero root with cubic nonlinearity is two and the unfolding is given by (6.110). On the other hand the codimension of (6.93) is *one* and the universally unfolded system is

$$\dot{r} = A_3 r^3 + ar . \tag{6.94}$$

We changed the notation and introduced $A_3 \equiv b^R$. This notation is used in the classification given in Section 6.5.1.

The bifurcation system (6.94) is case 2 (equation (6.109)), in the listing of Section 6.5.1 for the two codimension *one* cases.

If, for example, the third order term in the first equation of the normal form (6.92) vanishes, then a fifth order term will be the term of lowest order. In this case the codimension is *two* and the unfolding is given by (6.111) which is easy to understand because again only a symmetric deformation is possible.

We remark that the unfolding of the case of one zero root with cubic nonlinearity, which is given by the unfolding (6.110) with codimension *two*, is the same as in the static case (6.87).

These two cases exhaust the list of cases of codimension *two* in one variable. The remaining cases of codimension *two* in two variables are not easy to prove mathematically and we refer to the literature ([7], [44], [48], [50]). They are given in (6.112)–(6.114).

Finally, we consider as an example the equation

$$\ddot{x} + a\dot{x} + bx + Ax^3 + B\dot{x}^3 = 0 .$$

Such an equation appears, for example, in the field of planar flow-induced oscillations of slender rodlike continuous structures ([57]) after a discretization by a one mode Galerkin approximation. This equation has a degenerate linear part for $a = b = 0$, corresponding to $(4.12)_4$ or (6.112). However, the nonlinear equation given above does not represent a versal unfolding of this degenerate case $a = b = 0$ because its codimension is *three* and not *two*. This can be seen by inserting the coefficients A and B into (6.115). This results in a coefficient A_{21} in (6.115) which is zero. The consequence is that the normal form must now include fifth order terms and a third unfolding parameter must be introduced ([117]).

6.3   Determinacy

Already several times, we have implicitly used the concept of determinacy. Basically, determinacy answers the question up to which order terms in the Taylor expansion about the equilibrium position (which in our case is always the origin) must be retained in the bifurcation equations in order to completely describe the local stability behavior of the system. In other words, one can ask at which term the Taylor series can be truncated such that higher order terms do not qualitatively influence the local behavior of the system. Of course, it is important to point out that only the local behavior is investigated.

For one-dimensional bifurcation equations, no matter whether statical or dynamical, the problem of determinacy is easy to answer. This is, because the first non-zero term in the Taylor series expansion decides the behavior. Let us consider the simple example ([111] p. 59) $V(q) = (1/4)q^4 - (1/2)q^2$, whose graph is shown in Fig. 6.5c. At $q = 0$ the potential V has a local maximum of type q^2. Therefore, V is locally equivalent to $-q^2$. Finally the answer to: "how local is local?", is given by: "until another critical point is encountered" (see also p. 258).

We give now the definition of determinacy with respect to potentials. First, recall the definition of the k-jet of a function from p. 181 and that two functions V_1 and V_2 are of the same type (right-equivalent) if there exists a smooth change of coordinates $p = p(q)$ from which follows $V_1(q) = V_2(p(q))$ plus constant terms (Appendix M.1). Constant terms in the potential do not have any influence and can be ignored.

Definition 6.4 (Determinacy of potentials) *If the k-jet $j^k V$ of V is of a form that for any $\tilde{V}$ of degree $\geq k+1$, $V + \tilde{V}$ is locally equivalent to $j^k V$ by a smooth change of coordinates, then V is called k-determinate at 0.*

In less mathematical language, this means that adding terms of order $\geq k+1$ to a k-determinate function does not change the local qualitative behavior.

Unfortunately, for functions of several variables the first non-zero jet does not necessarily determine its determinacy as is the case for functions in one variable. This is shown by the example

$$V(q_1, q_2) = q_1^2 q_2 + q_2^\ell$$

where to the potential (6.74) (iii) a term q_2^ℓ, $\ell \geq 4$ has been added. The first non-vanishing jet is $q_1^2 q_2$, but this function is not locally equivalent to V under a smooth coordinate change, since the solutions of $q_1^2 q_2 = 0$ are the lines $q_1 = 0, q_2 = 0$, whereas the solution of $V(q_1, q_2) = 0$ is the line $q_2 = 0$. Clearly both solution sets are qualitatively different. Thus V is not equivalent to any jet $j^k V$ for $k < \ell$. On the other hand (6.74) (i) and (ii) are 3-determinate whereas (6.74) (iii) and (iv) are indeterminate.

Test for determinacy

We do not want to give a systematic discussion on how to check the determinacy of a function (see, for example, [111] p. 124 and the comments in Appendix M.1). Instead we present an algorithm (for more details and limitations see [42] p. 623) which allows to decide in four simple steps if a function $V(q)$ is determinate and, if so, how many terms of its Taylor series must be retained:

(i) Assume that $V(q_1, \ldots, q_n)$ is k-determinate.

(ii) Let $m_j(q)$ be the sequence of monomials in $q_1, \ldots, q_n$ of degree $2, 3, \ldots$:

$$m_j(q) = q_1^2, q_1 q_2, \ldots, q_n^2; q_1^3, q_1^2 q_2, \ldots .$$

that is, $m_1 = q_1^2$, $m_2 = q_1 q_2, \ldots$.

(iii) Calculate the set of polynomials $Q_{ij}(q)$ defined by

$$Q_{ij}(q) = j^{k+1} \left\{ \frac{\partial V}{\partial q_i} m_j(q) \right\} . \tag{6.95}$$

(iv) Can all monomials of degree $k + 1$ be written as linear superpositions of the Q_{ij} with constant coefficients? If this is the case then assumption (i) is correct.

We consider two examples:

Example 1: We consider the potential

$$V = q_1^2 + q_1 q_2^2 . \tag{6.96}$$

We assume (6.96) to be 3-determinate. From (6.96) follows

$$V_{,1} = 2q_1 + q_2^2, \qquad V_{,2} = 2q_1 q_2. \tag{6.97}$$

From (6.95) with (6.97) follows

$$
\begin{aligned}
Q_{11} &= 2q_1^3 + q_1^2 q_2^2, & Q_{12} &= 2q_1^2 q_2 + q_1 q_2^3, \\
Q_{13} &= 2q_1 q_2^2 + q_2^4, & Q_{21} &= 2q_1^3 q_2, \\
Q_{22} &= 2q_1^2 q_2^2, & Q_{23} &= 2q_1 q_2^3.
\end{aligned}
\tag{6.98}
$$

From the Q_{ij} listed in (6.98), which have been obtained from the quadratics of m_j, is clear that they are not sufficient for proving 3-determinacy because, for example, q_1^4 cannot be obtained. Therefore, also the following terms

$$
\begin{aligned}
Q_{14} &= 2q_1^4 + q_1^3 q_2^2, & Q_{15} &= 2q_1^3 q_2 + q_1^2 q_2^3, \\
Q_{16} &= 2q_1^2 q_2^2 + q_1 q_2^4, & Q_{17} &= 2q_1 q_2^3 + q_2^5, \\
Q_{24} &= 2q_1^4 q_2, & Q_{25} &= 2q_1^3 q_2^2, \\
Q_{26} &= 2q_1^2 q_2^3, & Q_{27} &= 2q_1 q_2^4
\end{aligned}
\tag{6.99}
$$

must be calculated. Now, we try again to express all monomials of degree 4 by a linear superposition of the 4-jets of Q_{ij}. These are

$$
q_1^4 = \frac{1}{2} Q_{14}, \quad q_1^3 q_2 = \frac{1}{2} Q_{21}, \quad q_1^2 q_2^2 = \frac{1}{2} Q_{22}, \quad q_1 q_2^3 = \frac{1}{2} Q_{23}.
$$

Obviously, it is not yet possible to obtain q_2^4. Hence, we assume (6.96) to be 4-determinate. To prove this, we need in addition the term

$$
Q_{18} = 2q_1^5 + q_1^4 q_2^2.
\tag{6.100}
$$

From (6.99) and (6.100) it is easy to see that (6.96) is 4-determinate, because all monomials of degree 5 can be obtained from a linear superposition of the 5-jets of Q_{ij}

$$
\begin{aligned}
q_1^5 &= \frac{1}{2} Q_{18}, & q_1^4 q_2 &= \frac{1}{2} Q_{24}, & q_1^3 q_2^2 &= \frac{1}{2} Q_{25}, \\
q_1^2 q_2^3 &= \frac{1}{2} Q_{26}, & q_1 q_2^4 &= \frac{1}{2} Q_{27}, & q_2^5 &= Q_{17} - Q_{23}.
\end{aligned}
$$

It is not too surprising that (6.96) is 4-determinate and not 3-determinate because it can be written as

$$
V = q_1^2 + q_1 q_2^2 = \left(q_1 + \frac{1}{2} q_2^2 \right)^2 - \frac{1}{4} q_2^4.
\tag{6.101}
$$

From (6.101) follows that on the q_1 axis ($q_2 = 0$) V is positive. On the parabola $q_1 + (1/2)q_2^2 = 0$, V is negative. At the origin V is zero. If one adds the term q_2^4, a strict minimum is obtained. Hence, V cannot be 3-determinate, because adding a term of fourth degree yields a qualitative change, but adding terms of fifth degree does not lead to any qualitative changes. Four-determinacy of (6.101) can also be shown by the change of coordinates $u = q_1 + (1/2)q_2^2$ and $v = q_2/\sqrt{2}$ which transforms (6.101) into $V(u,v) = u^2 - v^4$ which certainly is 4- and not 3-determinate ([111] p. 162).

Example 2: We consider the potential (6.89) describing the buckling problem of a rectangular plate at a double eigenvalue ([90]).

From (6.89) follows

$$V_{,1} = q_1^3 + \mu q_1 q_2^2 = q_1(q_1^2 + \mu q_2^2), \qquad V_{,2} = q_2(\mu q_1^2 + q_2^2). \qquad (6.102)$$

We want to show that (6.89) is 4-determinate. Hence, it must be possible to generate all monomials of degree 5 by a linear superposition of the Q_{ij}, which have to be calculated up to fifth degree:

$$\begin{array}{ll} Q_{11} = q_1^3(q_1^2 + \mu q_2^2), & Q_{12} = q_1^2 q_2(q_1^2 + \mu q_2^2), \\ Q_{13} = q_1 q_2^2(q_1^2 + \mu q_2^2), & Q_{21} = q_1^2 q_2(\mu q_1^2 + q_2^2), \\ Q_{22} = q_1 q_2^2(\mu q_1^2 + q_2^2), & Q_{23} = q_2^3(\mu q_1^2 + q_2^2). \end{array} \qquad (6.103)$$

To obtain all monomials of fifth degree, it is convenient to calculate first $q_1^2 q_2^3$ and $q_1^3 q_2^2$. From (6.103) follows

$$q_1^2 q_2^3 = c_1 Q_{12} + c_2 Q_{21} = (c_1 + c_2\mu)q_1^4 q_2 + (c_1\mu + c_2)q_1^2 q_2^3$$

with constants c_1, c_2 to be determined. The conditions $c_1 + c_2\mu = 0$ and $c_1\mu + c_2 = 1$ give the solution

$$c_1 = -\frac{\mu}{1 - \mu^2}, \qquad c_2 = \frac{1}{1 - \mu^2}.$$

Thus, we obtain

$$q_1^2 q_2^3 = \frac{-\mu}{1 - \mu^2} Q_{12} + \frac{1}{1 - \mu^2} Q_{21}.$$

Similarly we obtain

$$q_1^3 q_2^2 = \frac{1}{1 - \mu^2} Q_{13} - \frac{\mu}{1 - \mu^2} Q_{22}.$$

From these expressions, it is clear that $\mu = 1$ is a special value that leads to a degenerate case as is explained on p. 205. The remaining four monomials become

$$\begin{array}{rcl} q_1^5 & = & Q_{11} - \mu q_1^3 q_2^2 \\ q_1^4 q_2 & = & Q_{12} - \mu q_1^2 q_2^3 \\ q_1 q_2^4 & = & Q_{22} - \mu q_1^3 q_2^2 \\ q_2^5 & = & Q_{23} - \mu q_1^2 q_2^3, \end{array}$$

where for $q_1^3 q_2^2$ and $q_1^2 q_2^3$ the above expressions must be inserted.

Relation between codimension and determinacy

There exist some important connections between codimension and determinacy ([23]). Here, we only mention that codim $V < \infty$ if and only if V is finitely determinate. This can be expressed by the following inequality ([23])

$$k(V) \leq \text{codim } V + 2 \tag{6.104}$$

where $k(V)$ denotes that V is k-determinate.

We make a final comment concerning the case $\mu = \mu_0 = 1$ of (6.89). We showed above that this is a special case, which allowes no simple classification. In fact V_{μ_0} is indeterminate. From (6.104) results that its codimension is also infinite. This result was already obtained on p. 205. The indeterminacy of (6.89) for $\mu = \mu_0 = 1$ is not too difficult to understand from another point of view, because in this case

$$V_{\mu_0}(q_1, q_2) = \frac{1}{4}(q_1^2 + q_2^2)^2$$

is rotationally symmetric (Appendix G.5). If we introduce polar coordinates $q_1 = r \cos \varphi$, $q_2 = r \sin \varphi$ into V_{μ_0}, we obtain

$$V_{\mu_0}(r, \varphi) = \frac{1}{4}r^4 \ .$$

Now $V_{\mu_0}(r, \varphi)$ depends only on one variable r. Any perturbation of V_{μ_0} by a term $r^k \cos k\varphi$ with arbitrary k will change V_{μ_0} qualitatively. This is easy to understand if we compare the zero-sets of V_{μ_0} and of its perturbation. They are different!

If V is finitely (k-) determinate, one can draw the important conclusion that a local coordinate system near the origin can be introduced. In this local coordinate system the behavior of the system is given by the k-jet (see p. 181) of V, that is, by a finite segment of its Taylor series expansion.

6.4   Unfolding

If a system of bifurcation equations (or the corresponding degenerate potential) at the critical parameter value $\lambda = \lambda_c$, is of finite codimension then an unfolding of this bifurcation system is of great practical importance because the unfolded system of bifurcation equations allows to understand the influence of all possible perturbations of the system on its bifurcation behavior.

Definition 6.5 (Versal unfolding) *A versal unfolding is an embedding of the bifurcation system (degenerate potential) in a parametrized family that includes all qualitatively different cases of the class to which this degenerate system belongs.*

In other words a versal unfolding gives all essential perturbations of a bifurcation system. For example, in the case of static systems equation (6.85) is already a *universal* unfolding of the degenerate potential $V = \frac{1}{4}Aq^4$ obtained from (6.81) for $a_1 = 0$.

Definition 6.6 (Universal unfolding) *An unfolding is called universal if it represents the most general type of a smooth, versal perturbation which V can be subjected to with the minimum number of parameters.*

Theorem 6.2 (Unfolding of potentials) *For all degenerate potentials $V(q_1, \ldots, q_n)$ with finite codimension ℓ, one can find ℓ independent functions $u_j(j = 1, 2, \ldots, \ell)$, which cannot be generated by the $V_{,i}$ (see p. 203), such that*

$$V_u = V(q_1, \ldots, q_n) + \sum_{j=1}^{\ell} a_j u_j(q_1, \ldots, q_n) \qquad (6.105)$$

is a universal unfolding of V.

That the universal unfolding of a potential can be given in the form (6.105) as a linear relationship in the parameters $a_j(j = 1, 2, \ldots, \ell)$ is nontrivial and requires advanced mathematics for its proof ([111] ch. 8).

For example, for the potential $V(q_1, q_2) = q_1^3 + q_2^3$, it is shown in Section 6.2.1 that $q_1, q_2, q_1 q_2$ cannot be generated by $V_{,1}$ and $V_{,2}$. Therefore, the three parameter universal unfolding is given by

$$V_u = q_1^3 + q_2^3 + a q_1 q_2 + b q_1 + c q_2 .$$

Finally, we rewrite the two simplest cases of unfolded bifurcation equations, derived for static systems from potentials, namely equations (6.84) and (6.87), by indicating explicitly all those terms which can be neglected. One has for (6.84) ([53])

$$q^2 + a + O(|q|^3 + |aq| + |a|^2) = 0 \qquad (6.106)$$

and for (6.87)

$$q^3 + aq + b + O(|q|^4 + |aq^2| + |a^2 q| + |bq| + |b|^2) = 0. \qquad (6.107)$$

That the terms in the brackets in (6.106) and (6.107) can be neglected, follows from the order relations $a = O(|q|^2)$ and $b = O(|q|^3)$ which result immediately from the solutions of (6.106) and (6.107) (see also the discussion at the end of Section 4.3.1).

Equations (6.106) and (6.107) describe also the steady state bifurcation behavior of the dynamic bifurcations which are represented by (6.108) and (6.110) and discussed in Section 6.6.2.

For the one-dimensional dynamic bifurcation equation $(6.93)_1$, the universally unfolded system is given by (6.94). It is not too difficult to see from Fig. 1.7a and Fig. 1.7b that (6.94) contains all qualitatively possible perturbations of (6.93).

A list of universal deformations of bifurcation equations of low codimension is given in Section 6.5.

6.5 Classification

One of the main reasons for the great success of the application of bifurcation theory to nonlinear stability problems is the fact that for cases of low codimension only a limited number of qualitatively different cases of loss of stability exist. For systems with small aspect ratio which are frequently met in applications, cases of low codimension prevail. We recall that the codimension of a particular case depends upon the structure of the eigenvalues at loss of stability and the order of the nonlinear terms. In the sections to follow, a complete list of all cases of low codimension for dynamical (up to codimension two) and statical (up to codimension four) problems will be given.

6.5.1 Dynamic bifurcation

There exist only seven cases up to and including codimension *two* for the time-continuous case and three cases of codimension *one* for the time-discrete case. They are investigated in detail in [7], [50]. We discuss them in application to the examples of Chapter 4 in Sections 6.6.2 and 6.6.3. The unfolding parameters are designated by a and b. They represent those parameters that one wants to keep variable in the stability analysis. The coefficients resulting from the reduction process and the application of normal form theory are designated by A_i, A_{ij}, respectively. They contain all other parameters of the system which are kept fixed in the stability investigation.

(A) Time-continuous cases

There exist two cases of codimension one and five cases of codimension two.

Codimension one

1. One zero root with quadratic nonlinearity ($z \in \mathbb{R}$):

$$\dot{z} = A_2 z^2 + a + O(|z|^3) \,. \tag{6.108}$$

2. One pair of purely imaginary roots with cubic nonlinearity:

$$\begin{aligned} \dot{r} &= A_3 r^3 + ar + O(|r|^4) \\ \dot{\varphi} &= \omega + O(|r|^2) . \end{aligned}$$
(6.109)

Codimension two

3. One zero root with cubic nonlinearity:

$$\dot{z} = A_3 z^3 + az + b + O(|z|^4) .$$
(6.110)

4. One pair of purely imaginary roots with fifth order nonlinearity:

$$\begin{aligned} \dot{r} &= A_5 r^5 + br^3 + ar + O(|r|^7) \\ \dot{\varphi} &= \omega + O(|r|^2) . \end{aligned}$$
(6.111)

5. Two zero roots with cubic nonlinearities:

$$\begin{aligned} \dot{z}_1 &= z_2 \\ \dot{z}_2 &= A_{30} z_1^3 + A_{21} z_1^2 z_2 + az_1 + bz_2 + O(|z_1|^5 + |z_2|^5) . \end{aligned}$$
(6.112)

6. One zero root and a purely imaginary pair:

$$\begin{aligned} \dot{r} &= ar + A_{30} r^3 + A_{12} r z^2 + O(|r|^5 + |z|^5) \\ \dot{z} &= bz + A_{21} r^2 z + A_{03} z^3 + O(|r|^5 + |z|^5) \\ \dot{\varphi} &= \omega + A_{20} r^2 + A_{02} z^2 + O(|r|^4 + |z|^4) . \end{aligned}$$
(6.113)

7. Two pairs of imaginary roots without low-order resonance, that is, $m\omega_1 + n\omega_2 \neq 0$ with $|m| + |n| \leq 4$:

$$\begin{aligned} \dot{r}_1 &= ar_1 + A_{30} r_1^3 + A_{12} r_1 r_2^2 + O(|r_1|^5 + |r_2|^5) \\ \dot{r}_2 &= br_2 + A_{21} r_1^2 r_2 + A_{03} r_2^3 + O(|r_1|^5 + |r_2|^5) \\ \dot{\varphi}_1 &= \omega_1 + O(|r_1|^2 + |r_2|^2) \\ \dot{\varphi}_2 &= \omega_2 + O(|r_1|^2 + |r_2|^2) . \end{aligned}$$
(6.114)

For the cases (5) and (6) it is assumed that only odd functions appear which excludes also constant terms following from imperfections. This is, generally, the case for perfect reflectionally symmetric oscillatory systems (Chapter 5). If imperfections are present the codimension for theses cases is greater than two ([129]). For a more complete listing, where also nonlinear terms of even (second) order are present in the normal form, see [50] ch. 7 and [7] ch. 6.

In the one-dimensional cases (6.108) and (6.110), the calculation of the corresponding coefficients A_i is trivial, because they follow immediately from the center manifold reduction. However, in the two simple two-dimensional cases (6.109) and (6.111) and in (6.112)–(6.114), either *normal form theory* or *averaging* must be used to calculate the coefficients A_i or A_{ij}, respectively. For the derivation of (6.112) and (6.115) see Appendix M.3.

In the cases (6.108)–(6.111) where only the coefficient A_i is present, the stability behavior of the degenerate (singular) bifurcation point is easily obtained because it depends only on the sign of A_i (see p. 51). For the remaining cases, this question is more subtle and will be discussed in Section 6.6.2.

Finally, we present the relationship between the coefficients A_i, A_{ij} in the above classifications and the coefficients $a_{ijk}, a_{ijk\ell}, a_{ijk\ell m}$ and b_{ik} which appear in the bifurcation equations after having performed the reduction process.

For (6.108) and (6.110) only one coefficient of the second and third order term must be calculated after the reduction process. A_3 in (6.109) is given by the real part of (6.34). The fastest way to calculate A_5 in (6.111) is again the use of complex variables and normal form theory. We denote the nonlinear terms in the bifurcation equation by g_3 and g_5 as is done in (6.25) for the third order terms. From g_5 we need only one term

$$g_5 = b_{32}^\star w^3 \overline{w}^2 + \ldots$$

because all other terms of fifth order can be eliminated. The normal form calculation yields

$$C_{32} = b_{32}^\star + \frac{i}{2\omega}\left[2b_{30}b_{12} - b_{12}\overline{b}_{12} - \frac{3}{2}b_{03}\overline{b}_{03} \right]$$

where $b_{32}^\star$ is the coefficient of the term $w^3 \overline{w}^2$ in the bifurcation equation before the application of the normal form theory. Moreover it has been assumed that only odd order terms were present, hence, only third order terms are showing up in the expression of C_{32}. The normal form equation of fifth order is

$$\dot{v} = i\omega v + C_{32}v^3 \overline{v}^2 + O(|v,\overline{v}|^7) \ .$$

The real coefficient A_5 in the amplitude equation $(6.111)_1$ is the real part of C_{32}.

If the cubic nonlinear terms are given by $(6.14)_2$, then for (6.112) we obtain (Appendix M.3)

$$A_{30} = a_{230} \qquad A_{21} = 3a_{130} + a_{221} \ . \tag{6.115}$$

Equation (6.112) is called the Bogdanov-Takens normal form.

If a third equation is added to (6.14) by normal form transformation ([126]) one obtains for (6.113)

$$A_{30} = \frac{3}{8}(a_{1300} + a_{2030}) + \frac{1}{8}(a_{1120} + a_{2210})$$

$$A_{12} = \frac{1}{2}(a_{1102} + a_{2012})$$

$$A_{21} = \frac{1}{2}(a_{3201} + a_{3021})$$

$$A_{03} = a_{3003} \qquad\qquad (6.116)$$

$$A_{02} = \frac{1}{2}(a_{2102} - a_{1012})$$

$$A_{20} = \frac{3}{8}(a_{2300} - a_{1030}) + \frac{1}{8}(a_{2120} - a_{1210}) \ .$$

To calculate the coefficients in (6.114) we add two more equations to (6.13) and additionally, since, the second order nonlinear terms can be eliminated, we perform the calculations only for third order nonlinear terms. The system of bifurcation equations is

$$\dot{y}_1 = \qquad -\omega_1 y_2 + g_1(y_1, y_2, y_3, y_4)$$
$$\dot{y}_2 = \omega_1 y_1 \qquad\qquad + g_2(y_1, y_2, y_3, y_4)$$
$$\dot{y}_3 = \qquad -\omega_2 y_4 + g_3(y_1, y_2, y_3, y_4)$$
$$\dot{y}_4 = \omega_2 y_3 \qquad\qquad + g_4(y_1, y_2, y_3, y_4) \ .$$

Transformation to complex variables $w_1 = y_1 + iy_2$, $w_2 = y_3 + iy_4$, $F_1 = g_1 + ig_2$ and $F_2 = g_3 + ig_4$ yields

$$\dot{w} = Cw + F(w)$$

where

$$C = \begin{pmatrix} i\omega_1 & 0 & 0 & 0 \\ 0 & -i\omega_1 & 0 & 0 \\ 0 & 0 & i\omega_2 & 0 \\ 0 & 0 & 0 & -i\omega_2 \end{pmatrix}$$

and

$$w = (w_1, \overline{w}_1, w_2, \overline{w}_2)^T \ , \qquad F = (F_1, \overline{F}_1, F_2, \overline{F}_2)^T \ .$$

The *resonance condition* (6.11) supplies the following two equations ($|m| = 3$)

$$m_1(i\omega_1) + m_2(-i\omega_1) + m_3(i\omega_2) + m_4(-i\omega_2) - i\omega_1 = 0 \quad (6.117)$$
$$m_1(i\omega_1) + m_2(-i\omega_1) + m_3(i\omega_2) + m_4(-i\omega_2) - i\omega_2 = 0 \ . (6.118)$$

The solutions of (6.117) are

$$m_1 = 2, \quad m_2 = 1, \quad m_3 = 0, \quad m_4 = 0$$
$$m_1 = 1, \quad m_2 = 0, \quad m_3 = 1, \quad m_4 = 1.$$

Similarly follows for (6.118)

$$m_1 = 0, \quad m_2 = 0, \quad m_3 = 2, \quad m_4 = 1$$
$$m_1 = 1, \quad m_2 = 1, \quad m_3 = 1, \quad m_4 = 0.$$

That is, the following cubic terms remain in the *normal form* equations

$$\dot{w}_1 = i\omega_1 w_1 + b_{12100} w_1^2 \overline{w}_1 + b_{11011} w_1 w_2 \overline{w}_2$$
$$\dot{w}_2 = i\omega_2 w_2 + b_{21110} w_1 \overline{w}_1 w_2 + b_{20021} w_2^2 \overline{w}_2 .$$

The coefficients can be calculated in the following way

$$b_{12100} = \frac{1}{2} \frac{\partial^3 F_1}{\partial w_1^2 \partial \overline{w}_1} , \qquad\qquad b_{11011} = \frac{\partial^3 F_1}{\partial w_1 \partial w_2 \partial \overline{w}_2}$$

$$b_{21110} = \frac{\partial^3 F_2}{\partial w_1 \partial \overline{w}_1 \partial w_2} , \qquad\qquad b_{20021} = \frac{1}{2} \frac{\partial^3 F_2}{\partial w_2^2 \partial \overline{w}_2}$$

Returning to the real variables we obtain

$$b_{12100} = \frac{1}{16} \left(\frac{\partial^3 g_1}{\partial y_1^3} + \frac{\partial^3 g_1}{\partial y_1 \partial y_2^2} + \frac{\partial^3 g_2}{\partial y_1^2 \partial y_2} + \frac{\partial^3 g_2}{\partial y_2^3} - \right.$$
$$\left. -i \left(\frac{\partial^3 g_2}{\partial y_1^3} + \frac{\partial^3 g_2}{\partial y_1 \partial y_2^2} - \frac{\partial^3 g_1}{\partial y_1^2 \partial y_2} - \frac{\partial^3 g_1}{\partial y_2^3} \right) \right)$$

$$b_{11011} = \frac{1}{8} \left(\frac{\partial^3 g_1}{\partial y_1 \partial y_3^2} + \frac{\partial^3 g_1}{\partial y_1 \partial y_4^2} + \frac{\partial^3 g_2}{\partial y_2 \partial y_3^2} + \frac{\partial^3 g_2}{\partial y_2 \partial y_4^2} - \right.$$
$$\left. -i \left(\frac{\partial^3 g_2}{\partial y_1 \partial y_3^2} + \frac{\partial^3 g_2}{\partial y_1 \partial y_4^2} - \frac{\partial^3 g_1}{\partial y_2 \partial y_3^2} - \frac{\partial^3 g_1}{\partial y_2 \partial y_4^2} \right) \right)$$

$$\text{(6.119)}$$

$$b_{21110} = \frac{1}{8} \left(\frac{\partial^3 g_3}{\partial y_1^2 \partial y_3} + \frac{\partial^3 g_3}{\partial y_2^2 \partial y_3} + \frac{\partial^3 g_4}{\partial y_1^2 \partial y_4} + \frac{\partial^3 g_4}{\partial y_2^2 \partial y_4} - \right.$$
$$\left. -i \left(\frac{\partial^3 g_4}{\partial y_1^2 \partial y_3} + \frac{\partial^3 g_4}{\partial y_2^2 \partial y_3} - \frac{\partial^3 g_3}{\partial y_1^2 \partial y_4} - \frac{\partial^3 g_3}{\partial y_2^2 \partial y_4} \right) \right)$$

$$b_{20021} = \frac{1}{16} \left(\frac{\partial^3 g_3}{\partial y_3^3} + \frac{\partial^3 g_3}{\partial y_3 \partial y_4^2} + \frac{\partial^3 g_4}{\partial y_3^2 \partial y_4} + \frac{\partial^3 g_4}{\partial y_4^3} - \right.$$
$$\left. -i \left(\frac{\partial^3 g_4}{\partial y_3^3} + \frac{\partial^3 g_4}{\partial y_3 \partial y_4^2} - \frac{\partial^3 g_3}{\partial y_3^2 \partial y_4} - \frac{\partial^3 g_3}{\partial y_4^3} \right) \right) .$$

The coefficients in the amplitude equations are the real parts in (6.119).

In the notation of (6.14) they take the form

$$A_{30} = \frac{3}{8}(a_{13000} + a_{20300}) + \frac{1}{8}(a_{11200} + a_{22100})$$

$$A_{12} = \frac{1}{4}(a_{11020} + a_{11002} + a_{20120} + a_{20102}) \qquad (6.120)$$

$$A_{21} = \frac{1}{4}(a_{32010} + a_{30210} + a_{42001} + a_{40201})$$

$$A_{03} = \frac{3}{8}(a_{30030} + a_{40003}) + \frac{1}{8}(a_{30012} + a_{40021}) .$$

The relationship between the unfolding parameters a, b and the physical parameters is explained in Section 6.6.2.

(B)   Time-discrete cases

Codimension one

1. One root $\mu = +1$:

$$u_{t+1} = a + u_t + A_2 u_t^2 + O(|u_t|^3) . \qquad (6.121)$$

2. One root $\mu = -1$:

$$u_{t+1} = (-1 + a)u_t + A_3 u_t^3 + O(|u_t|^5) . \qquad (6.122)$$

3. A pair of complex roots $\mu_{1,2} = \nu \pm i\eta$ with $|\mu_{1,2}| = 1$:

$$w_{t+1} = (\mu + c + C_3 w_t \overline{w}_t)w_t \qquad (6.123)$$

with $w_t = u_t + iv_t$.

The coefficients A_2, A_3 and C_3 in the equations above follow from the reduction process and the normal form theory calculations of Section 6.1.2 where $A_2 = a_2$ in (6.59), A_3 is given by (6.61) and $C_3 = A_{21}$ is given by (6.70). As before a and c are the unfolding parameters.

6.5.2   Static bifurcation: elementary catastrophe theory

The bifurcation equations of statical systems (Section 2.2) can be derived from a potential function. For potentials up to codimension *four* (parameters a, b, c, d) there exist only *seven* qualitatively different types, if we use as equivalence relation *right-equivalence* (Appendix M.1). These seven types form the famous list of *elementary catastrophes* (see also the table in Section 6.2.1). In this list, we write down the potentials rather than the bifurcation equations and we also include their names.

For a simple eigenvalue ($n_c = 1$, one variable q), the following table results ([111], [121], [42])

$$V = \frac{1}{3}q^3 + aq \qquad \text{Fold} \qquad (6.124/1)$$

$$V = \frac{1}{4}q^4 + \frac{1}{2}aq^2 + bq \qquad \text{Cusp} \qquad (6.124/2)$$

$$V = \frac{1}{5}q^5 + \frac{1}{3}aq^3 + \frac{1}{2}bq^2 + cq \qquad \text{Swallowtail} \quad (6.124/3)$$

$$V = \frac{1}{6}q^6 + \frac{1}{4}aq^4 + \frac{1}{3}bq^3 + \frac{1}{2}cq^2 + dq \qquad \text{Butterfly.} \quad (6.124/4)$$

For double eigenvalues ($n_c = 2$, two variables q_1, q_2), the catastrophes are called umbilics and are as follows

$$V = q_1^3 - 3q_2q_1^2 + a(q_1^2 + q_2^2) + bq_1 + cq_2 \qquad \text{elliptic} \qquad (6.124/5)$$
$$V = q_1^3 + q_2^3 + aq_1q_2 + bq_1 + cq_2 \qquad \text{hyperbolic} (6.124/6)$$
$$V = q_2^2q_1 + q_1^4 + aq_1^2 + bq_2^2 + cq_1 + dq_2 \qquad \text{parabolic.} \quad (6.124/7)$$

Whereas the classification for bifurcation equations in one variable can easily be continued, this is more difficult for the case of double or multiple critical eigenvalues. In [3], the classification process is continued far beyond the list (6.124) of the *elementary catastrophes*. Though a classification as given in [3] is important mathematically, from the point of view of applications, singularities with high codimension ℓ are difficult to analyse. Moreover, the complete bifurcation set of these problems is not easy to understand. However, we must note that for (4.173), the equations obtained for buckling of a rectangular plate at a double eigenvalue, we have codimension $\ell = 8$. Therefore, the bifurcation behavior must be studied in a 10-dimensional space analogous to the three-dimensional space for (6.87) as presented in Fig. 6.7. The corresponding bifurcation diagram must be represented in an eight-dimensional parameter space corresponding to the two-dimensional plane for (6.87).

For example, the calculation of the unfolding parameter values a, b in (6.124/2) is given in Section 6.5.3.

6.5.3 The unfolding theory of Golubitsky and Schaeffer

We mentioned in the Introduction the unfolded potential (6.124/2) in connection with Fig. 1.5. This unfolding includes all possible, qualitatively different imperfections for a bifurcation equation at a simple eigenvalue where the third order term does not vanish. However, it is shown

below that by transforming a physical problem into the canonical form
(6.124/2) a mixing between the distinguished parameter and the imperfection parameters can occur. This mixing is generally undesirable when
trying to interpret the solutions of the problem physically.

Let us explain this by means of a simple example, taken from [44]. It
is the system depicted in Fig. 6.8. It consists of two rigid rods of length

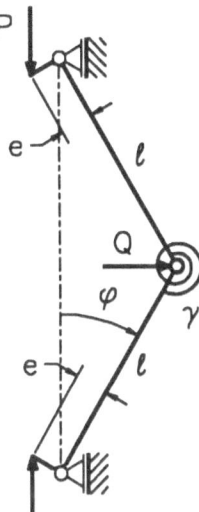

Figure 6.8. Conservative system of two rigid rods with
a torsional elastic hinge, having one degree of freedom

ℓ, connected at the inner hinge by a linearly elastic torsional spring with
stiffness γ. The system has one degree of freedom expressed by the angle
φ. The loading is given by the axial load P, with an excentricity e
and the transversal load Q. The axial load P will be the distinguished
parameter. It is further assumed that the spring is imperfect, that is, its
torque does not vanish for $\varphi = 0$ but for $\varphi = \varphi_0 \neq 0$. This static system
is conservative and can be analysed by writing down the potential energy
V, which reads

$$V(\varphi, P, Q, e, \varphi_0) = \frac{1}{2}\gamma(2(\varphi - \varphi_0))^2 - Q\ell(\sin\varphi - \sin\varphi_0)$$
$$- P[2\ell(\cos\varphi_0 - \cos\varphi) + 2e\sin\varphi - 2e\sin\varphi_0] \tag{6.125}$$

where the imperfections Q, e, φ_0 are assumed to be small. From (6.125)
we obtain the equilibrium equation

$$G = \frac{\partial V}{\partial \varphi} = 2\gamma(2\varphi - 2\varphi_0) -$$
$$- Q\ell\cos\varphi - 2P\ell\sin\varphi - 2Pe\cos\varphi = 0 . \tag{6.126}$$

Since the imperfections are small we expand the trigonometric functions up to third order about $\varphi = 0$, to obtain the 3-jet of G

$$
j^3 G = \frac{P\ell}{3}\varphi^3 + \frac{1}{2}(Q\ell + 2Pe)\varphi^2 + \\
+ (4\gamma - 2P\ell)\varphi - (Q\ell + 2Pe + 4\gamma\varphi_0) .
\tag{6.127}
$$

Setting the imperfection parameters $Q = e = \varphi_0 = 0$, we obtain the critical parameter value (buckling load of the perfect system) $P_c = 2\gamma/\ell$. Now we divide (6.127) by $P_c\ell/3$. This gives from $j^3 G = 0$

$$
\frac{P}{P_c}\varphi^3 + \frac{3}{2}\frac{Q\ell + 2Pe}{P_c\ell}\varphi^2 + \frac{6(2\gamma - P\ell)}{P_c\ell}\varphi - 3\frac{Q\ell + 2Pe + 4\gamma\varphi_0}{P_c\ell} = 0 .
\tag{6.128}
$$

Next we introduce a new parameter μ by

$$
P = P_c + \frac{P_c}{6}\mu = \frac{2\gamma}{\ell} + \frac{1}{3}\frac{\gamma}{\ell}\mu .
\tag{6.129}
$$

Inserting (6.129) into (6.128) yields

$$
\left(1 + \frac{\mu}{6}\right)\varphi^3 + \left(\frac{3}{4}\frac{Q\ell}{\gamma} + 3\frac{e}{\ell} + \frac{1}{2}\frac{e}{\ell}\mu\right)\varphi^2 - \\
- \mu\varphi - \left(\frac{3}{2}\frac{Q\ell}{\gamma} + 6\frac{e}{\ell} + 6\varphi_0\right) - \frac{e}{\ell}\mu = 0 .
\tag{6.130}
$$

In (6.130) we can neglect the terms $\frac{\mu}{6}\varphi^3$ and $\frac{1}{2}\frac{e}{\ell}\mu\varphi^2$ because they are at least of fourth order in φ, that is, $O(|\varphi|^4)$ (see Section 6.4 and p. 101). If we set the imperfection parameters in (6.130) to zero we obtain

$$
\varphi^3 - \mu\varphi = 0 .
\tag{6.131}
$$

The universal unfolding of (6.131) in the theory of Golubitsky and Schaeffer has codimension two and can have one of the two following forms ([48] vol. 1 p. 130)

$$
\begin{aligned}
G_1(q, \lambda, \alpha, \beta) &= q^3 - \lambda q + \alpha + \beta q^2 &= 0 \\
G_2(q, \lambda, \alpha, \beta) &= q^3 - \lambda q + \alpha + \beta\lambda &= 0 .
\end{aligned}
\tag{6.132}
$$

The transformation of (6.130) to (6.132) can be performed by inserting

$$
\varphi = q + \delta
\tag{6.133}
$$

into (6.130). Using the abbreviations for the imperfection parameters

$$
A = \frac{3}{4}\frac{Q\ell}{\gamma} + 3\frac{e}{\ell} , \quad B = -\left(\frac{3}{2}\frac{Q\ell}{\gamma} + 6\frac{e}{\ell} + 6\varphi_0\right) , \quad \text{and} \quad \varepsilon = \frac{e}{\ell}
\tag{6.134}
$$

we obtain

$$q^3 + (3\delta + A)q^2 + (3\delta^2 + 2A\delta - \mu)q + \delta^3 + A\delta^2 + B - \mu(\delta + \varepsilon) = 0 . \quad (6.135)$$

In (6.135) we have now the possibility to remove either the term $\mu(\delta + \varepsilon)$ and to obtain G_1 in (6.132) or two remove the term $(3\delta + A)q^2$ and to obtain G_2 in (6.132). In the first case we use $\delta = -\varepsilon$ and the parameters are

$$\alpha = -\varepsilon^3 + A\varepsilon^2 + B , \qquad \beta = -3\varepsilon + A , \qquad \lambda = \mu - 3\varepsilon^2 + 2A\varepsilon . \quad (6.136)$$

For the second case we have $\delta = -A/3$ and the parameters are

$$\alpha = -\frac{A^3}{27} + B + \frac{A^2}{3}\varepsilon , \qquad \beta = \left(\frac{A}{3} - \varepsilon\right) , \qquad \lambda = \mu + \frac{A^2}{3} . \quad (6.137)$$

The necessary transformations changed the variable by equation (6.133) and the parameters by (6.136) or (6.137). It is essential to note that the distinguished parameter μ appears only in λ. The practically important result of this transformation is that the mathematical imperfection parameters α, β are expressed only by physical imperfection quantities. This is in agreement with the definition of *bifurcation equivalence* given in Appendix M.4.

Now let us compare this with the approach used in *elementary catastrophe theory*. Again we start with (6.130). Neglecting *h.o.t.* according to (6.107) we rewrite (6.130) using (6.134) in the form

$$\varphi^3 + A\varphi^2 - \mu\varphi + B - \varepsilon\mu = 0 . \quad (6.138)$$

With the transformation

$$\psi = \varphi + \frac{A}{3}$$

we obtain

$$\psi^3 + a\psi + b = 0 \quad (6.139)$$

where

$$a = -\frac{1}{3}A^2 - \mu , \qquad b = \frac{2A^3}{27} + \frac{A\mu}{3} + B - \varepsilon\mu . \quad (6.140)$$

The major disadvantage of the new parameters a, b in (6.140) is that b is a combination of imperfection parameters and the distinguished parameter μ.

Summing up, we can say that by a variation of the distinguished parameter μ (that is, P according to (6.129) in the *Golubitsky and Schaeffer theory*) only λ in (6.136) and (6.137) is affected and the imperfection parameters α, β remain unchanged. On the other hand, in the unfolding of the *elementary catastrophe theory* a variation of the distinguished parameter affects, in general, both parameters a, b. That means, if one wants to see how the system behaves under a variation of only the distinguished parameter one must vary both a and b simultaneously if one makes use of (6.139). This will result in a loading path in the bifurcation diagram of Fig. 6.7 which is neither parallel to the a-axis nor to the b-axis.

This becomes also quite clear from the comparison of two bifurcation graphs. If we set $B - \varepsilon\mu = 0$ in (6.138) we obtain for φ the graph number 2 in Fig. 6.15. However, even if $B - \varepsilon\mu = 0$ we can obtain from (6.140) $b \neq 0$ and for ψ according to (6.139) the graph number 4 in Fig. 6.15. These two bifurcation graphs are qualitatively different. However, we may obtain graph 2 in Fig. 6.15 from (6.139) if we vary a and b simultaneously. Therefore, in applications the concept of *contact* or *bifurcation equivalence* seems to be better suited to the study of bifurcation problems than the concept of right equivalence (see Appendix M).

The bifurcation diagram of G_1 in (6.132) is given in Fig. 6.15. It gives a stratification of the α, β plane into domains of qualitatively similar behavior. This similar behavior is expressed by qualitatively similar graphs of the bifurcation solutions.

The calculation of the bifurcation diagram of Fig. 6.15 is explained in Section 6.6.1.

A listing of the universal unfoldings with respect to *bifurcation equivalence* of several practically important cases is given in [48] vol. 1 ch. 4. These unfoldings are performed keeping the difference between the distinguished parameter and the imperfection parameters.

In order to show that this theory supplies quite interesting new results we give a listing of the cases of codimension one in one variable q. We denote the distinguished parameter by λ and the imperfection parameter by α. There are three cases:

$$
\begin{aligned}
(1) \qquad & q^3 + \lambda + \alpha q &&= 0 \\
(2) \qquad & q^2 - \lambda q + \alpha &&= 0 \qquad\qquad (6.141) \\
(3) \qquad & q^2 \pm \lambda^2 + \alpha &&= 0 \,.
\end{aligned}
$$

Case (1) does not present anything new compared to Fig. 6.10 where we must identify $b = \lambda$ and $a = \alpha$. In the terminology of Golubitsky and Schaeffer this is a *hysteresis*.

Case (2) is the perturbation of the transcritical bifurcation $q^2 - \lambda q = 0$ which is discussed in connection with the frame of Fig. 6.12, and the shell of Fig. 6.19 and appears again in Fig. 6.39 where we discuss the motion of a robot. The imperfect case is shown in Fig. 6.13, Fig. 6.14 and Fig. 6.20.

Case (3) is new and must be treated separately for the two different signs:

(a) $q^2 + \lambda^2 + \alpha = 0$. This bifurcation equation leads to the *isola* bifurcation which together with the bifurcation diagram is depicted in Fig. 6.9a.

(b) $q^2 - \lambda^2 + \alpha = 0$. This bifurcation equation leads to a simple bifurcation which together with the bifurcation diagram is depicted in Fig. 6.9b.

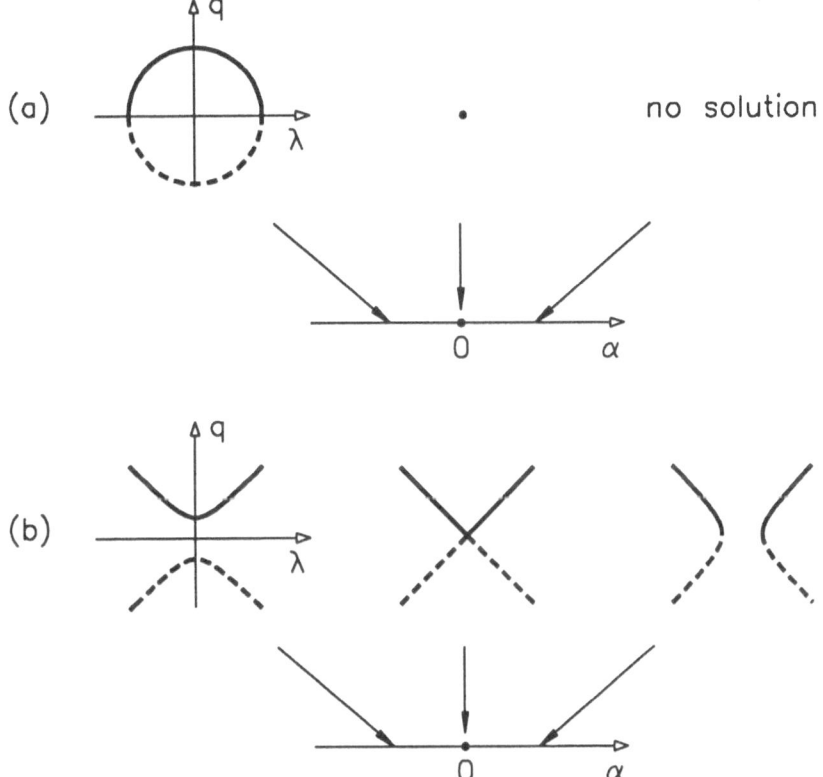

Figure 6.9. Bifurcation diagrams (below) and corresponding solutions (above) for $(6.141)_3$: (a) positive, (b) negative sign

In Section 6.6.1 we present a simple example where we explain the appearence of $(6.141)_3$ in a slightly modified way. In fact we find for the problem considered that the normal form is

$$q(q^2 \pm \lambda^2) = 0 \qquad (6.142)$$

instead of $(6.141)_3$ if we set $\alpha = 0$.

6.5.4 Restricted generic bifurcation

In (6.124) universal unfoldings of potentials, up to four parameters, are listed. However, as mentioned in Section 6.2.1 on p. 204 the universal unfolding of the buckling problem of the rectangular plate at a double eigenvalue requires an eight-parameter family. In [111] p. 320 such an unfolding with physically interpretable parameters is discussed. However, as already mentioned, from a practical point of view, such a high-dimensional parameter family presents two essential difficulties. First, it is difficult to find physically relevant quantities as parameters. Second, it is a difficult task to analyse the bifurcation set in an eight-dimensional parameter space.

From an engineering point of view, a different approach proposed by *J. Hale* is more appropriate for practical needs. For technical problems, generally a set of physically relevant parameters is known. Choosing some (r) of these parameters for the unfolding one wants to know whether in a r-parameter family, some sort of genericity or robustness can be achieved. We note that for this r-parameter family $r \geq 1$ but that r is in general smaller than the number ℓ of parameters necessary for the universal unfolding.

In [54], such a concept is worked out. In this approach, genericity refers to the change in the number of solutions as parameters are varied.

Definition 6.7 (Restricted generic bifurcation) *If under a variation of parameters in a r-parameter family $(r < \ell)$ the number of solutions changes in almost all cases only by two then the unfolding is of restricted generic type.*

As examples, let us consider the buckling problems of the rectangular plate of Section 4.3.3 and of the annular plate of Section 5.2. In both cases, the following parameters are of technical relevance:

(1)   the inplane thrust λ,

(2)   the deviation σ of the length ℓ for the rectangular plate or the radius b for the annular plate from their critical values ℓ_c or b_c for which multiple eigenvalues exist,

(3), (4) the contributions γ_1 and γ_2 of the transversal loading t obtained by the projection on the two adjacent buckling modes, respectively.

The corresponding unfolded system follows from (4.188) or from (5.67), respectively, to

$$q_1^3 + \mu_1 q_1 q_2^2 - \lambda q_1 + \gamma_1 = 0$$
$$q_2^3 + \mu_2 q_2 q_1^2 - (\lambda + \sigma)q_2 + \gamma_2 = 0 .$$

$$(6.143)$$

The derivation of the parameters will be explained in Section 6.6.1. In [91] it is shown that for (6.143) the change in the number of solutions is always *two* except at a finite number of special points. Hence, (6.143) is robust in a certain sense. Whereas, it is shown in [147] that if we have, for example, $\gamma_1 = 0$, this type of robustness is lost.

We return to the concept of *restricted generic bifurcation* in Section 6.6.1 when we discuss Fig. 6.15. There we try to explain this concept intuitively in connection with the two parameter unfoldings of equation (6.136) from a slightly different point of view.

6.6 Bifurcation diagrams

A complete nonlinear stability or bifurcation analysis consists of several steps as indicated at the end of the Introduction. As already mentioned in Section 6.2.1 for the potential (6.87), the final step is the calculation of the bifurcation diagram.

Definition 6.8 (Bifurcation diagram) *A bifurcation diagram is a partition of the parameter space $\Lambda \in R^\ell$ into domains of qualitatively different types of system behavior.*

In this partition is ℓ, generally, equal to the codimension of the problem. The practical use of a bifurcation diagram stems from the fact that from the location of the parameter value in the diagram the qualitative behavior of the system can be estimated. Also the changes of the states under a variation of the parameters can be seen. We will not give a systematic treatment of how to calculate bifurcation diagrams because this is done for statical systems (the potentials (6.124)) in [111], [121], [42], for dynamical systems in [7], [2], [50] and for statical and dynamical systems in [48]. Rather, we wish to explain how to use bifurcation diagrams to analyse the examples of Chapters 4 and 5.

6.6.1 Statical systems

We recall that $V : \mathbb{R}^n \times \mathbb{R}^\ell \to \mathbb{R}$ is a parametrized family of smooth functions $V(q_1, \ldots, q_n, a_1, \ldots, a_\ell)$. The *equilibrium set* M is defined by (see also (2.19))

$$M = \left\{ (q, a) \middle| \frac{\partial V}{\partial q_i} = 0 , \quad i = 1, \ldots, n \right\} . \qquad (6.144)$$

The equilibrium set M consists of the critical points of V. Next, we calculate the set of the *degenerate critical points* of V. This is the *singularity* *set S defined by*

$$S = \left\{ (q, a) \left| \frac{\partial V}{\partial q_i} = 0 \right., \quad \det\left[\frac{\partial^2 V}{\partial q_i \partial q_j}\right] = 0 \right\} . \tag{6.145}$$

If we project S down into the parameter space $\{a\}$ we obtain the *bifurcation set B*. The calculation of this projection can be performed by eliminating the state variables from M and S.

A)   One-dimensional bifurcation equations

The cusp and the fold

For the potential (6.124/2) the set of equilibrium positions obtained from (6.87) forms the set M given by (6.144) which is a smooth surface as is shown in Fig. 6.7. From Fig. 6.7 follows that a domain in the parameter plane (hatched) exists where there are three possible equilibria (q_1, q_2, q_3) for each parameter value (a, b). Two equilibria (q_4, q_5) exist for parameter values on the cusp shaped curve which is the bifurcation set B. Only one equilibrium exists for parameter values outside the hatched domain. The corresponding potentials are also sketched in Fig. 6.7. To obtain the equation for the cusp shaped curve we differentiate (6.87) with respect to q and set this expression to zero. That is

$$\frac{\partial^2 V}{\partial q^2} = 3q^2 + a = 0 . \tag{6.146}$$

This is the condition for the stability boundary. Equations (6.87) and (6.146) define the singularity set S according to (6.145). The projection of the singularity set in the parameter space is performed by eliminating q from (6.87) and (6.146). We obtain

$$4a^3 + 27b^2 = 0 . \tag{6.147}$$

The semicubical parabola (6.147) represents the *bifurcation set B* (Fig. 6.7) in the parameter space. The name bifurcation set becomes understandable because, if under variation of parameters the parameter value crosses (6.147), a qualitative change in the behavior of the system can occur. The parameter plane together with the semicubical parabola represents the *bifurcation diagram* for this example. This diagram gives a partition of the parameter space into domains of qualitatively similar behavior, expressed by the number of equilibrium states.

The bifurcation diagram for the potential (6.83) is given by the parameter line a and the origin which is the bifurcation set. In Fig. 6.4 both the bifurcation graph and the bifurcation diagram are shown. The graph possesses a *fold* point.

We note that to the semicubical parabola in Fig. 6.7 correspond *fold points* because here a stable and an unstable equilibrium meet. This relates exactly to the diagram of Fig. 6.4. In Section 6.6.2, this will be called a saddle node bifurcation.

In general, one can say that in a bifurcation diagram there is one point that is most degenerate (in Fig. 6.7 the *cusp* point $a = b = 0$). All other points represent either parameter values leading to degeneracies of lower order (in Fig. 6.7 the fold points corresponding to (6.147)) or to a regular behavior.

From the bifurcation diagram, those systems which are not *structurally stable* or *robust* can be easily recognized. In example (6.87), these are the systems for which $a = b = 0$ or $a = 0, b \neq 0$ or $a \neq 0, b = 0$ or $a \neq 0, b \neq 0$ but still satisfying (6.147).

Since we already discussed the case $a \neq 0, b = 0$ (Fig. 1.5a changes for a small perturbation $b \neq 0$ to Fig. 1.5b), we consider the case $a = 0, b \neq 0$. From Fig. 6.7, the three graphs in Fig. 6.10 are obtained, giving the

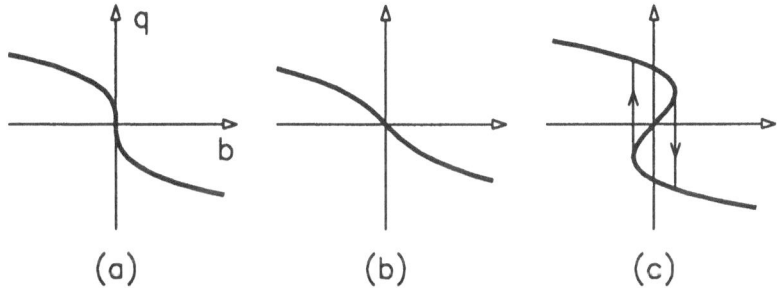

Figure 6.10. Three sections of the surface of Fig. 6.7 for (a) $a = 0$, (b) $a > 0$, and (c) $a < 0$

perturbation of $a = 0$ due to small nonzero values of a. From Fig. 6.10, it is again obvious that a system with $a = 0$ is not robust. In the case $a \neq 0, b \neq 0$ and a, b fulfilling (6.147) we obtain the situation depicted in Fig. 1.5b, at the critical parameter value a_c. Here a small perturbation of the parameters of the system results either in a system with one or three equilibrium positions.

The potential $(6.124)_2$ is of great importance in applications to model jump effects. To explain this, let us consider Fig. 6.7 with a fixed at a negative value. We assume that the state of the system corresponds to q_1 on the upper surface. If b is increased, position q_4 will be reached. Then the system can move into position q_5 only by a jump. Hence, the hysteresis diagram of Fig. 6.10c is obtained. Such a behavior, for example, can be displayed by the rod of Fig. 1.6a if $P > P_c$. Then $-a$ is proportional to $P - P_c$ and b is proportional to Q. That is, we are able to explain the jump behavior of the buckled rod under a varying transversal load Q.

Also, the two diagrams of Fig. 1.5 a and b follow from Fig. 6.7. They are sections of the surface cut by two vertical planes, one along the a-axis and the other parallel to the a-axis at a distance b.

The nonlinear stability behavior at a simple eigenvalue of any three-determinate statical problem is described by the one-dimensional bifurcation equation (6.87), which corresponds to the cusp catastrophe, and therefore, the solutions are given by Fig. 6.7. For example, the cusp catastrophe includes rod and plate buckling and buckling of the spherical shell (4.251) at a simple eigenvalue. However, if the quadratic term in (4.251) does not vanish the stability problem is described by the simpler fold catastrophe. This one follows from (6.124/1) with the bifurcation graph and diagram given in Fig. 6.4.

Considering these results, a comment seems to be appropriate. In practical problems, the case is sometimes met where the second order term does not vanish in a bifurcation equation. However, its coefficient is some orders of magnitude smaller than the coefficient of the next higher order. For example, such a case can be found in the buckling problem of the spherical shell (4.251) or for the stability problem of a tractor-semitrailer with a slightly unsymmetrically loaded semitrailer (Section 6.6.2 no. (3)). We discuss the latter case. In [166], it is shown that the steady state bifurcation equation is given by

$$a_0 q + A_2 q^2 + A_3 q^3 = 0 . \tag{6.148}$$

In a strict mathematical treatment, since $A_2 \neq 0$, the third order term

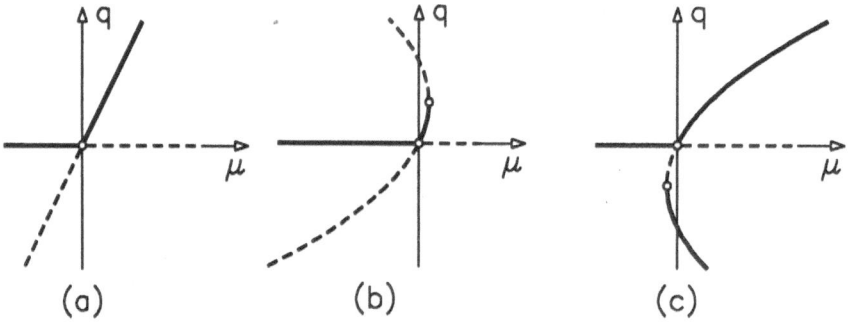

Figure 6.11. Bifurcation graphs of (6.148) for (a) only a quadratic term and (b) and (c) also with cubic terms of different sign

could be neglected. If $a_0 = -c\mu$ and $c \gg A_2$ then the solution may have the form depicted in Fig. 6.11a which gives an almost vertical bifurcation branch. To obtain more information about the behavior of the system after loss of stability of the basic state, the next higher order non-vanishing term must be considered. In this case, if $A_3 \neq 0$, the third order term should be included to obtain Fig. 6.11b,c. The difference in these two figures follows from the sign of A_3. From Fig. 6.11b,c it is obvious that due to the presence of the third order term the significance of the results obtained from the quadratic bifurcation system can be strongly diminished.

This example demonstrates that in applications the concepts of *determinacy* and of *locality* must also be considered from an application oriented point of view rather than from a purely mathematical one.

The cusp catastrophe with distinguished parameter

In the calculation of the parameters a and b in (6.87), the inconvenience of the mixing of distinguished and imperfection parameters can occur as is explained in Section 6.5.3. Hence, from a physical point of view it is often much more appropriate to use the unfolding discussed in Section 6.5.3. Engineers ([151]) have already done this for simple cases, without making

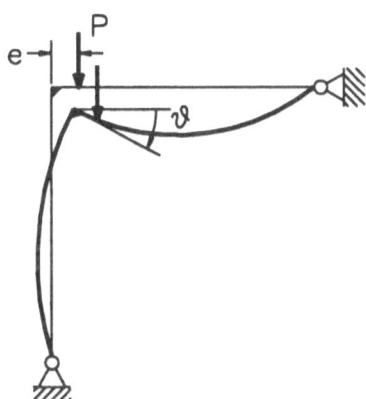

Figure 6.12. An engineering system with the nonsymmetric bifurcation behavior shown in Fig. 6.14

use of the general theory developed in [48]. This can be seen in connection with Fig. 6.4, which is related to the bifurcation equation (6.84). The analysis of the frame depicted in Fig. 6.12 yields a bifurcation equation in the form ([76] eq. 4.8)

$$A_2\vartheta^2 + a_1\vartheta + a_0 = 0 . \tag{6.149}$$

The angle ϑ describes the rotation of the corner of the frame and the parameter a_1 depends only on the load P and the parameter a_0 only on the excentricity e (Fig. 6.12). For these two parameters, the surface in Fig. 6.13 can be calculated. Intersections of this surface with vertical planes $e \approx a_0$ =const. results in the technically more meaningful bifurcation graphs of Fig. 6.14 ([151]). The reader should verify the geometrically obvious facts of Fig. 6.13 by performing the transformation from (6.149)

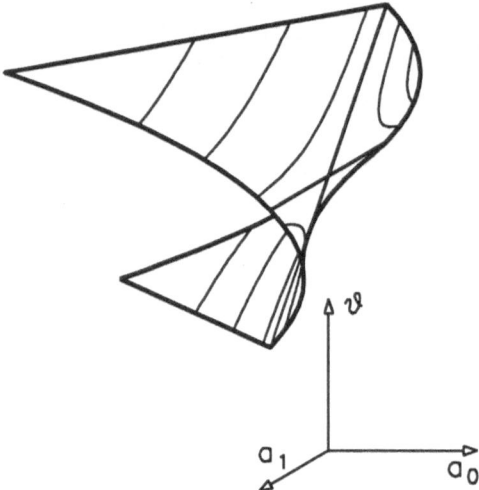

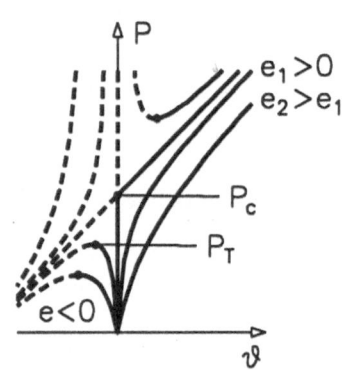

Figure 6.13. Fold surface corresponding to (6.149) with the intersection curves a_0 =const.

Figure 6.14. Nonsymmetric bifurcation point yielding an imperfection sensitive structural behavior

to (6.84) similarly to the calculation in Section 6.5.3. Further we note that the frame of Fig. 6.12 is an imperfection sensitive structure because already a small imperfection $e < 0$ results in a considerable decrease of the critical buckling load P_c to the limit load P_T.

Let us now return to the cubic bifurcation equation (6.131) and calculate the bifurcation diagram for an unfolding in the sense of the theory of Golubitsky and Schaeffer with a distinguished parameter λ. Such an unfolding is given by (6.132)$_1$

$$G(q, \lambda, \alpha, \beta) = q^3 - \lambda q + \alpha + \beta q^2 = 0 \ . \tag{6.150}$$

We differentiate (6.150) twice to obtain

$$G' = 3q^2 - \lambda + 2\beta q = 0 \tag{6.151}$$
$$G'' = 6q + 2\beta = 0 \ . \tag{6.152}$$

To identify the location of the turning points in the bifurcation diagram (plates 4 and 8 in Fig. 6.15) we form the expression $qG' - G = 0$ to eliminate λ and then we insert from (6.152) into this expression to find

$$\alpha = \frac{\beta^3}{27} \ . \tag{6.153}$$

Equation (6.153) together with the line of bifurcation points $\alpha = 0$ gives the stratification of the α, β parameter plane. Both lines correspond

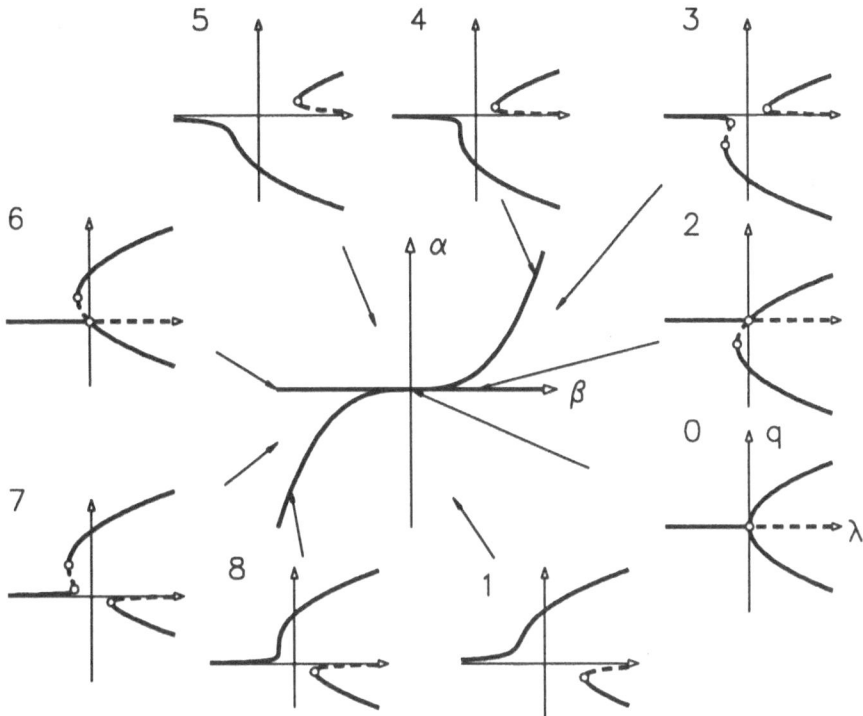

Figure 6.15. Bifurcation diagram in the (α, β) plane with the corresponding eight qualitatively different bifurcation graphs in plates 1–8 which are obtained from the unfolding of plate 0

to structurally unstable cases as is evident from the bifurcation graphs in Fig. 6.15. This is because a small perturbation away from this lines changes the graphs qualitatively. We see that more complicated graphs displaying hysteresis effects are obtained for this case as compared to the solutions shown in Fig. 1.5a and b. However, these graphs can be also obtained from (6.87) and from Fig. 6.7 because (6.87) is a universal unfolding. In fact, if we cut the surface of Fig. 6.7 by properly chosen vertical planes whose projections into the a, b parameter plane are not parallel to the a-axis we can obtain all bifurcation graphs of Fig. 6.15 (see also the remarks in Section 6.5.3).

Restricted generic bifurcation

In connection with Fig. 6.15 and the unfolding (6.150), we are able to give an intuitive explanation of the meaning of a *restricted generic unfolding*. In [51] the following alternative definition of restricted generic bifurcation in given:

Definition 6.9 (Restricted generic bifurcation) *An unfolding is called restricted generic if for sufficiently large parameter values a perturbation does not lead to qualitative changes in the bifurcation graphs.*

To explain this definition, we consider (6.150) where either $\alpha \neq 0$ and $\beta = 0$ or $\alpha = 0$ and $\beta \neq 0$ are chosen. In the first case we obtain

$$q^3 - \lambda q + \alpha = 0 \ .$$

If α is not equal to zero and is large enough, we obtain the graphs in plates 1 or 5 of Fig. 6.15. Obviously, a small perturbation by the term βq^2 does not lead to a qualitative change. Hence, the one parameter unfolding of $q^3 - \lambda q = 0$ with α is of restricted generic type.

Now let us consider the second case, which is given by

$$q^3 - \lambda q + \beta q^2 = 0 \ .$$

Again, we select the nonzero unfolding parameter β large enough. This results in the bifurcation graphs depicted in plates 2 or 6 of Fig. 6.15. However, these are not robust, since for a perturbation by arbitrarily small values of α the bifurcation points vanish. Hence, the one parameter unfolding of $q^3 - \lambda q = 0$ with βq^2 is not of the restricted generic type.

An example for an isola bifurcation

We reconsider the double pendulum of Fig. 6.8. However, we make several changes. First, we restrict ourselves to the perfect problem. Second, we assume the rods to be axially linearly elastic (Fig. 6.16) with stiffness c ([44]). Finally, the torsional spring is assumed to have a nonlinear point

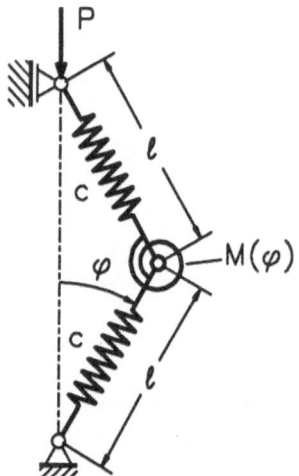

Figure 6.16. Two bar system of axially elastic rods having two degrees of freedom

symmetric characteristic, that is

$$M(\psi) = \gamma_1\psi + \gamma_3\psi^3 + \ldots . \tag{6.154}$$

Under these assumptions the potential energy $V(\varphi, \ell)$ up to at least fourth order terms is

$$V = 2P\ell\cos\varphi + 2\frac{1}{2}c(\ell - \ell_0)^2 + \frac{1}{2}\gamma_1(2\varphi)^2 + \frac{1}{4}\gamma_3(2\varphi)^4 + \ldots . \tag{6.155}$$

In (6.155) denotes ℓ_0 the length of the rods for the unloaded system. The equilibrium equations for the two degrees of freedom system are

$$\frac{\partial V}{\partial \ell} = 2P\cos\varphi + 2c(\ell - \ell_0) = 0$$

$$\tag{6.156}$$

$$\frac{\partial V}{\partial \varphi} = -2P\ell\sin\varphi + 4\gamma_1\varphi + 16\gamma_3\varphi^3 + \ldots = 0 .$$

The structure of (6.156) allows to eliminate the variable ℓ measuring the axial compression of the rods by expressing ℓ from the first equation and inserting into the second. If we further restrict to terms of third order we obtain one equation for φ in the form

$$\left[\frac{P\ell_0}{3} - \frac{4}{3}\frac{P^2}{c} + 16\gamma_3\right]\varphi^3 + \left[\frac{2P^2}{c} - 2P\ell_0 + 4\gamma_1\right]\varphi + h.o.t. = 0 . \tag{6.157}$$

Bifurcation from the solution $\varphi = 0$ can occur only if

$$P^2 - c\ell_0 P + 2c\gamma_1 = 0 ,$$

or

$$P_{1,2} = \frac{c\ell_0}{2} \pm \sqrt{\frac{(c\ell_0)^2}{4} - 2c\gamma_1} . \tag{6.158}$$

Thus, for $c\ell_0^2 > 8\gamma_1$ there are two distinct bifurcations and for $c\ell_0^2 < 8\gamma_1$ there is none. This latter result is easy to understand physically. In this case the axial springs are so soft that the shortening of the two rods compensates the increase of the load that much that buckling never occurs.

We are now interested in the case where the two bifurcation loads coincide. Then $c\ell_0^2 = 8\gamma_1$ and $P_{1,2} = P_c = c\ell_0/2$. Furthermore, we introduce two parameters λ and μ by

$$P = P_c + \lambda = \frac{c\ell_0}{2} + \lambda , \qquad \gamma_1 = \frac{c\ell_0^2}{8} + \frac{\mu}{2c} . \tag{6.159}$$

Inserting (6.159) into (6.157) yields

$$A_3\varphi^3 + (\lambda^2 + \mu)\varphi = 0 \qquad\qquad (6.160)$$

where

$$A_3 = \frac{c}{2}\left(16\gamma_3 - \frac{c\ell_0^2}{6} - \lambda\ell_0 - \frac{4\lambda^2}{3c}\right). \qquad\qquad (6.161)$$

From (6.161) we can conclude the following:

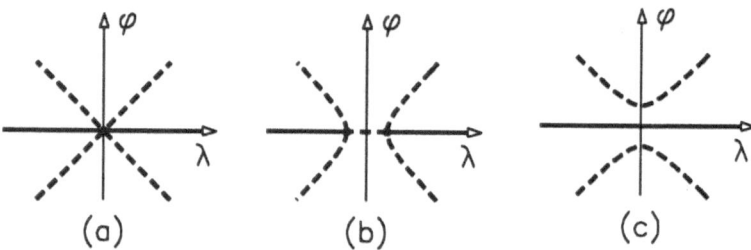

Figure 6.17. Bifurcation solutions of the two bar system of Fig. 6.16 with a linear torsional spring: (a) critical case, (b) the axial spring is stiffer than in the critical case, (c) the axial spring is softer than in the critical case

(i) If in (6.154) $\gamma_i = 0$ for all $i \geq 3$, that is, the characteristic of the torsional spring is linear, we obtain $A_3 < 0$ and introduce $-A = A_3$. (Again the terms $\lambda\varphi^3$ and $\lambda^2\varphi^3$ can be neglected.) In this case the perfect bifurcation problem is described by

$$A\varphi^3 - \lambda^2\varphi = 0$$

with the solutions:

$$\varphi = 0, \qquad \varphi = \pm\frac{\lambda}{\sqrt{A}}. \qquad\qquad (6.162)$$

The solutions (6.162) are drawn in Fig. 6.17a. Moreover in Fig. 6.17b the case is shown if $\mu < 0$, that is, if the two bar system buckles and no adjacent stable equilibrium exists. In Fig. 6.17c the case $\mu > 0$ is shown, that is, the stiffness of the axial springs, compared to the torsional spring, is so small that no instability occurs (see [44] for a more detailed discussion of this case).

(ii) We assume now that γ_3 in (6.161) is large enough such that $A_3 > 0$. In this case we obtain from (6.160) the following solutions:

$$\varphi = 0, \qquad A_3\varphi^2 + \lambda^2 = -\mu. \qquad\qquad (6.163)$$

In Fig. 6.18 the solutions of (6.163) are depicted for (a) $\mu = 0$, (b) $\mu < 0$ and (c) $\mu > 0$. The physical interpretation of the isola bifurcation is quite clear. The progressive characteristic of the torsional spring on one hand and the shortening of the lengths of the rods on the other hand force the system back into the stable straight equilibrium position.

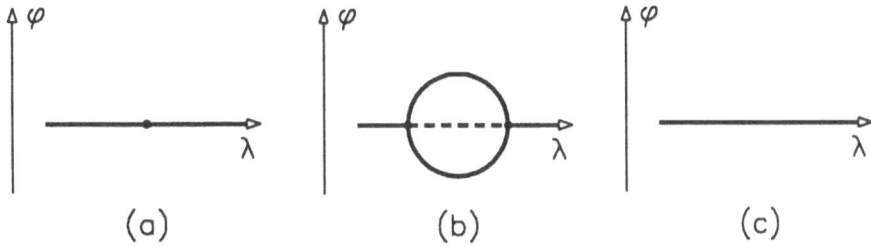

Figure 6.18. Bifurcation solutions of the two bar system of Fig. 6.16 with a progressively nonlinear torsional spring: (a) critical case, (b) the axial spring is slightly stiffer than in the critical case, (c) the axial spring is softer than in the critical case

Finally we remark that (6.142) in the framework of contact equivalence (Appendix M) has codimension five ([44]). However, if we require the problem to be reflectional symmetric the codimension is one and the unfolding is given by (6.160) ([45]).

Shell buckling

Here we discuss those cases of shell buckling from Section 4.3.5 which are given by the one-dimensional bifurcation equations (4.251). We consider two distinct cases.

First, the case of a two-determinate bifurcation equation is found if in (4.251) $r_{ccc} \neq 0$. From (4.241)$_1$ follows that $r_{ccc} \neq 0$ if $c = 2n$, that is, for c even. In this case, we have one bifurcation equation with quadratic nonlinearity. The theory of *Golubitsky and Schaeffer* supplies an unfolding in the form of (6.149), given by

$$A_2 \beta_c^2 + a_1 \beta_c + a_0 = 0 \qquad (6.164)$$

where

$$A_2 = -\frac{1}{8} r_{ccc} \frac{3}{2\mu_c}$$

and

$$a_1 = -\frac{\partial w_c}{\partial \lambda}\bigg|_{\lambda=\lambda_c} (\lambda - \lambda_c) = -w_c , \qquad (6.165)$$

where w_c is given by (4.242). The third coefficient a_0 may be calculated making the assumption that the thickness of the shell varies with the angle ϑ ([125], [124]). Theoretically, the behavior of (6.164) is completely described by Fig. 6.14. Physically, the imperfection behavior can be explained as follows. If the shell thickness varies as depicted in Fig. 6.19,

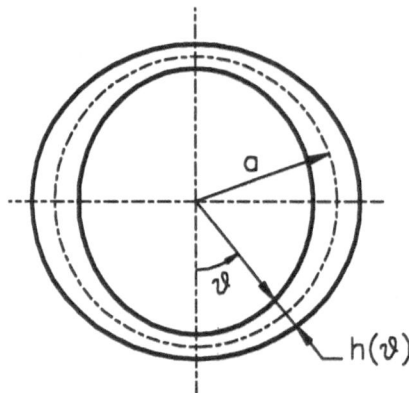

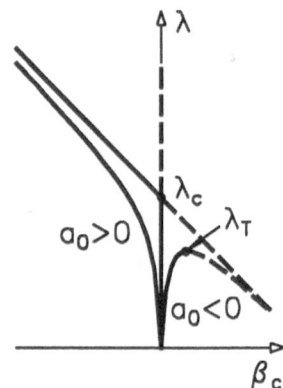

Figure 6.19. Imperfect shell due to variable shell thickness having a non-symmetric bifurcation point according to Fig. 6.20

Figure 6.20. Postbuckling behavior of the shell of Fig. 6.19 for quadratic terms in the bifurcation equation

that is, the shell is thicker at the equator and thinner at the poles the coefficient a_1 is positive and this leads to the behavior shown in the left part of Fig. 6.20. A coefficient a_0 less than zero is obtained if the shell is thicker at the poles and thinner at the equator. Then the behavior sketched in the right part of Fig. 6.20 is obtained. This is understandable physically, because in both cases the imperfection has a preference for the even buckling mode shown in Fig. 4.25. However, for $a_0 > 0$ the deflection at the poles is outward, whereas for $a_0 < 0$ the deflection at the poles is inward. Increasing the load the former stays stable whereas the latter becomes unstable at λ_T. Anyway, the third order terms calculated below will have a strong influence on Fig. 6.20 as is shown in Fig. 6.21. Hence, even for strong imperfections ($a_1 > 0$), no stable behavior for load

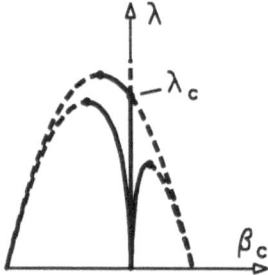

Figure 6.21. Postbuckling behavior of the shell of Fig. 6.19 taking into accout quadratic and cubic terms in the bifurcation equation

values above the buckling load λ_c will be found for imperfect problems.

Second, we consider bifurcation at a simple eigenvalue, where we have $r_{ccc} = 0$. Such a case follows from (4.251), if $c = 2n + 1$ (that is, c is odd). Then a single bifurcation equation with a third order nonlinearity is obtained. This equation is

$$A_3 \beta_c^3 + a_1 \beta_c + a_0 = 0 \; , \tag{6.166}$$

where A_3 can be calculated from (4.251). First, we note that the sums in (4.251) are finite, because r_{ccj} and r_{jcc} which are given by $(4.241)_1$ are zero for $j > 2c$. This follows from the properties of the Legendre polynomials ([124]). In order to make an important point we rewrite the third-order coefficient from (4.251) in the form

$$
A_3 = -r_{cccc} \left(\frac{2}{3\delta \mu_c} - \frac{\lambda}{3} \right) + \left\{ \frac{1}{2} \sum_j \frac{r_{ccj} r_{jcc}}{\mu_j} + \right.
$$
$$
\left. + \sum_{\substack{j \\ j \neq c}} \left[\frac{r_{ccj} r_{jcc}}{\delta^2 \omega_j} \left(\frac{1}{2\mu_j} + \frac{1}{\mu_c} \right) \left(\frac{2}{\mu_c} + \frac{1}{\mu_j} \right) \right] \right\} \; . \tag{6.167}
$$

The coefficient A_3 is positive because the value of the curly bracket is positive and much larger than the negative term. The negative term is the direct contribution of the third order terms in the shell equations to the bifurcation equations. The terms in the curly bracket are the contribution to the bifurcation equation coming from the elimination of the passive variables due to the *Ljapunov-Schmidt* reduction. Hence, we have encountered a case where a simple *Galerkin* reduction with an ansatz in one buckling mode would result in a leading term of third order in the bifurcation equation with a negative coefficient A_3. However, the bifurcation equation obtained by the Ljapunov-Schmidt reduction has a cubic term with a positive coefficient. Hence, the stability behavior of these two bifurcation systems is completely different. For the Galerkin system a supercritical bifurcation with stable non-trivial solutions would be obtained (Fig. 1.4a) whereas the correct bifurcation equations have subcritical bifurcating solutions (Fig. 1.4b). Therefore, for this problem, the Ljapunov-Schmidt reduction must not be replaced by the simpler Galerkin method.

The coefficients a_0 and a_1 are as before. The behavior is the same as is depicted in Fig. 6.21 but without the slight asymmetry.

B) Two-dimensional bifurcation equations

Here we discuss the plate and shell buckling problems at double eigenvalues of Chapter 4.

(a) Plate buckling

Before getting specific, it is important to point out that the buckling behavior of an elastic system governed by a potential at a simple or double eigenvalue with codimension ≤ 4 is described by one of the seven cases of (6.124). However, we know that buckling of a rectangular plate at a double eigenvalue has codimension 8, and hence, is not included in this listing.

Therefore, we use the theory of *restricted generic bifurcation*. First, we explain how to obtain the unfolding parameters $\lambda, \sigma, \gamma_1$ and γ_2 in (6.143). The distinguished parameter λ is given by (4.167). Further, it is assumed that a transversal loading $\bar{t}(X, Y)$ in z-direction acts on the rectangular plate of Fig. 4.17. In the non-dimensionalized plate equations (4.143) the term $t(x, y)$ must be added at the right hand side of $(4.143)_2$. This term $t(x, y)$ is related to $\bar{t}$ by

$$ t = \frac{\bar{t}b^4\sqrt{12(1 - \nu^2)}^3}{Eh^4} \, . $$

The parameters $\bar{\gamma}_1$ and $\bar{\gamma}_2$ follow from the imperfect plate equation (4.143) by projection on the buckling modes w_m and w_{m+1} to

$$ \bar{\gamma}_1 = -(t, w_m), \qquad \bar{\gamma}_2 = -(t, w_{m+1}) \, . \tag{6.168} $$

The parentheses represent the inner product defined by (4.170).

The unfolding parameter σ is more difficult to obtain as is shown in [123]. A simplified analysis is presented in [90]. According to Fig. 6.22, the slopes d_m, d_{m+1} of the two intersecting curves at the critical value

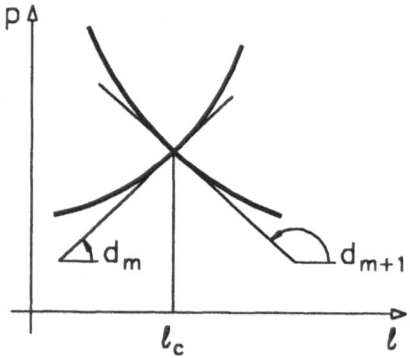

Figure 6.22. Local approximation of the eigenvalue curves at a double eigenvalue by their tangents

ℓ_c are calculated. The curved branches of the stability boundary are replaced by the linear part of their Taylor series expansions. Setting $\ell = \ell_c + \Delta\ell$, we obtain for the bifurcation equations (4.173)

$$ \begin{aligned} Br^3 + Crs^2 - \alpha(\tilde{\lambda} - d_m\Delta\ell)r + \bar{\gamma}_1 &= 0 \\ Cr^2s + Ds^3 - \beta(\tilde{\lambda} - d_{m+1}\Delta\ell)s + \bar{\gamma}_2 &= 0 \end{aligned} \tag{6.169} $$

with $\tilde{\lambda}$ given by (4.167). To transform (6.169), into (6.143) we introduce

$$\lambda = \tilde{\lambda} - d_m \Delta \ell , \quad \tilde{\lambda} - d_{m+1} \Delta \ell = \lambda + (d_m - d_{m+1}) \Delta \ell = \lambda + \sigma . \quad (6.170)$$

The transformation of (6.169) to (6.143) considering (6.170) is given by the change of variables (4.187) where μ_1 and μ_2 are given by (4.189) and $\gamma_1 = \overline{\gamma}_1 \sqrt{B/\alpha^3}$ and $\gamma_2 = \overline{\gamma}_2 \sqrt{D/\beta^3}$. We proceed now with (6.143).

The first question to be asked is: how many solutions can a system like (6.143) have? If we denote the set of equations (6.143) by $\mathbf{G} = 0$, where $\mathbf{G} = (G_1, G_2, \ldots, G_{n_c})^T$ and the components G_i are polynomials of degree k_i in n_c variables, then the number N of solutions is given by

$$N = \prod_{i=1}^{n_c} k_i . \quad (6.171)$$

For example, we obtain for (6.87) with $n_c = 1$, $k_1 = 3$: $N = 3$ and for (6.143) with $n_c = 2$, $k_1 = 3$, $k_2 = 3$: $N = 9$.

If in (6.143) $\gamma_1 = \gamma_2 = 0$, it is possible to give all nine solutions of (6.143) explicitely. They are (we recall that q_1 and q_2 are the amplitudes of the modes w_m and w_{m+1}, respectively)

(1) $q_1 = q_2 = 0$ (trivial solution, unbuckled state)

(2,3) $q_1 = \pm\sqrt{\lambda} , \quad q_2 = 0 ; \quad \lambda \geq 0 ;$

(buckled state in mode w_m)

(4,5) $q_1 = 0, \quad q_2 = \pm\sqrt{\lambda + \sigma} ; \quad \lambda \geq -\sigma ; \quad\quad (6.172)$

(buckled state in mode w_{m+1})

(6,7,8,9) $q_1 = \pm\sqrt{\dfrac{-\sigma\mu_1 + \lambda(1 - \mu_1)}{1 - \mu_1\mu_2}} , \quad q_2 = \pm\sqrt{\dfrac{\sigma + \lambda(1 - \mu_2)}{1 - \mu_1\mu_2}} ;$

(the condition for the existence of these solutions is that the radicands are nonnegative. The corresponding buckled states are a superposition of the modes w_m and w_{m+1}).

A bifurcation diagram (Fig. 6.23a) for the solutions (6.172) can be drawn giving a partition of the (λ, σ) parameter plane into domains with the same number of solutions. The qualitative behavior of the solutions (6.172) for increasing values of λ depends upon the modal parameters μ_1 and μ_2 in (6.143). In [123] it is shown that in the μ_1, μ_2 plane (Fig. 6.24) six regions with qualitatively different solutions exist. The partition follows from (6–9) in (6.172) and is given by the curve $\mu_1\mu_2 = 1$ and by the lines $\mu_1 = 1$ and $\mu_2 = 1$.

In general, for a plate buckling problem the modal parameters μ_1 and μ_2 depend upon the shape of the plate, the boundary conditions and the inplane loading.

Mode jumping

We apply the preceeding results to study a physically very interesting phenomenon called *mode jumping*. The first complete theoretical explanation of this phenomenon is given in [123].

Performing buckling experiments with rectangular plates (Fig. 4.17) without any transversal loading ($t \equiv 0$), the following findings are reported in [135]: "A plate of given length ℓ buckles at the critical thrust λ_c in a buckling pattern with m half waves. However, if the load λ is increased beyond λ_c a *secondary bifurcation* occurs at λ_s and a sudden change from the buckling pattern with m to a pattern with $m + 1$ half waves is observed."

It is intuitively clear that such a phenomenon can occur in the elastic range without plastic deformation only if ℓ is close to (in fact it must be slightly smaller) the critical length ℓ_c for a double eigenvalue of the linear buckling problem with modes w_m and w_{m+1}.

Specifying simply supported boundaries at all edges of the plate as in Fig. 4.17, the mode jumping phenomenon cannot be explained. This is because in the case of simply supported edges the modal parameters belong to region 1 in Fig. 6.24. The corresponding bifurcation diagram is shown in Fig. 6.23a and the solutions (6.172) are sketched in Fig. 6.23b

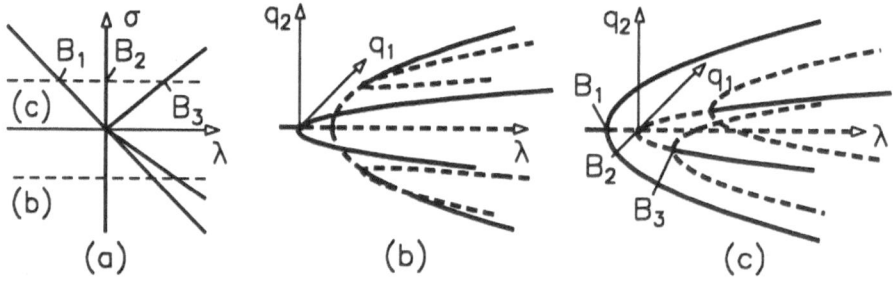

Figure 6.23. (a) Bifurcation diagram and three-dimensional representation of the bifurcation solutions for (b) the lower and (c) the upper dotted line in (a). For example, in case (c) for increasing values of λ the w_{m+1} mode with amplitude q_2 which bifurcates at B_1 is stable (full line) whereas the w_m mode with amplitude q_1 which bifurcates at B_2 is initially unstable (dotted line) but becomes stable after a secondary bifurcation at B_3. At B_1 the number of solutions given by (6.172) changes from 1 to 3 at B_2 from 3 to 5 and at B_3 form 5 to 9

and Fig. 6.23c. The practical use of the diagrams in Fig. 6.23 is the following. For a given small value of $\sigma \neq 0$, we draw in Fig. 6.23a a line parallel to the λ-axis which represents a loading path. If we increase λ starting

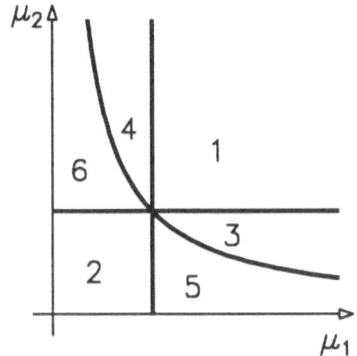

Figure 6.24. Stratification of the μ_1, μ_2 parameter plane into six regions with qualitatively different solutions. The solutions in Fig. 6.23 correspond to domain 1 and those in Fig. 6.25 to domain 3

from negative values, three boundaries are crossed where the number of solutions changes. The corresponding solutions are plotted in Fig. 6.23b for $\sigma < 0$ and in Fig. 6.23c for $\sigma > 0$. In both cases if λ is increased beyond λ_c (bifurcation point B_1 in Fig. 6.23c), it is clear that the plate stays in the mode shape (w_m with amplitude q_1 in Fig. 6.23b and w_{m+1} with amplitude q_2 in Fig. 6.23c), it buckled initially, and no secondary bifurcation is possible. Only the application of transversal loads could result in a change in the buckling pattern. (A study of the behavior of the plate in the postbuckling regime for the case of Fig. 6.23 under a variation of the transversal loads γ_1 or γ_2, but keeping the thrust λ at a fixed value is given in [147].) Hence, for a plate with simply supported boundaries and a pure thrust the mode jumping phenomenon cannot occur.

However, if one assumes clamped boundaries at the shorter edges $x = 0$ and $x = \ell$ and simply supported boundaries at the longer edges $y = 0$ and $y = 1$ (Fig. 4.17), the bifurcation diagram of Fig. 6.25a with the corresponding solutions depicted in Fig. 6.25b and c is obtained. In

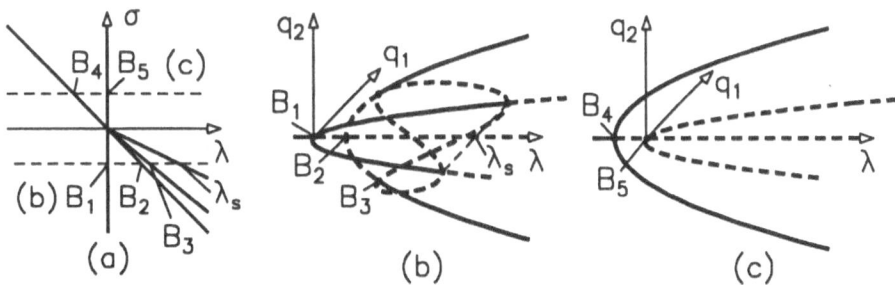

Figure 6.25. (a) Bifurcation diagram and the corresponding bifurcation solutions: (b) for the lower and (c) for the upper dotted line in (a)

the case $\sigma < 0$ (that is, $\ell < \ell_c$) the plate buckles for $\lambda = \lambda_c = 0$ in the w_m mode with amplitude q_1. Increasing λ beyond λ_c now, contrary to the behavior shown in Fig. 6.23, on the initially stable branch a secondary bifurcation occurs at $\lambda = \lambda_s$. For values $\lambda \geq \lambda_s$ the state w_m with amplitude q_1 is unstable. However, the state w_{m+1} with amplitude q_2, which after bifucation at B_2 was initially unstable, exchanged its stability at the secondary bifurcation B_3. Therefore, at $\lambda = \lambda_s$ the system moves from the state with m half waves and amplitude q_1 into the adjacent stable state with $m + 1$ half waves and amplitude q_2.

Furthermore, we note that the mode jumping phenomenon is robust with respect to small perturbations. These may be due to imperfections $\gamma_1 \neq 0, \gamma_2 \neq 0$ as is shown in [139].

Finally we remark that program packages ([130], [36]) are available which allow to calculate solutions as they are depicted in Fig. 6.25b, treating the original differential equations provided these are ordinary differential equations.

We return to the plate buckling problem in Section 6.6.5.

(b) Shell buckling

The mathematically most complicated bifurcation case of the axisymmetric shell buckling problem studied in Section 4.3.5 is that of a double eigenvalue. We set $c = 2n$ and $d = 2n + 1$ in (4.252). Then it follows from (4.241) that

$$r_{ccc} \neq 0, \ r_{cdd} \neq 0, \ r_{dcd} = r_{ddc} \neq 0,$$

$$\text{and all other} \quad r_{ijk} = 0, \quad i, j, k = c, d. \tag{6.173}$$

Thus, (4.252) simplifies to

$$-\frac{3}{2}\frac{1}{\delta\mu_c}r_{ccc}\beta_c^2 - \frac{3}{2}\frac{1}{\delta\mu_c}r_{cdd}\left(\frac{1}{3} + \frac{2}{3}\frac{\mu_c}{\mu_d}\right)\beta_d^2 = 0$$

$$-\frac{3}{2}\frac{2}{\delta\mu_c}r_{dcd}\left(\frac{2}{3}\frac{\mu_c}{\mu_d} + \frac{1}{3}\right)\beta_c\beta_d = 0 . \tag{6.174}$$

Further, it is shown in [124] that the following assumptions can be made

$$\frac{\mu_c}{\mu_d} \approx 1 , \qquad \frac{r_{cdd}}{r_{ccc}} \approx \frac{r_{dcd}}{r_{ccc}} \approx 1 \qquad \text{and} \qquad \delta\mu_c \approx 1 . \tag{6.175}$$

With (6.175) follows from (6.174) that

$$\beta_c^2 + \beta_d^2 = 0$$

$$\beta_c\beta_d = 0 . \tag{6.176}$$

From (6.176) we can derive by a scaling of the variables: $x = \beta_c$, $y = \beta_d/\sqrt{3}$ the following normal form of the degenerate bifurcation equations

$$x^2 + 3y^2 = 0$$
$$xy = 0 .$$

These bifurcation equations can be obtained from the potential

$$V = x^2 y + y^3 . \tag{6.177}$$

The potential (6.177) coincides with the second case of (6.74) and is a *hyperbolic umbilic*. It is three-determinate and its universal unfolding can be given by

$$V_u = x^2 y + y^3 + a_0(x^2 + y^2) + a_1 x + a_2 y . \tag{6.178}$$

The unfolding parameters a_0, a_1 and a_2 result from the following arguments. The calculation of a_0 is analogous to (6.165) because of $(6.175)_1$. The two other coefficients a_1 and a_2 can be calculated from an imperfection in the thickness of the shell. For example, we assume that the thickness of the spherical shell varies with the angle ϑ (Fig. 6.19). Then a_1 and a_2 are the contributions of the projection of the thickness variation on the buckling modes. For details, see [125]. A complete representation of (6.178) requires a five-dimensional space, two variables and three parameters. Hence, it is only meaningful to draw either the bifurcation diagram in three-dimensional parameter space (Fig. 6.26) or alternatively the solutions depending on a single parameter. We do not go into detail,

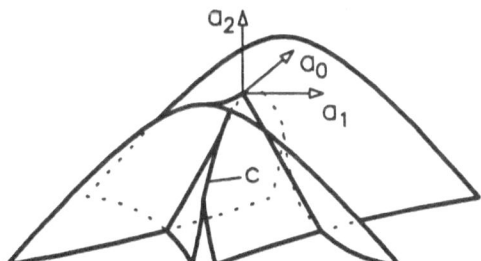

Figure 6.26. Bifurcation diagram of the hyperbolic umbilic (6.178) in a_0, a_1, a_2 parameter space

because the solutions are extensively studied in [151]. It is important to remark in connection with the surfaces of Fig. 6.26 that the most degenerate point is the origin. Through the origin passes a line which corresponds to cusp points marked by c in Fig. 6.26. All other bifurcation points, which are obtained if one crosses through a surface, are fold points.

Concerning the practical significance of a local bifurcation analysis of shell buckling problems, we refer to the comment on p. 138 where it is indicated that shell buckling problems are characterized by boundary layer solutions, and hence, the solutions for the shell buckling problem obtained in this section are only of very limited practical importance.

6.6.2   Time-continuous dynamical systems

In this section, we are going to discuss in varying amount of detail the significance of the cases (6.108)–(6.114) for the mechanical or technical problems investigated in Chapter 4. First, it is important to recall that all obtained diagrams are robust because they belong to versal cases. In fact they are universal. Second, we study only the steady state solutions.

1) The saddle-node bifurcation (6.108)

The steady state solution of (6.108) $\dot{z} = A_2 z^2 + a$ corresponds to the bifurcation graph given in Fig. 6.4 (q replaced by z, $A_2 > 0$). The name saddle-node bifurcation stems from the fact that varying the parameter value a from positive to negative values the three phase diagrams of Fig. 6.27 are obtained. To obtain the planar flow in Fig. 6.27, the equation

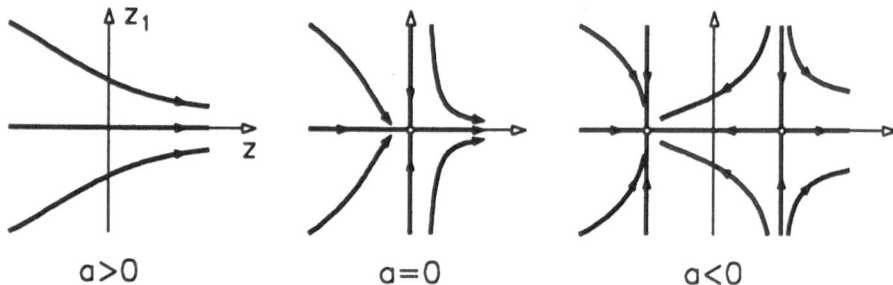

$$a>0 \qquad\qquad a=0 \qquad\qquad a<0$$

Figure 6.27. Saddle-node bifurcation in a planar system corresponding to (6.108) for $A_2 > 0$

$\dot{z}_1 = -z_1$ according to (3.20) is added to (6.108). This gives the complete qualitative behavior of a two-dimensional system under variation of the parameter a. The z-axis represents the center manifold. Note, however, that the stability of the nontrivial steady state solutions is just opposite to that of the potential by comparing Fig. 6.4 with Fig. 6.27. The same is true in case (3) below.

2) The Hopf bifurcation (6.109)

This practically very important case has already been considered several times. The bifurcation diagram is the parameter line a together with the bifurcation point $a = 0$.

The parameter a is given by the expression (see also p. 248)

$$a = \frac{d}{d\lambda}(\Re\mu(\lambda))\Big|_{\lambda=\lambda_c} (\lambda - \lambda_c) . \tag{6.179}$$

For a simple bifurcation, it must be required that the derivative of the real part of $\mu(\lambda)$ denoted by $\Re\mu(\lambda)$ is not equal to zero in order to obtain from

$$A_3 r^3 + ar = 0$$

besides the trivial solution $r = 0$ the nontrivial solution $r = \sqrt{-a/A_3}$.

Depending on the sign of A_3 two qualitatively different cases are obtained which are drawn in Fig. 1.7a and b. In the case $A_3 < 0$, a stable bifurcation point (Fig. 1.7a) is found from which a family of stable limit cycles bifurcates for increasing values a. Physically, a *soft* generation of self-sustained oscillations occurs. In the case $A_3 > 0$, the bifurcation point is unstable (Fig. 1.7b). From this point a family of unstable limit cycles bifurcates. Physically, a *hard self-excitation* is given since for values below the critical parameter value ($a < 0$) an instability can occur provided the perturbation is large enough. We recall that the stability of the two cases $A_3 > 0$ and $A_3 < 0$ at the critical parameter value $a = 0$ can be calculated as on p. 51.

With this case, we can explain another interesting feature, by comparing the solutions of (6.109) with the solutions of its linear part $\dot{r} = ar$. The latter are shown in Fig. 6.28 for $a < 0, a = 0$ and $a > 0$. If we com-

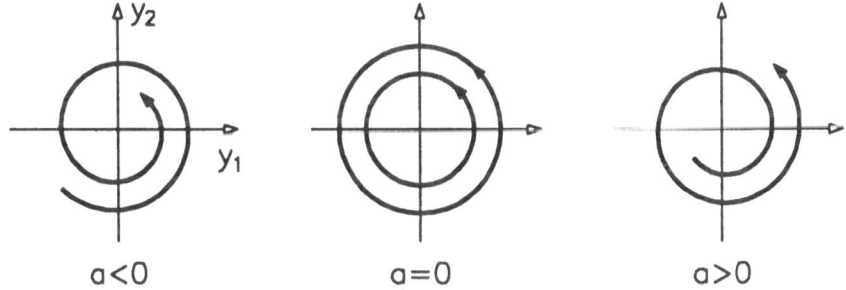

$$a<0 \qquad\qquad a=0 \qquad\qquad a>0$$

Figure 6.28. Solutions of the linear equations $\dot{r} = ar$ and $\dot{\varphi} = \omega$ for $a < 0, a = 0$ and $a > 0$

pare the cases $a = 0$ in Fig. 6.28 and in Fig. 1.7 we see that they are qualitatively different. For the linear case a family of closed orbits is obtained which is a highly non-robust situation, whereas from the nonlinear case a decision about the stability of the equilibrium position follows as is shown in Fig. 1.7.

If third order nonlinear terms determine the behavior, then the Hopf bifurcation is one of the two generic ways of loss of stability in one parameter families, and hence, occurs quite frequently. As examples we found Hopf bifurcations in Chapter 4 for various types of double pendula, the tractor-semitrailer and the railway vehicle and in Chapter 5 for the fluid conveying tube.

A more general view on the calculation of the unfolding parameter a

We start with (3.4) to which, making use of the suspension trick, we add the equation $\dot{\varepsilon} = 0$, where $\varepsilon = \lambda - \lambda_c$. This results in

$$\dot{\xi} = \mathbf{A}(\varepsilon)\xi + f(\xi, \varepsilon)$$
$$\dot{\varepsilon} = 0 . \tag{6.180}$$

Next we expand $\mathbf{A}(\varepsilon)$ into a series in ε

$$\mathbf{A}(\varepsilon) = \mathbf{A}_0 + \varepsilon_1 \frac{\partial \mathbf{A}}{\partial \varepsilon_1} + \varepsilon_2 \frac{\partial \mathbf{A}}{\partial \varepsilon_2} + \ldots$$

where $\varepsilon^T = (\varepsilon_1, \ldots, \varepsilon_\ell)$, $\mathbf{A}_0 = \mathbf{A}(0)$ and $\partial \mathbf{A}/\partial \varepsilon_i$ is evaluated at $\varepsilon = 0$. The next step is the transformation of (6.180) by means of (3.7) yielding

$$\dot{y} = \mathbf{B}^{-1}\left(\mathbf{A}_0 + \varepsilon^T \left(\frac{\partial \mathbf{A}}{\partial \varepsilon}\right) + \ldots\right)\mathbf{B}y + h.o.t.$$
$$\dot{\varepsilon} = 0$$

or in the usual notation

$$\dot{y} = \mathbf{J}y + \mathbf{B}^{-1}\varepsilon^T \left(\frac{\partial \mathbf{A}}{\partial \varepsilon}\right)\mathbf{B}y + h.o.t.$$
$$\dot{\varepsilon} = 0 . \tag{6.181}$$

Since ε is considered to be a (trivial) variable the linear part is unchanged and we focus on the quadratic terms containing ε and y.

Now we apply center manifold theory. First, we decompose (6.181) into

$$\dot{y}_c = \mathbf{J}_c y_c + \mathbf{P}_c \left(\varepsilon_1 \mathbf{B}^{-1}\frac{\partial \mathbf{A}}{\partial \varepsilon_1}\mathbf{B} + \varepsilon_2 \mathbf{B}^{-1}\frac{\partial \mathbf{A}}{\partial \varepsilon_2}\mathbf{B} + \ldots\right)y + \ldots$$
$$\dot{y}_s = \mathbf{J}_s y_s + \mathbf{P}_s \left(\varepsilon_1 \mathbf{B}^{-1}\frac{\partial \mathbf{A}}{\partial \varepsilon_1}\mathbf{B} + \varepsilon_2 \mathbf{B}^{-1}\frac{\partial \mathbf{A}}{\partial \varepsilon_2}\mathbf{B} + \ldots\right)y + \ldots \tag{6.182}$$
$$\dot{\varepsilon} = 0$$

where $\mathbf{P}_c$ and $\mathbf{P}_s$ are projection matrices given by $\mathbf{P}_c = (\mathbf{E}_{n_c}, 0)$ and $\mathbf{P}_s = (0, \mathbf{E}_{n_s})$ where $\mathbf{E}_{n_c}$ and $\mathbf{E}_{n_s}$ are the n_c- and n_s-dimensional unit matrices, respectively. Second, we eliminate y_s from (6.182)$_1$ by expressing

$$y_s = h(y_c, \varepsilon) \tag{6.183}$$

and inserting (6.183) into (6.182). To do this we calculate

$$\dot{y}_s = \frac{\partial h}{\partial y_c}\mathbf{J}_c y_c + \frac{\partial h}{\partial \varepsilon}0 + h.o.t.$$

and insert this expression into $(6.182)_2$ to obtain

$$\frac{\partial h}{\partial y_c} \mathbf{J}_c y_c = \mathbf{J}_s h + \mathbf{P}_s \left(\varepsilon_1 \mathbf{B}^{-1} \frac{\partial \mathbf{A}}{\partial \varepsilon_1} \mathbf{B} + \right.$$
$$\left. + \varepsilon_2 \mathbf{B}^{-1} \frac{\partial \mathbf{A}}{\partial \varepsilon_2} \mathbf{B} + \dots \right) \left(\begin{matrix} y_c \\ h(y_c) \end{matrix} \right) + h.o.t.$$

Solving this equation for h by a series expansion as we did for (3.25) we can eliminate y_s from $(6.182)_1$. This supplies the following system on the center manifold

$$\begin{aligned} \dot{y}_c &= \mathbf{J}_c y_c + (\varepsilon_1 \mathbf{A}_1 + \varepsilon_2 \mathbf{A}_2 + \dots) y_c + h.o.t. \\ \dot{\varepsilon} &= 0 \end{aligned} \tag{6.184}$$

where $\mathbf{A}_1, \mathbf{A}_2, \dots$ result from projecting $(6.182)_1$ onto the y_c components.

Now we can apply the normal form transformation to (6.184). We do this for the case of a Hopf bifurcation. As in (6.23) we work with complex variables and restrict to a scalar parameter $\varepsilon = \varepsilon_1$. Then we obtain

$$\begin{aligned} \dot{w} &= i\omega_c w + \varepsilon(\alpha w + \beta \overline{w}) + h.o.t. \\ \dot{\overline{w}} &= -i\omega_c \overline{w} + \varepsilon(\overline{\alpha} \overline{w} + \overline{\beta} w) + h.o.t. \\ \dot{\varepsilon} &= 0 \end{aligned} \tag{6.185}$$

where $\alpha = \alpha_1 + i\alpha_2$ and $\beta = \beta_1 + i\beta_2$ result from the matrix $\mathbf{A}_1$ in (6.184) as it is shown below. The resonance condition (6.11) for $(6.185)_1$ after inserting the three eigenvalues $(i\omega_c, -i\omega_c, 0)$

$$m_1(i\omega_c) + m_2(-i\omega_c) + m_3(0) - i\omega_c = 0$$

gives the solution $m_1 = 1$, $m_2 = 0$, $m_3 = 1$. Hence, the second order term $\alpha w^{m_1} \overline{w}^{m_2} \varepsilon^{m_3} = \alpha \varepsilon w$ cannot be removed and $(6.185)_1$ can only be simplified to the normal form

$$\dot{v} = (i\omega_c + \alpha\varepsilon)v + b_{21} v^2 \overline{v} , \tag{6.186}$$

where the coefficient b_{21} of the third order terms is the same as calculated in Section 6.1.1. The coefficient α can be calculated by rewriting $(6.185)_1$ in real coordinates. From

$$\begin{aligned} (y_1 + iy_2)^{\cdot} &= i\omega_c(y_1 + iy_2) + \\ &\quad + \varepsilon \left((\alpha_1 + i\alpha_2)(y_1 + iy_2) + (\beta_1 + i\beta_2)(y_1 - iy_2) \right) + h.o.t. \end{aligned}$$

we obtain

$$\begin{pmatrix} y_1 \\ y_2 \end{pmatrix}^{\cdot} = \begin{pmatrix} 0 & -\omega_c \\ \omega_c & 0 \end{pmatrix} \begin{pmatrix} y_1 \\ y_2 \end{pmatrix} +$$

$$+ \varepsilon \begin{pmatrix} \alpha_1 + \beta_1 & -\alpha_2 + \beta_2 \\ \alpha_2 + \beta_2 & \alpha_1 - \beta_1 \end{pmatrix} \begin{pmatrix} y_1 \\ y_2 \end{pmatrix} + h.o.t.$$

$$= \mathbf{J}_c \begin{pmatrix} y_1 \\ y_2 \end{pmatrix} + \varepsilon \mathbf{A}_1 \begin{pmatrix} y_1 \\ y_2 \end{pmatrix} + h.o.t.$$

The entries $\alpha_1, \alpha_2, \beta_1, \beta_2$ follow from a comparison with the elements a_{ij} of $\mathbf{A}_1$ given in (6.184). Since in (6.186) only α is needed we have

$$\alpha_1 = \frac{1}{2}(a_{11} + a_{22})$$

$$\alpha_2 = -\frac{1}{2}(a_{12} - a_{21}) .$$

(6.187)

In polar coordinates the unfolded equations are

$$\dot{r} = \alpha_1 \varepsilon r + A_3 r^3 \quad + O(|r|^5)$$
$$\dot{\varphi} = \omega_c + \varepsilon \omega_1 + A_2 r^2 \quad + O(|r|^4)$$

where setting $a = \alpha_1 \varepsilon$ results in the same unfolding parameter as it is given in (6.179) because

$$\alpha = \frac{\partial \mu}{\partial \varepsilon} ,$$

(6.188)

where $\mu = \sigma \pm i\omega_c$ is the critical eigenvalue.

To show that (6.188) is correct we calculate the eigenvalues of $\mathbf{J}_c + \varepsilon \mathbf{A}_1$ obtained from (6.184). Let $P(\mu, \varepsilon) = 0$ be the characteristic polynomial. For $\varepsilon = 0$, $P(\mu, 0) = \mu^2 + \omega_c^2 = 0$ has the solution $\mu_c = \pm i\omega_c$ and $\partial P(\mu, 0)/\partial \mu$ is regular. Hence, the equation $P(\mu, \varepsilon) = 0$ can be solved locally in the neighborhood of $\varepsilon = 0$ to give $\mu = \mu(\varepsilon)$. That is (for $\mu_c = i\omega_c$)

$$\left(\frac{\partial P}{\partial \mu} \frac{\partial \mu}{\partial \varepsilon} + \frac{\partial P}{\partial \varepsilon} \right) \bigg|_{\varepsilon=0} = 2i\omega_c \frac{\partial \mu}{\partial \varepsilon} + \frac{\partial P}{\partial \varepsilon} \bigg|_{\varepsilon=0} = 0 .$$

(6.189)

From the characteristic polynomial of $\mathbf{J}_c + \varepsilon \mathbf{A}$

$$P(\mu, \varepsilon) = \mu^2 + \omega_c^2 + \varepsilon[(-a_{11} - a_{22})\mu + \\ + \omega_c(a_{21} - a_{12})] + \varepsilon^2(a_{11}a_{22} - a_{21}a_{12})$$

we obtain

$$\frac{\partial P}{\partial \varepsilon} \bigg|_{\varepsilon=0} = -\mu_c(a_{11} + a_{22}) + \omega_c(a_{21} - a_{12})$$

$$= -i\omega_c(a_{11} + a_{22}) + \omega_c(a_{21} - a_{12}) .$$

Inserting into (6.189) yields

$$2i\omega_c \frac{\partial \mu}{\partial \varepsilon} - i\omega_c(a_{11} + a_{22}) + \omega_c(a_{21} - a_{12}) = 0$$

with the solution

$$\frac{\partial \mu}{\partial \varepsilon} = \frac{1}{2}(a_{11} + a_{22}) - \frac{i}{2}(a_{12} - a_{21})$$

which is exactly $\alpha = \alpha_1 + i\alpha_2$ given by (6.187) and (6.188).

Return to physical coordinates

Now it is important to point out how from the solution of (6.109), given in mathematical variables for a two-dimensional system, we can calculate the system behavior in its physical variables. This can be done by proceeding in backward direction through those steps we performed to obtain the bifurcation equations listed in Section 6.5.1. The first step is to return from the polar coordinates r, φ to the cartesian coordinates $z_1 = r\cos\omega t$, $z_2 = r\sin\omega t$. The transformation from z_1, z_2 to y_1, y_2 according to (6.2) is trivial, because, $y_1 = z_1$, $y_2 = z_2$, up to terms of order $O(|z_1|^3 + |z_2|^3)$. Next the transformation from $y_1, y_2 \to \xi_1, \ldots, \xi_n$ must be performed. Recalling that, $y_i = O(|y_1|^2 + |y_2|^2)$ for $i = 3, \ldots, n$, we obtain from (3.7)

$$\begin{pmatrix} \xi_1 \\ \vdots \\ \vdots \\ \xi_n \end{pmatrix} = \mathbf{B} \begin{pmatrix} y_1 \\ y_2 \\ 0 \\ \vdots \\ 0 \end{pmatrix}.$$

Hence, after a Hopf bifurcation, all physical components perform limit cycle oscillations with the same frequency but with amplitudes depending on the matrix $\mathbf{B}$ given by (3.7).

Cases (1) and (2) above are the only bifurcations that can occur generically in one-parameter families. They are either the birth or annihilation of a pair of singular points or the birth or annihilation of a limit cycle from a singular point.

Finally, we mention that with the book [56] also a computer program is supplied which allows to calculate all those quantities that determine the Hopf bifurcation according to (6.109) including the center manifold reduction.

3) One zero root with additional degeneracy (6.110)

The study of the steady state solutions of (6.110) is completely analogous to the study of the potential (6.124/2). Hence, the bifurcation set in the parameter plane (a, b) is given in the lower part of Fig. 6.7 and the singular points are given by the surface in the upper part of Fig. 6.7. If we assume $A_3 < 0$ in (6.110), the corresponding vector field drawn for a section with a vertical plane at $a < 0$ is shown in Fig. 6.29.

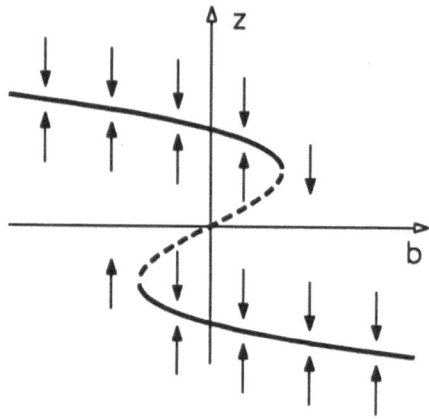

Figure 6.29. Equilibria and their stability for $A_3 < 0$ and $a < 0$ of (6.110)

The case (6.110) explains all the divergent bifurcations in Section 4.1. For example, the double pendulum of Fig. 4.3 with a stiff spring (c large) is governed by (6.110). The motion of the tractor-semitrailer of Fig. 4.8 is also governed by (6.110) if the distribution of the loading is such that the location of the center of gravity is moved towards the front of the trailer, that is, if d is large.

For the double pendulum $A_3 < 0$ is found, and as second parameter either a transversal load or an imperfect spring could be selected. Since nothing special compared to the static case in Section 6.6.1 will be found, we rather discuss the behavior of the tractor-semitrailer at loss of stability at a zero eigenvalue.

Before we treat this problem some general remarks should be made how to relate the unfolding parameters a and b in the normal form (6.110) to the physical or technical parameters. We already know that a depends on the distinguished parameter λ. Similar to (6.179) we obtain

$$a = \frac{d}{d\lambda}\mu_1(\lambda)\bigg|_{\lambda=\lambda_c} (\lambda - \lambda_c) . \tag{6.190}$$

Consideration of imperfections

To calculate the imperfection parameter b we select one or several parameters $e_1, \ldots, e_m$ which seem to be of physical or technical relevance. Hence, the equations of motion (3.4) take the form

$$\dot{\boldsymbol{\xi}} = \boldsymbol{F}(\boldsymbol{\xi}, \lambda, e_1, \ldots, e_m) = \boldsymbol{A}(\lambda, \boldsymbol{e})\boldsymbol{\xi} + \boldsymbol{f}(\boldsymbol{\xi}, \lambda, \boldsymbol{e}) . \tag{6.191}$$

We expand the right-hand side of (6.191) into a power series in e. This yields

$$\dot{\xi} = A(\lambda, 0)\xi + A_1(\lambda, 0)e\xi + f(\xi, \lambda, 0) + \sum_{j=1}^{m} \frac{\partial f}{\partial e_j}\bigg|_{e=0} e_j \,, \quad (6.192)$$

where $A_1(\lambda, 0)$ is the linear part in the Taylor expansion of $A(\lambda, e) = A(\lambda, 0) + A_1(\lambda, 0)e + \dots$. Performing the change of variables (3.7), (6.192) can be transformed into a form which is an extension of (3.8)

$$\begin{aligned} \dot{y} &= B^{-1}A(\lambda_c)By + B^{-1}A_1(\lambda_c, 0)eBy + \\ &+ B^{-1}f(By, \lambda_c) + \sum_{j=1}^{m} B^{-1}\frac{\partial f}{\partial e_j}\bigg|_{e=0} e_j \,. \end{aligned} \quad (6.193)$$

Only the constant terms in e_j in that line in (6.193) that corresponds to the zero eigenvalue of the linear operator must be retained for the evaluation of the imperfections. The imperfection terms multiplied by y are of higher order and can be neglected (Section 6.4).

We want to explain the type of analysis used above a little bit more general ([129]). For example, we rewrite (6.192) by introducing a small parameter ε following from a scaling of the variables and the parameters (see, for example, p. 263). This results in a system of the form

$$\dot{\xi} = \varepsilon e + A(\varepsilon)\xi + f(\xi, \varepsilon) \,. \quad (6.194)$$

Transformation to Jordan form of A by $\xi = By$ according to (3.7) results in

$$\dot{y} = B^{-1}\varepsilon e + J_0 y + \varepsilon J_1 y + B^{-1}f(By, \varepsilon) \quad (6.195)$$

where

$$\begin{aligned} J_0 &= J(0) = B^{-1}A(0)B \\ A(\varepsilon) &= A(0) + \varepsilon A_1 + O(\varepsilon^2) \qquad \text{and} \\ J_1 &= B^{-1}A_1 B \,. \end{aligned}$$

System (6.195) can be rewritten in the form

$$\dot{y} = \varepsilon\delta + J_0 y + \varepsilon J_1 y + g(y, \varepsilon) \quad (6.196)$$

which is nothing else than (6.193).

Now the question arises: can we eliminate all the components of the imperfection vector δ by a coordinate transformation of order ε?

This means that we have to check whether all components of δ are in the range of J_0. For example, if we have a zero eigenvalue in J_0 in the first row then the component δ_1 of δ cannot be eliminated by a coordinate shift of order ε (see also the discussion in Section 6.2.2). Therefore, it is preferable to treat a problem in the form of (6.196) in the following way. First, for $\varepsilon = 0$ the perfect bifurcation problem is treated, and second, the influence of small imperfections is taken care of in the unfolding of the bifurcation equations.

Tractor-semitrailer dynamics

For the tractor-semitrailer the speed V of the vehicle is one important parameter. Quite naturally, the steering angle δ can be selected as second parameter. Calculations yield $A_3 > 0$. Hence, the solution graph for $\delta \equiv 0$ is that of Fig. 1.5a, which is drawn again in Fig. 6.30a with V as the distinguished parameter. The solutions which bifurcate at $V = V_c$ are

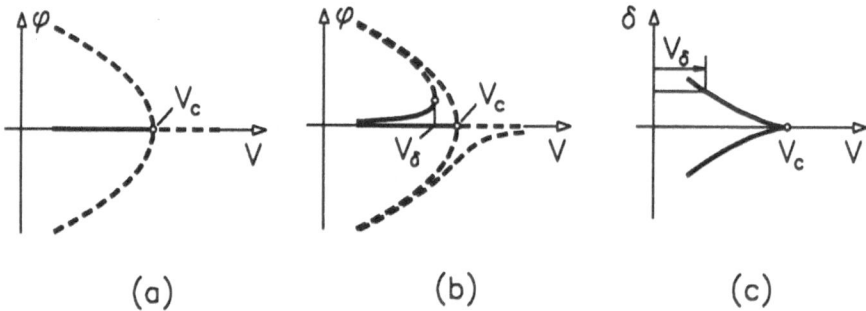

(a)   (b)   (c)

Figure 6.30. Steady state solutions for the tractor-semitrailer: (a) steering angle $\delta = 0$, (b) $\delta \neq 0$, (c) critical speed $V = V_\delta$ depending on δ

subcritical, and hence, unstable. The corresponding mode of instability is called *jack-knifing*. It is expressed by a monotonous increase of the angle φ between the axis of the trailer and the tractor (Fig. 4.8). This instability leads to a very dangerous driving state. Fig. 6.30b corresponds to Fig. 1.5b and shows the influence of a small steering angle δ on the driving behavior of the tractor-semitrailer. We note that the critical speed is reduced, which is intuitively quite comprehensible, because one expects the critical speed of the vehicle to be smaller in a cornering motion than in a straight line motion. The strong reduction of the critical speed depending on δ can be seen from the bifurcation diagram Fig. 6.30c.

For this problem, we also discuss the behavior of the system if we distinguish in the unfolding between the main parameter V and the imperfection parameters. Besides δ, we select as second imperfection y_s, which is a measure of the excentricity of the center of gravity of the trailer (Fig. 4.8). Such an excentricity could have resulted from improper loading. Now the bifurcation diagram Fig. 6.31 analoguous to Fig. 6.15 is obtained, but with different stability properties. From this diagram and

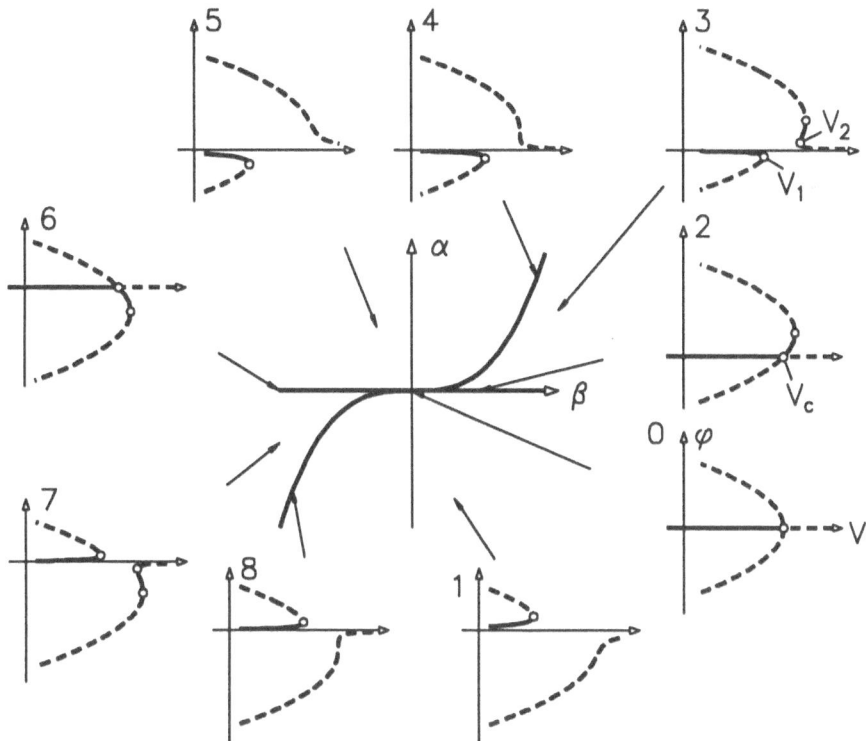

Figure 6.31. Bifurcation diagram and corresponding steady state solutions for the excentrically loaded tractor-semitrailer of Fig. 4.8. The plates 0 and 1 correspond to Fig. 6.30a and b

the corresponding bifurcation graphs, interesting and unexpected driving states of the tractor-semitrailer vehicle can be predicted, which, of course, do not correspond to typical driving conditions. The parameters λ, α, β in (6.150) are related to the physical quantities by

$$\lambda \sim V_c - V , \qquad \alpha = m_1\delta - m_2 y_s , \qquad \beta \sim y_s , \qquad (6.197)$$

where m_1 and m_2 are positive constants ([67]). The interesting information to be obtained from Fig. 6.31 is that there are still stable driving states for speeds $V > V_c$ (see also Fig. 6.21). Let us focus now on plates 2 and 3 in Fig. 6.31. The graph in plate 2 corresponds to a non-robust case ($\alpha = 0$) which according to (6.197)$_2$ even for a speed below V_c requires a steering angle $\delta \neq 0$ depending on the excentricity y_s of the loading. The corresponding jack-knifing angle φ is zero. After reaching the bifurcation point $V = V_c$, a stable supercritical branch exists for increasing speed V. However, it occurs with $\varphi \neq 0$ ($\varphi > 0$ in plate 2). If $\alpha \neq 0$, the graph in plate 3 of Fig. 6.31 is valid. Now, there exist two isolated stable branches for a steering value $\delta > 0$, one with $\varphi < 0$ and the other with $\varphi > 0$. Further, it follows from (6.197) that y_s must be positive. The state $\varphi < 0$ corresponds to a regular driving state for a range of speeds $0 < V \leq V_1$. We note from Fig. 4.8 that for the regular driving state a negative φ corresponds to a positive δ. The state $\varphi > 0$ represents a state of driving best characterized by the term "power-slide" from auto racing, and exists as a stable state only in a very limited domain of speeds above $V = V_c$. The configuration of the tractor-semitrailer depicted in the lower part of Fig. 4.8 is just in this state of power slide.

4) A purely imaginary pair with additional degeneracy (6.111)

The calculation of the unfolding parameters a, b is given by

$$a = \left.\frac{\partial \Re\mu}{\partial \varepsilon_1}\right|_{\substack{\varepsilon_1=0\\\varepsilon_2=0}} \varepsilon_1 + O(\varepsilon_1^2) , \qquad b = \left.\frac{\partial A_3}{\partial \varepsilon_2}\right|_{\substack{\varepsilon_1=0\\\varepsilon_2=0}} \varepsilon_2 + O(\varepsilon_2^2) . \qquad (6.198)$$

In the above expressions, $\Re\mu$ and A_3 are the real part of the complex eigenvalue and the coefficient of the third order term in the normal form expressed in polar coordinates, respectively. The directions ε_1 and ε_2 are shown in Fig. 4.2. In the direction of ε_1 $\Re\mu$ is changed keeping A_3 unchanged, whereas in the direction of ε_2 A_3 is changed keeping $\Re\mu$ unchanged. The calculation of the differential quotient $\partial A_3/\partial \varepsilon_2$ has been performed numerically making use of a symmetric difference quotient.

The steady state solutions of (6.111) follow from

$$A_5 r^5 + b r^3 + ar = 0 \qquad (6.199)$$

to

$$r = 0 , \qquad r_{1,2}^2 = -\frac{b}{2A_5} \pm \sqrt{\left(\frac{b}{2A_5}\right)^2 - \frac{a}{A_5}} .$$

These are the bifurcation graphs, shown in Fig. 6.33 for $A_5 < 0$. The value r_n of the turning point can be calculated from (6.199) searching for the maximum values of $a(r)$. That is, we calculate

$$\frac{da}{dr} = -4A_5 r^3 - 2br = 0 , \qquad b = -2A_5 r_n^2 . \qquad (6.200)$$

The corresponding value $a = a_n$ is

$$a_n = A_5 r_n^4 . \qquad (6.201)$$

Elimination of r_n from (6.200)$_2$ and (6.201) yields

$$b^2 = 4A_5 a$$

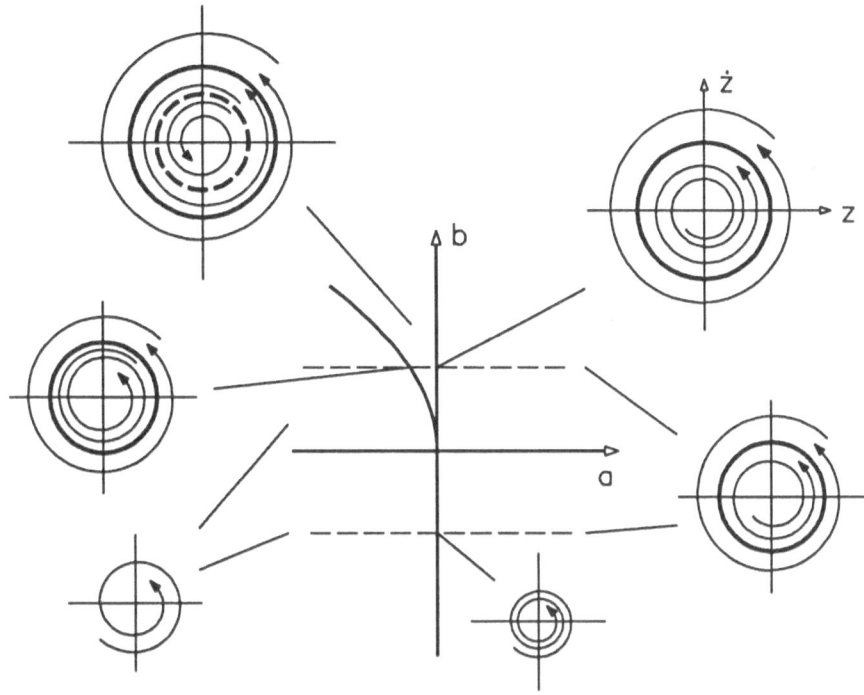

Figure 6.32. Bifurcation diagram and phase portraits for (6.111) for $A_5 < 0$

which gives the corresponding bifurcation set in the a, b parameter plane. The other bifurcation sets are the lines $a = 0$ and $b = 0$. Hence, we obtain the bifurcation diagram ([7]) in Fig. 6.32. The three bifurcation graphs of Fig. 6.33 correspond to the horizontal lines $b < 0$, $b = 0$ and $b > 0$ in the bifurcation diagram of Fig. 6.32.

Exchange of stability of a periodic solution by a steady state analysis

Two problems of Chapter 4 are determined by (6.111). These are the double pendulum with axially elastic rods (Fig. 4.1) and the railway bogy (Fig. 4.11). Both problems, especially railway vehicles are technically important applications.

In order to solve these problems we introduce a second parameter and by a proper choice of its value, we can create a more degenerate bifurcation. This more degenerate case allows us to treat the problem of exchange of stability of a periodic solution in a simple way. As already mentioned in Sections 4.1.1 and 4.1.5, these two problems possess one purely imaginary pair of eigenvalues at loss of stability. However, varying a second parameter (the stiffness $\bar{\gamma}$ for the pendulum and the angle δ of conicity of the wheels for the rail vehicle) we are able to annihilate the coefficient A_3 in the normal form (6.109) of the Hopf bifurcation. Hence, for these critical parameter values it is necessary to include fifth order terms in the bifurcation equations which, after the normal form transformation, are represented by the coefficient A_5 in (6.111). The codimension of (6.111) is *two* compared to *one* for (6.109). The dependence of the two mathematical unfolding parameters ε_1 and ε_2, which determine by (6.198) a and b, on the two physical unfolding parameters $\lambda_1 = F - F_c$ and $\lambda_2 = \overline{\gamma} - \overline{\gamma}_c$ can best be seen from Fig. 4.2. A similar situation holds for the railway axle. Fig. 6.33b corresponds to the system

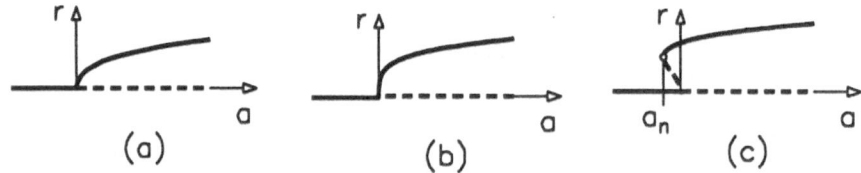

Figure 6.33. Amplitude of the limit cycles of (6.111) for three different values of b corresponding to Fig. 6.32 and for $A_5 < 0$: (a) $b < 0$, (b) $b = 0$ and (c) $b > 0$

for which $A_3 = 0$. Unfolding of this state gives us the particularly important graph of Fig. 6.33c which allows us to calculate the *stability of periodic solutions* by investigating a steady state solution.

Let us summarize this important point. For the railway vehicle, at $a = 0$, which corresponds to $V = V_c$, a family of limit cycles bifurcates off the steady state (Fig. 6.33), the stability of which is easy to determine by a linearized analysis. They are unstable if the bifurcation at $a = 0$ ($V = V_c$) is subcritical. However, the amplitude curve of the limit cycles has a *turning point* at $a = a_n$. Without making use of the methods of bifurcation theory the calculation of this turning point would be rather complicated. This is, because the stability of an oscillating solution for a nonlinear system which, generally, has a large number of degrees of freedom must be investigated. However, the application of the bifurcation approach, described above, allows us to calculate the limit point of the amplitude-parameter relationship of the periodic solution from the simple analysis of (6.111), that is, studying the steady state of a two-dimensional system.

In connection with this problem we are also able to comment again on the notion of *locality* of the bifurcation analysis. In fact the meaning of the definition we gave on p. 208 can be well explained with this problem. We see from Fig. 6.33c that the range of a can be quite restricted if we work with third order terms only, because the next bifurcation at a_n due to the fifth order terms can be very close to $a = 0$. However, in the fifth order model under the correct assumption that no additional bifurcation appears due to terms of higher (seventh) order the parameter variation can be of substantial amount. In this case both local bifurcations at

$a = 0$ and $a = a_n$ are incorporated in a "global" bifurcation problem.

Practical applications include the calculation of the critical speed V_n (Fig. 1.3) for a railway vehicle or even a whole train with many degrees of freedom from a two-dimensional bifurcation system ([162]). Here, the problem must be analyzed for that conicity angle δ for which A_3 becomes zero. Then, in the unfolding of the higher degenerate, that is, the fifth order system, the conicity angle which actually occurs for a real wheel must be introduced. This approach yields also the change in the stability behavior of a railway vehicle during the lifetime of its wheelset due to the certainly quasistatic change of the angle δ due to wear.

For the double pendulum depicted in Fig. 4.1, the higher degenerate system means that depending on the stiffness of the axial spring either a soft or a hard excitation of self-sustained oscillations may occur at loss of stability ([146]).

5) Two zero roots (6.112)

This case, which is called Bogdanov-Takens bifurcation, applies to planar oscillations of the tube in Section 5.3 (Fig. 5.9 and Fig. 5.10) and to the double pendulum with end support of Fig. 4.3, if the damping is large enough, because then the matrix $\mathbf{J}_c$ takes the form of case 4 in (4.12). For this latter case the unfolding parameters a and b in (6.112) are linear combinations of the two mathematical unfolding parameters $\varepsilon_1, \varepsilon_2$ which depend linearly on the two physical parameters $P - P_c$ and $c - c_c$ (Fig. 4.4). For the unfolding (6.112), the calculation of the stratification of the parameter plane which is presented in Fig. 6.34 is not difficult except for one bifurcation sequence. We give a brief presentation and for a more detailed description we refer the reader [7] p. 300, [21] and [50] p. 364.

To be specific, we assume

$$A_{30} < 0 \qquad \text{and} \qquad A_{21} < 0 . \tag{6.202}$$

For the steady state solutions of (6.112), the following equations

$$z_2 = 0 , \qquad A_{30}z_1^3 + A_{21}z_1^2 z_2 + az_1 + bz_2 = 0 \tag{6.203}$$

must be solved. The solutions to equation $(6.203)_2$ are

$$z_{10} = 0 , \qquad z_{11}, z_{12} = \pm\sqrt{-\frac{a}{A_{30}}} . \tag{6.204}$$

From the assumption $A_{30} < 0$ follows that for $a > 0$ there are three equilibria. However, for $a < 0$, there is only one, namely the origin. Hence, the influence of a in the bifurcation diagram will be that either phase diagrams with one or three equilibria are obtained. Now we study

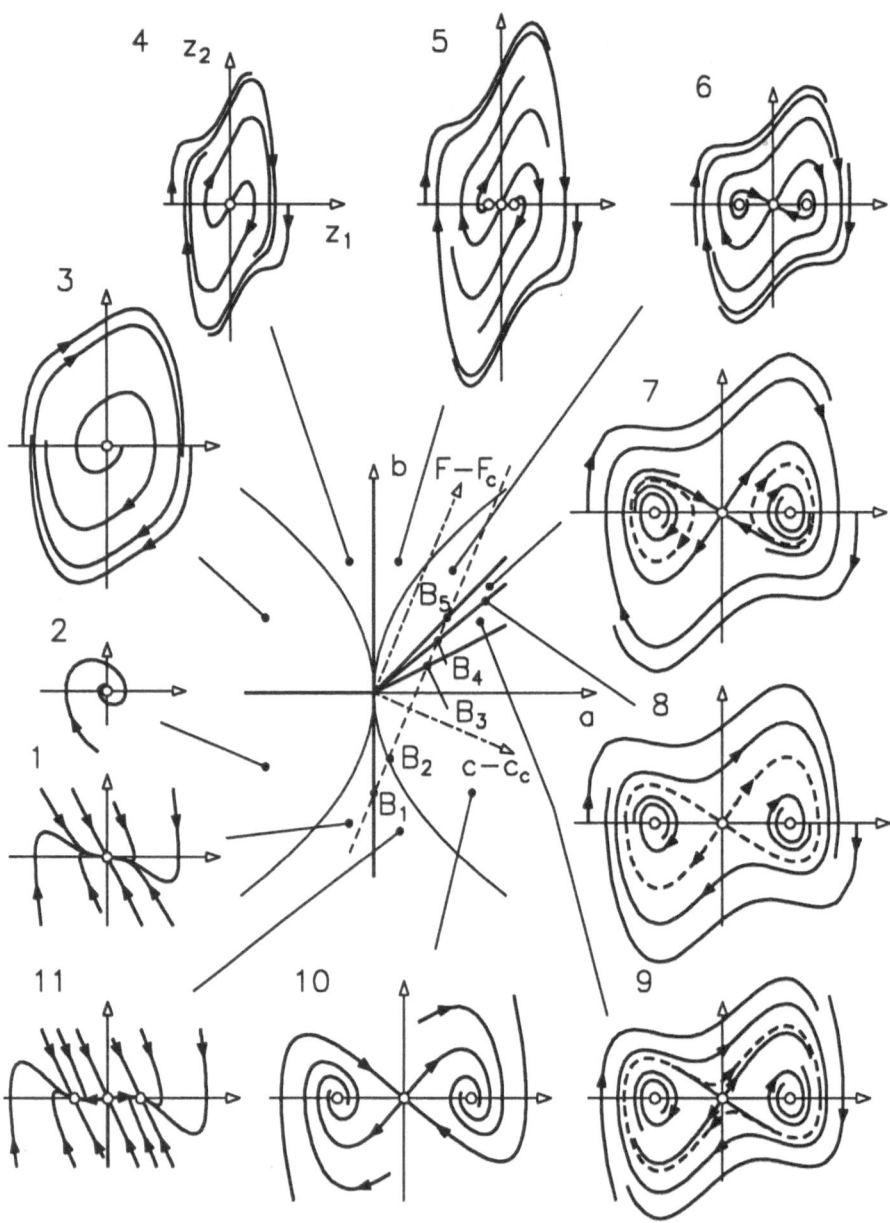

Figure 6.34. Bifurcation diagram in the a, b parameter plane and phase portraits of the double pendulum of Fig. 4.3 with large damping corresponding to (6.112) with $A_{30} < 0$ and $A_{21} < 0$. The broken line parallel to the F-axis represents a loading path

the influence of the second parameter b. The parameter b corresponds to the (negative) linear damping coefficient in the equation of a single degree of freedom oscillator

$$\ddot{z}_1 - az_1 - b\dot{z}_1 - A_{30}z_1^3 - A_{21}z_1^2\dot{z}_1 = 0 . \tag{6.205}$$

Equation (6.205) follows immediately from (6.112). Hence, in general the variation of b will influence the stability behavior of the equilibria. However, we shall see that the nonlinear terms still have an important influence. We study now the change of the phase plane diagrams under variation of the parameters. We rotate around the origin of the bifurcation diagram in the center of Fig. 6.34 in a clockwise direction.

In the quadrant $a < 0$, $b < 0$ there is only one equilibrium. It is asymptotically stable. This follows from (6.205) because an oscillator with a hard spring is given and both linear and nonlinear damping have positive coefficients. From the linear stability analysis of the equilibrium position we obtain the characteristic equation

$$s^2 - bs - a = 0$$

with the solutions

$$s_{1,2} = \frac{b}{2} \pm \sqrt{\frac{b^2}{4} + a} . \tag{6.206}$$

We obtain for $b^2/4 + a > 0$ real roots and for $b^2/4 + a < 0$ complex conjugate roots. Therefore, in the quadrant $a < 0$, $b < 0$ two domains are obtained. The equilibria are as follows

$$(1) \quad \frac{b^2}{4} + a > 0 \quad \text{a stable node}$$

$$(2) \quad \frac{b^2}{4} + a < 0 \quad \text{a stable focus.}$$

Keeping a fixed and changing b from negative to positive values, the stable focus turns locally unstable but the nonlinear damping term has a positive coefficient. We have already studied this case in (2.8). It gives rise to a stable limit cycle, whose amplitude can be calculated as is done in (2.16). The quadrant $a < 0$, $b > 0$ is again divided into two domains. The phase portraits are as follows

$$(3) \quad \frac{b^2}{4} + a < 0 \quad \text{an unstable focus and a stable limit cycle}$$

$$(4) \quad \frac{b^2}{4} + a > 0 \quad \text{an unstable node and a stable limit cycle.}$$

Now, we keep $b > 0$ fixed and vary a from a negative to a positive value. This results in the birth of two additional equilibrium positions z_{11}, z_{12}

(6.204) inside the limit cycle. In order to perform a linearized stability analysis of the equilibria z_{11} and z_{12} we insert $z_1 = \pm\sqrt{a/-A_{30}} + \xi$ into (6.205) and retain only linear terms in ξ to obtain

$$\ddot{\xi} - \left(b - \frac{A_{21}}{A_{30}}a\right)\dot{\xi} + 2a\xi = 0 . \tag{6.207}$$

The origin is unstable since in (6.206) $b > 0$. Since $a > 0$, the origin is a saddle because (6.206) has one positive and one negative root. Now, we keep a fixed and greater than zero, and decrease b. The roots of the characteristic equation of (6.207) are

$$s_{1,2} = \frac{1}{2}\left(b - \frac{A_{21}}{A_{30}}a\right) \pm \sqrt{\frac{1}{4}\left(b - \frac{A_{21}}{A_{30}}a\right)^2 - 2a} . \tag{6.208}$$

If $b - A_{21}a/A_{30} > 0$, the non-trivial equilibria are unstable and depending on the sign of the expression under the square root in (6.208) they are nodes for large values of b and, finally, unstable foci for smaller values of b. The results are given in plates (5) and (6).

A bifurcation from unstable to stable non-trivial equilibria occurs at $b - A_{21}a/A_{30} = 0$ and we enter domain (7). Locally about each non-trivial equilibrium an unstable limit cycle (dotted closed curves) is created. In addition one global limit cycle (full line) exists. Decreasing b further, the amplitude of the two unstable limit cycles increases until a homoclinic orbit as depicted in plate (8) is obtained. The analytical calculation of the corresponding parameter values is non-trivial. In order to determine these parameter values, one must make use of the Hamiltonian structure of (6.112) and evaluate elliptic integrals ([7] p. 303, [21]). In drawing Fig. 6.34 we have calculated boundary 8, that is, the parameter values which lead to the *homoclinic orbit* numerically. This can be done by solving a boundary value problem, requiring that the solution starting very close to the origin in the direction of the linearized unstable manifold ends up very close to the origin again in the direction of the linearized stable manifold ([136]). Moving into domain (9), only two limit cycles exist (the inner one given by the dotted closed curve is unstable and the outer one given by the full line closed curve is stable) which coalesce at the border between domains (9) and (10) and annihilate each other. The analytical calculation of this value is given in [21] and in [50] p. 375. Again we have calculated it numerically by locating a saddle node bifurcation point along the branch of periodic solutions. That is, one has to look for an eigenvalue $+1$ of the linearized Poincaré mapping. Interpretation of the remaining two plates (10) and (11) is obvious and does not require further comments.

An application of the bifurcation diagram of Fig. 6.34 can be given for the double pendulum of Fig. 4.3. For this purpose, in Fig. 6.34 the F-

and c-axis obtained from Fig. 4.4 are drawn. For a given value of $c \neq c_c$ we introduce a loading path in Fig. 6.34 parallel to the F-axis. Along this line, for increasing values of F, different types of behavior, which depend on the corresponding parameter values are obtained. That is, first a stable node is found which changes to a saddle and two adjacent stable nodes at the bifurcation point B_1. At B_2, the nodes change to foci. In both cases the double pendulum is in a buckled stable state. Increasing F further, nothing happens until we reach B_5 where the buckled equilibrium becomes unstable and a limit cycle oscillation is excited which winds around all three equilibria.

Clearly, local and global motions about the straight and deflected positions of the pendulum are possible in this bifurcation case.

6) One zero root and a purely imaginary pair (6.113)

This is the classical coupling case of flutter and divergence. For the pendulum of Fig. 4.3 this case can occur if the absolute value of damping is small.

The analysis of the steady state solutions of (6.113) is simplified by the fact that the third equation decouples from the other two, and hence, (6.113) reduces to a planar system. However, we will find below one case where this reduction of the dimension is not admissible.

Sometimes, it is convenient to introduce polar coordinates for the parameters as follows

$$a = \varepsilon \sin \alpha , \qquad b = \varepsilon \cos \alpha . \tag{6.209}$$

The advantage of (6.209) is that if we keep ε fixed and not equal to zero, only α must be varied to obtain the complete bifurcation behavior in the neighborhood of the critical parameter value $a = b = 0$. Further, we scale the variables by

$$r \to \sqrt{\varepsilon} r , \qquad z \to \sqrt{\varepsilon} z .$$

With these expressions the terms of up to and including third order in the first two equations of (6.113) are

$$\begin{aligned}
\dot{r} &= \varepsilon r(\sin \alpha + A_{30} r^2 + A_{12} z^2) = P(r, z) \\
\dot{z} &= \varepsilon z(\cos \alpha + A_{21} r^2 + A_{03} z^2) = Q(r, z) .
\end{aligned} \tag{6.210}$$

From the conditions given in Theorem 2.2, the complete stratification of the parameter plane can be determined.

First, we note that four different sets of equilibrium positions for (r, z) can be obtained from (6.210):

(1) $(0, 0)$ for all α

(2) $(0, \pm\sqrt{-\dfrac{\cos \alpha}{A_{03}}})$ for $-\dfrac{\cos \alpha}{A_{03}} > 0$

(3) $\left(\sqrt{-\dfrac{\sin \alpha}{A_{30}}}, 0\right)$ for $-\dfrac{\sin \alpha}{A_{30}} > 0$ (6.211)

(4) $\left(\sqrt{\dfrac{(A_{03} \sin \alpha - A_{12} \cos \alpha)}{(A_{12}A_{21} - A_{30}A_{03})}}, \ \pm\sqrt{\dfrac{(A_{30} \cos \alpha - A_{21} \sin \alpha)}{(A_{12}A_{21} - A_{30}A_{03})}}\right)$.

From (2.43) follows the determinant Δ

$$\Delta = (\sin \alpha + A_{12}z_0^2 + 3A_{30}r_0^2)(\cos \alpha + 3A_{03}z_0^2 + A_{21}r_0^2) -$$
$$- 4r_0^2 z_0^2 A_{12}A_{21} .$$
(6.212)

For case (1) of (6.211), we have $\Delta = \sin \alpha \cos \alpha$ and the structurally unstable cases are $\alpha = 0, \pi/2, \pi, 3\pi/2$. Hence, in each quadrant of the bifurcation diagram of Fig. 6.35 the origin is an equilibrium of a different stability type. Now we assume $A_{03} < 0$, $A_{30} > 0$, $A_{12} > 0$ and $A_{21} < 0$. Then two more bifurcation sets can be obtained. From $(6.211)_4$ we obtain the lines L_1 and L_2 from the requirement that the radicands must be nonnegative (see also p. 267). The line L_3, which corresponds to a homoclinic orbit, follows from inserting $(6.211)_4$ into (2.44). The stability of the singular points can be determined from a linear analysis making use of (2.43) and the trace σ. It is easy to check that for a stable singular point $\Delta > 0$ and $\sigma < 0$ must hold.

Now, the plates in Fig. 6.35 can be determined. In the plates, an equilibrium on the r-axis corresponds to an oscillation and an equilibrium on the z-axis corresponds to a statically buckled configuration.

For a quasistatically increasing load the physical behavior of the double pendulum of Fig. 4.3 can be best understood from the loading path drawn in the bifurcation diagram of Fig. 6.35. In domain (2) the straight pendulum position is stable (at least for small perturbations). At the bifurcation point B_1 entering into domain (3), a static bifurcation occurs to a stable buckled position. Increasing F further, point B_2 is reached where an oscillation about the stable buckled state sets in. For increasing values of F, line L_3 is reached. Line L_3 corresponds to plate (5), where a saddle point connection exists. Such a separatrix is also called a *heteroclinic orbit*. As mentioned in Theorem 2.2, a separatrix joining saddle points is a structurally unstable structure. Therefore, it is not admissible to study the behavior of the three-dimensional system (6.113) by a two-dimensional approximation for parameter values in the neighborhood of L_3. Here, the full three-dimensional system (6.113) must be considered.

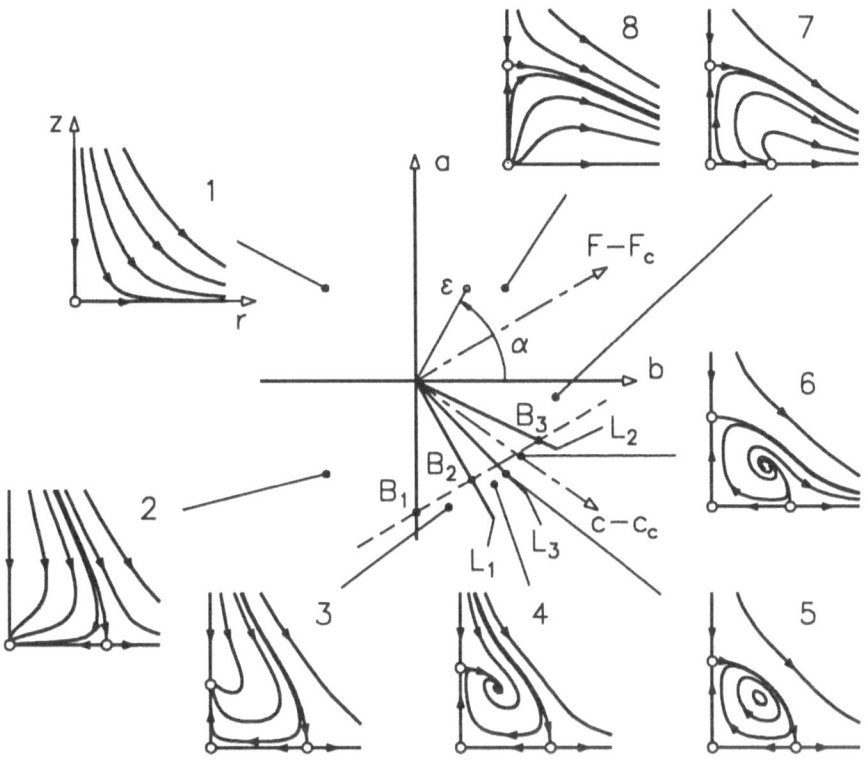

Figure 6.35. Bifurcation diagram in the a, b parameter plane and corresponding r, z plates of the double pendulum of Fig. 4.3 with small damping according to (6.113)

The reason is that under a perturbation (and the third equation can be interpreted as perturbation of the planar system) such a heteroclinic structure can take the form depicted in Fig. 6.36 leading to transversal heteroclinic points. However, transversal heteroclinic points are a strong indication that chaotic motions might occur for those parameter values and proper initial conditions. Using a numerical method, the existence of transverse heteroclinic points for the double pendulum has been shown in [126]. Thus, for increasing F after a sequence of three bifurcations chaotic motion can occur. In the literature this is sometimes called the Arnold-Ruelle-Takens scenario of transition to chaos in contrast to the period-doubling route to chaos found for the logistic map or the Duffing oscillator ([50]).

An experiment demonstrating this behavior is easy to perform using a fluid-carrying tube with an elastic support (Fig. 5.7). If we restrict ourselves to planar motions the system can be approximately modelled by one having the properties of the double pendulum of Fig. 4.3. Increasing

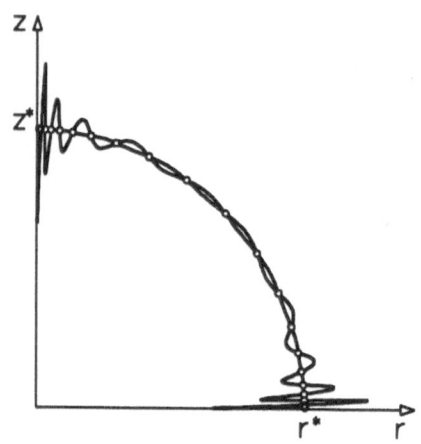

Figure 6.36. Transversal heteroclinic points following from a perturbation of the heteroclinic orbit of
plate 5 in Fig. 6.35, due to the three-
dimensional system (6.113)

the parameter F corresponds to an increase of the flow rate.

Another example is the tractor-semitrailer of Fig. 4.8 loaded such that
the center of gravity is at $d = d_c$ (Fig. 4.9), which, in fact, is the optimal
loading condition ([67]) concerning the linearized stability limit. However, for this problem no non-trivial stable states exist after a loss of
stability ([67]).

In [50] a complete classification of this type of coupling between divergence and flutter is given. It is shown that seven qualitatively different
cases exist.

7) Two purely imaginary pairs of eigenvalues (6.114)

Before we proceed as under number (6), we shortly explain how to calculate the mathematical unfolding parameters a, b from the physical parameters U and c_2 as they are shown in Fig. 4.7 ([66]). For this purpose
we introduce the two physical parameters λ_1 and λ_2 defined by

$$\lambda_1 = U - U_c , \qquad \lambda_2 = c_2 - c_{2c} .$$

Now we proceed as described in case (2) of the current Section for two
parameters λ_1 and λ_2. This results in a suspended system analogous to
(6.184) in the form

$$\begin{aligned} \dot{y}_c &= \ \mathbf{J}_c y_c + (\lambda_1 \mathbf{A}_1 + \lambda_2 \mathbf{A}_2) y_c + h.o.t. \\ \dot{\lambda} &= \ 0 \end{aligned} \qquad (6.213)$$

where $y_c \in \mathbb{R}^4$, $\lambda \in \mathbb{R}^2$ and $\mathbf{J}_c$ is given by (4.20).

We apply normal form theory to the six-dimensional suspended system (6.213) as we did in the case of the Hopf bifurcation in case (2) of

the current Section. The result of these calculations is

$$
\begin{aligned}
\dot{r}_1 &= (\alpha_{11}\lambda_1 + \alpha_{12}\lambda_2)r_1 + A_{30}r_1^3 + A_{12}r_1r_2^2 + O(|r_1|^5 + |r_2|^5) \\
\dot{r}_2 &= (\alpha_{21}\lambda_1 + \alpha_{22}\lambda_2)r_2 + A_{21}r_1^2r_2 + A_{03}r_2^3 + O(|r_1|^5 + |r_2|^5) \\
\dot{\varphi}_1 &= \omega_{1c} + \omega_{11}\lambda_1 + \omega_{12}\lambda_2 + O(|r_1|^2 + |r_2|^2) \\
\dot{\varphi}_2 &= \omega_{2c} + \omega_{21}\lambda_1 + \omega_{22}\lambda_2 + O(|r_1|^2 + |r_2|^2)
\end{aligned}
$$

where

$$
\begin{aligned}
\alpha_{11} &= \frac{\partial\sigma_1}{\partial\lambda_1}, & \omega_{11} &= \frac{\partial\omega_1}{\partial\lambda_1}, & \alpha_{12} &= \frac{\partial\sigma_1}{\partial\lambda_2}, & \omega_{12} &= \frac{\partial\omega_1}{\partial\lambda_2}, \\
\alpha_{21} &= \frac{\partial\sigma_2}{\partial\lambda_1}, & \omega_{21} &= \frac{\partial\omega_2}{\partial\lambda_1}, & \alpha_{22} &= \frac{\partial\sigma_2}{\partial\lambda_2}, & \omega_{22} &= \frac{\partial\omega_2}{\partial\lambda_2}.
\end{aligned}
$$

Here $\mu_1 = \sigma_1 \pm i\omega_1$ and $\mu_2 = \sigma_2 \pm i\omega_2$ are the two eigenvalues passing through the imaginary axis.

The final step to obtain (6.114) is to introduce by $(\varepsilon_1, \varepsilon_2)$ the mathematical parameter vector which is related to (λ_1, λ_2) by

$$
\begin{pmatrix} \varepsilon_1 \\ \varepsilon_2 \end{pmatrix} = \begin{pmatrix} \alpha_{11} & \alpha_{12} \\ \alpha_{21} & \alpha_{22} \end{pmatrix} \begin{pmatrix} \lambda_1 \\ \lambda_2 \end{pmatrix}
$$

and setting $a = \varepsilon_1$ and $b = \varepsilon_2$.

For the further treatment, we introduce by means of (6.209) the parameters ε, α instead of a, b. Also a scaling $r_1 \to \sqrt{\varepsilon}r_1$ and $r_2 \to \sqrt{\varepsilon}r_2$ of the variables is introduced. This results in amplitude equations up to terms of third order as follows

$$
\begin{aligned}
\dot{r}_1 &= \varepsilon r_1(\sin\alpha + A_{30}r_1^2 + A_{12}r_2^2) \\
\dot{r}_2 &= \varepsilon r_2(\cos\alpha + A_{21}r_1^2 + A_{03}r_2^2) \, .
\end{aligned}
\tag{6.214}
$$

Again, we find four different types of steady state (equilibrium) solutions (r_1, r_2) of (6.214) which have the same structure as those in (6.211). However, now only the positive values of the square roots are relevant. The bifurcation diagram of (6.214) is calculated for the flow-excited double pendulum (Fig. 4.6) of Section 4.1.3. In the plates a nontrivial equilibrium on the r_1 or r_2 axes denotes an oscillation with amplitude r_1 or r_2 and frequences ω_1 or ω_2, respectively. The simple non-trivial solutions (2) and (3) of (6.211) correspond to the domains (3) and (5) in the bifurcation diagram Fig. 6.37. If one enters domains (2) or (6) the stable oscillations that existed in domains (3) and (5) are still present, and in addition, saddle points on the other axis are created. Between the domains (2) and (6) is domain (1) which corresponds to solution (4) of

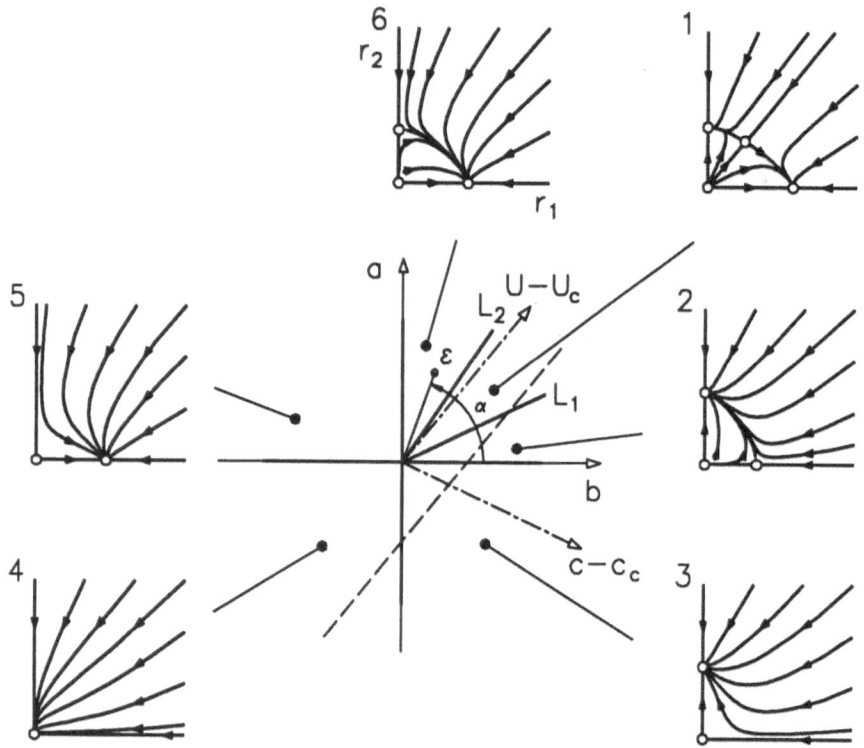

Figure 6.37. Bifurcation diagram in the a, b parameter plane with the corresponding r_1, r_2 plates according to (6.114) for the fluid excited double pendulum of Fig. 4.6

(6.211), that is to the sector where the expressions under the roots are nonnegative. To be more specific the border lines L_1 and L_2 are given by

$$L_1 : a = \frac{A_{12}}{A_{03}} b \qquad \text{and} \qquad L_2 : a = \frac{A_{30}}{A_{21}} b .$$

Mathematically, a coupling of the two simple one-frequency oscillations r_1, ω_1 and r_2, ω_2 occurs, and hence, we obtain a motion on a torus. However, in this problem this two frequency motion is unstable. Therefore, depending on the initial conditions, only one of the basic motions with a single frequency will exist. The stability of the steady state solutions can be checked by calculating the eigenvalues of the linearized system

$$\begin{pmatrix} \xi_1 \\ \xi_2 \end{pmatrix}^{\cdot} = \begin{pmatrix} a + 3A_{30}r_1^2 + A_{12}r_2^2 & 2A_{12}r_1r_2 \\ 2A_{21}r_1r_2 & b + 3A_{03}r_2^2 + A_{21}r_1^2 \end{pmatrix} \Bigg|_{\substack{r_1 = r_{10} \\ r_2 = r_{20}}} \begin{pmatrix} \xi_1 \\ \xi_2 \end{pmatrix}$$

which is obtained from inserting $r_1 = r_{10} + \xi_1$ and $r_2 = r_{20} + \xi_2$ into (6.214) and linearization of the resulting equation with respect to $\boldsymbol{\xi}$.

For a physical interpretation, we proceed as we did in the cases before. We start by drawing the axes system of the physical parameters U and c in the bifurcation diagram of Fig. 6.37. Next, we recall from Section 4.1.3 and Fig. 4.6 the physical meaning of the two parameters. U is the velocity of the fluid flow acting at m_1 which is responsible for a galloping instability. $c = c_2$ is the stiffness of the torsional spring located at the support of the pendulum. The system has two eigenmodes with amplitudes r_1 and r_2. They are depicted in Fig. 6.38. They show that

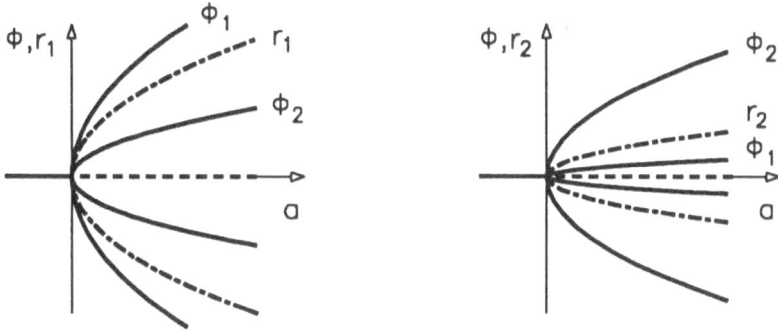

Figure 6.38. Eigenmodes ϕ_1 and ϕ_2 for (a) $c_2 < c_{2c}$ and (b) $c_2 > c_{2c}$ corresponding to domains (6) and (2) of Fig. 6.37

for a spring constant less than critical ($c_2 < c_{2c}$) the amplitude $\phi_1 > \phi_2$ and for a spring constant greater than critical ($c_2 > c_{2c}$) the amplitude $\phi_2 > \phi_1$. In these cases, both angles oscillate either with frequency ω_1 or ω_2, respectively. For loading paths as the one shown in Fig. 6.37 by the broken line, nothing special happens since the motion on the torus is unstable. Hence, only motions with one frequency are found for this example.

An application of this bifurcation case to a technical system is given in [66] where the loss of stability of the motion of a truck-trailer system is investigated.

In [50], it is shown that the complete classification of (6.114) contains twelve different cases.

6.6.3   Time-discrete dynamical systems

For time-discrete dynamical systems, exists a complete classification for codimension *one* ([7]). These are the three cases (6.121)–(6.123). We recall that (6.121)–(6.123) are the unfolded bifurcation equations which follow from application of center manifold reduction and normal form theory to the different types of loss of stabilty of the periodic motion of a robot. Hence, we shall discuss their physical significance by means of the robot example presented in Section 4.2.

1) $\mu = +1$: Saddle-node bifurcation (6.121)

The discussion of (6.121) is similar to that of (6.108) on p. 246. The details present no problems and will not be repeated here, because for the physical problem of the robot a different type of bifurcation called *transcritical bifurcation* applies. It is a bifurcation from the trivial solution as long as the perfect system is considered. Inclusion of imperfections leads to a higher degenerate (codimension *two*) case ([48]). The bifurcation equation for the perfect system is

$$u_{t+1} = (1+a)u_t + A_2 u_t^2 + O(|u_t|^3) . \tag{6.215}$$

The coefficient A_2 follows from the reduction process. The unfolding parameter a is given by

$$a = (\omega_0 - \omega_{0c})\frac{\partial \mu}{\partial \omega_0}\bigg|_{\omega_0 = \omega_{0c}} . \tag{6.216}$$

Here ω_0 is the parameter whose quasistatic increase leads to the loss of stability of the basic periodic motion at ω_{0c}.

The fixed points of the mapping (6.215) follow from inserting $u_{t+1} = u_t$ into (6.215) to

$$u_{t0} = 0 , \qquad u_{t0} = -\frac{a}{A_2} . \tag{6.217}$$

In Fig. 6.39, these solutions are shown for $A_2 < 0$. The stability of the solutions in Fig. 6.39 can be determined from a linearized stability ana-

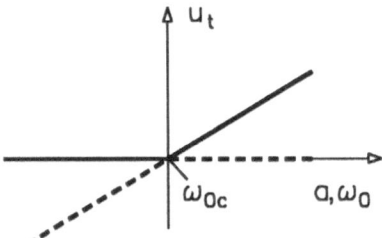

Figure 6.39. Transcritical bifurcation according to (6.215)

lysis. That is, one has to check whether the eigenvalues of the linearized map are smaller or larger than one. From (6.215) we obtain

$$f'(u_{t0}) = 1 + a + 2A_2 u_{t0} . \tag{6.218}$$

The eigenvalues of the two solutions (6.217) follow from (6.218) to

$$f'(0) = 1 + a , \qquad f'\left(-\frac{a}{A_2}\right) = 1 - a .$$

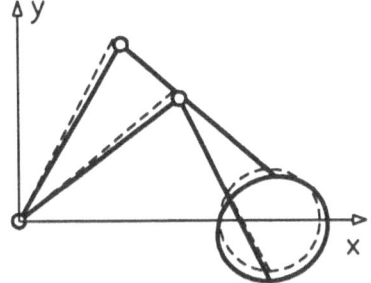

Figure 6.40. The full line gives the motion of the endpoint of the robot after a transcritical bifurcation in comparison to the motion before bifurcation given by the dotted line

Hence, the trivial solution $(6.217)_1$ is stable for $a < 0$ and unstable for $a > 0$. The bifurcated solution $(6.217)_2$ is stable for $a > 0$ and unstable for $a < 0$. To understand the behavior of the robot after loss of stability, the motion of the endpoint is depicted in Fig. 6.40. It can be seen that the bifurcated solution is still periodic moving along a closed curve which is similar to a circle but shifted away from the prescribed original circular path.

2) $\mu = -1$: Flip bifurcation (6.122)

The unfolding parameter a is as in (6.216). In Fig. 6.41, u_{t+1} is depicted as function of u_t for $A_3 > 0$. We observe that there are no non-trivial

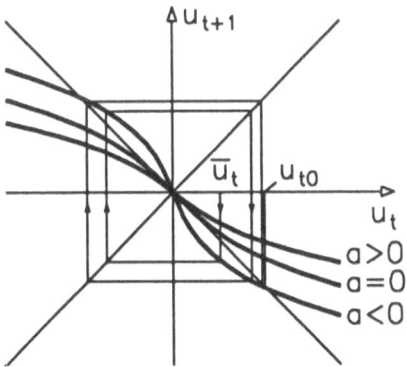

Figure 6.41. One-dimensional map (6.122) for three different values of a, possessing a two-periodic cycle u_{t0} for $a < 0$

fixed points bifurcating off $u_t = 0$ at $a = 0$ for $a \neq 0$ and a small (compare this figure with Fig. 2.4 which corresponds to the case $\mu = +1$). However, we notice that if for $a < 0$ an initial value $\bar{u}_t \neq 0$ is iterated, the orbit approaches a two-periodic cycle u_{t0}. Hence, it is natural to calculate the second iterate u_{t+2} to obtain

$$u_{t+2} = f(u_{t+1}) \;=\; (-1+a)u_{t+1} + A_3 u_{t+1}^3$$
$$= (1-2a)u_t - 2A_3 u_t^3 + O(|u_t|^5) \;. \tag{6.219}$$

Inserting $u_{t+2} = u_t$ into (6.219), the fixed points

$$u_{t0} = 0 \;, \qquad u_{t0} = \pm\sqrt{-\frac{a}{A_3}} \tag{6.220}$$

of the second iterate are obtained. That is, at $a = 0$ a bifurcation to a two-periodic motion occurs. In Fig. 6.42, the graph u_{t+2} is depicted

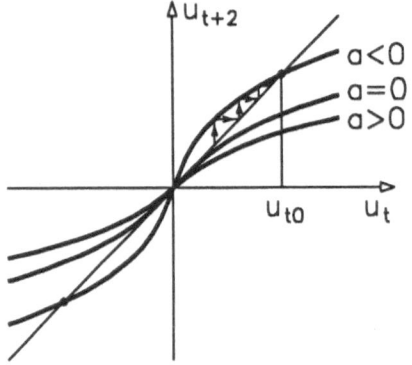

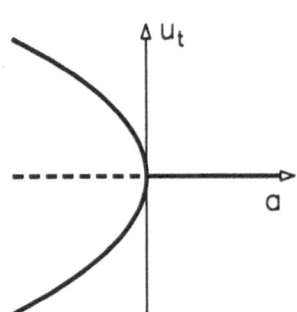

Figure 6.42. Second iterate u_{t+2} over u_t for the map (6.122)

Figure 6.43. Flip bifurcation at $a = 0$ to an orbit of period two, for a variation of a from positive to negative values

as function of u_t. For $a < 0$, two nontrivial fixed points exist which correspond to period-two points of the original mapping. The bifurcation graph is shown in Fig. 6.43. Contrary to Fig. 6.39, we vary now a from positive to negative values. This is because as usually we start from a locally stable trivial solution, for which the modulus of the eigenvalue of the linear map must be smaller than one. The result of decreasing a is that the eigenvalue crosses the unit circle (Fig. 4.15) from inside to outside. In the theory of nonlinear oscillations the *period doubling* bifurcation is called *subharmonic bifurcation*. This is well understood from Fig. 6.44. The stability of the solutions in Fig. 6.43 can be similarly determined as it was done for the transcritical bifurcation. Finally, the motion of the endpoint of the robot is shown in Fig. 6.45. It is of the same type as depicted in Fig. 6.44.

3) $\mu_{1,2} = \nu \pm i\eta$, $|\mu_{1,2}| = 1$: **Hopf bifurcation (6.123)**

In this case the bifurcation equations take their simplest form in complex variables $w_t = u_t + iv_t$. First, we introduce polar coordinates both for

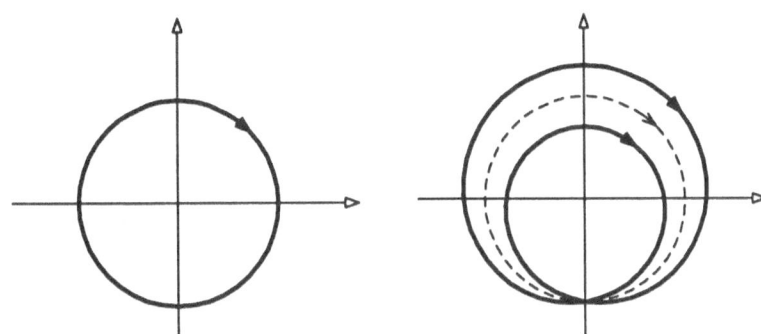

Figure 6.44. Transition from a periodic orbit to a two-periodic orbit due to a flip bifurcation

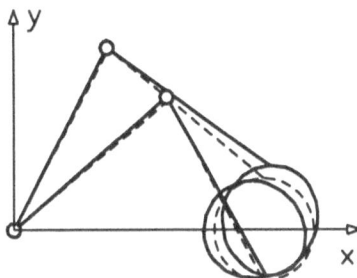

Figure 6.45. As in Fig. 6.40 but after a flip bifurcation

the variables and the coefficients

$$w_{t+1} = r_{t+1}e^{i\varphi_{t+1}}, \ w_t = r_t e^{i\varphi_t}, \ \mu = e^{i\alpha}, \ c = ae^{i\gamma}, \ C_3 = A_3 e^{i\beta}. \quad (6.221)$$

Inserting (6.221) into (6.123) yields

$$
\begin{aligned}
r_{t+1}e^{i\varphi_{t+1}} &= \left[(1 + ae^{i(\gamma-\alpha)})\, e^{i\alpha} + A_3 e^{i\beta} r_t^2\right] r_t e^{i\varphi_t} \\
&= \left[1 + ae^{i(\gamma-\alpha)} + A_3 e^{i(\beta-\alpha)} r_t^2\right] r_t e^{i(\varphi_t + \alpha)} \ .
\end{aligned}
\quad (6.222)
$$

Setting $\chi = \gamma - \alpha$ and $\psi = \beta - \alpha$ and taking the absolute value of each side of (6.222), we obtain

$$
\begin{aligned}
r_{t+1} &= \left|1 + a(\cos\chi + i\sin\chi) + A_3(\cos\psi + i\sin\psi)r_t^2\right| r_t \\
&= \sqrt{(1 + a\cos\chi + A_3\cos\psi\, r_t^2)^2 + (a\sin\chi + A_3\sin\psi\, r_t^2)^2}\, r_t \quad (6.223) \\
&= (1 + a\cos\chi + A_3\cos\psi\, r_t^2)r_t + O(|r_t|^4 + |ar_t^2| + |a|^2) \ .
\end{aligned}
$$

Here, the square root has been expanded into its power series. From (6.223) and setting $r_{t+1} = r_t$ to obtain the steady state solutions follows

$$(a\cos\chi + A_3\cos\psi\, r_t^2)r_t = 0$$

with the steady state amplitudes

$$r_{t0} = 0, \qquad r_{t0} = \sqrt{-\frac{a \cos \chi}{A_3 \cos \psi}} \qquad (6.224)$$

where

$$a = (\omega_0 - \omega_{0c})\frac{d}{d\omega_0}(\Re\mu)\bigg|_{\omega_0=\omega_{0c}} \qquad \text{and} \qquad \gamma = (\omega_0 - \omega_{0c})\frac{d}{d\omega_0}(\Im\mu)\bigg|_{\omega_0=\omega_{0c}}.$$

The nontrivial solution (6.224) represents the motion on a torus. As before the stability can be determined from a linearized analysis. If we have

$$f'(r_{t0}) = 1 + a \cos \chi + 3A_3 \cos \psi r_{t0}^2 < 1 \qquad (6.225)$$

then the solution is stable. Inserting (6.224) into (6.225) yields the stability behavior giving one or the other case shown in Fig. 6.46.

Finally, we describe the motion of the robot after the prescribed pe-

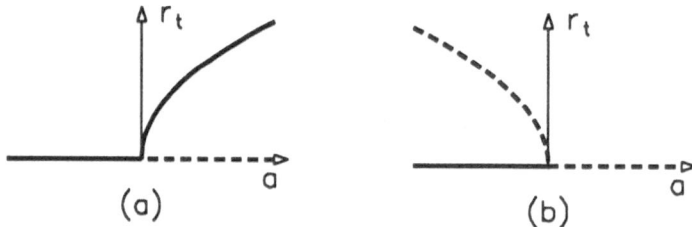

Figure 6.46. Hopf bifurcation of the mapping (6.123): (a) supercritical, (b) sub-critical

riodic motion loses stability by a Hopf bifurcation. For this case, the scenario of Fig. 4.14 applies. The original motion is given by a closed curve and corresponds to the fixed point $x_0 = 0$. After a supercritical Hopf bifurcation we obtain the motion on a torus which is represented by an asymptotically stable limit cycle in the Poincaré map. The cross-section of the torus is depicted in Fig. 6.47, where several iterated paths (transients) are marked. The motion of the endpoint of the robot is depicted in Fig. 6.48 and is shown to stay within two limiting circles. The motion on the torus can be either periodic or quasiperiodic depending whether the ratio of the frequency ω_0 of the basic periodic motion and the frequency η obtained from the Hopf bifurcation are rational or not. The orbit of the periodic motion is closed whereas the orbit of the quasi-periodic motion densely (uniformly) covers the whole torus ([5] p. 287).

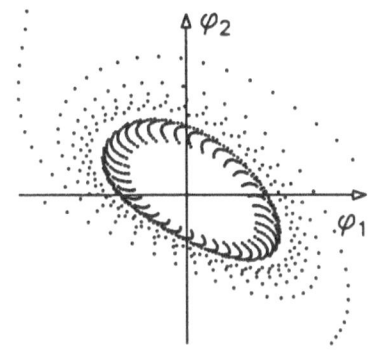

Figure 6.47. Poincaré section of the torus and nearby orbits after a Hopf bifurcation. The motion shown is quasiperiodic

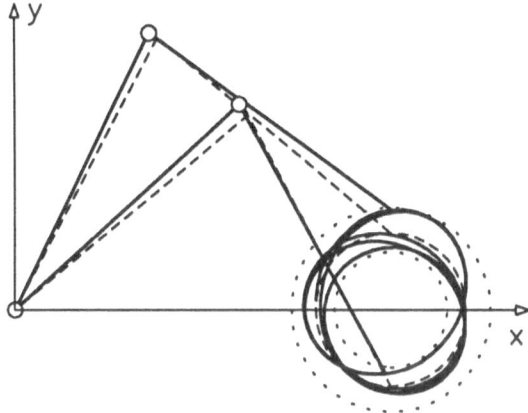

Figure 6.48. As in Fig. 6.40 but after a Hopf bifurcation. The motion is confined by two limiting circles

6.6.4 Symmetric dynamical systems

In this section, we discuss briefly the simple losses of stability of the trivial equilibrium position for the spherical double pendulum with the follower force loading described described in Section 5.1 and the fluid-carrying tube discussed in Section 5.3. Both systems are $O(2)$-symmetric, that is, they are equivariant both under a rotation and a reflection. Studying these systems will allow us to gain insight into the interesting phenomenon of symmetry breaking. Furthermore, we are able to treat both problems simultaneously because the structure of the bifurcation equations depends only on the matrix $\mathbf{J}_c$ given by (5.30)–(5.33). Partly they coincide for both cases. We will treat only the two simplest cases given by (5.30).

1) Two zero roots

In the corresponding bifurcation equations (5.40), the coefficient A_3 results from the center manifold reduction. The unfolding parameter a is given by

$$a = \frac{d\mu(\lambda)}{d\lambda}\bigg|_{\lambda=\lambda_c} (\lambda - \lambda_c) . \qquad (6.226)$$

In (6.226) is μ the eigenvalue and λ is either proportional to the follower force or the flow rate, respectively, for the two physical systems given by Fig. 5.2 or Fig. 5.7. Again we look for the steady state solutions of (5.40). There are three different cases:

(i)

$$v_1 = v_2 = 0 . \qquad (6.227)$$

This is the trivial solution representing the straight double pendulum or the straight downhanging tube.

(ii)

$$v_1 \neq 0, \ v_2 = 0 \quad \text{or} \quad v_1 = 0, \ v_2 \neq 0. \qquad (6.228)$$

Inserting (6.228) into (5.40) we find

$$av_1 + A_3 v_1^3 = 0 \quad \text{or} \quad av_2 + A_3 v_2^3 = 0$$

with the solutions

$$v_1 = 0, \ v_1 = \pm\sqrt{-\frac{a}{A_3}}; \quad v_2 = 0, \ v_2 = \pm\sqrt{-\frac{a}{A_3}} . \qquad (6.229)$$

(iii)

$$v_1 = v_2 = v . \qquad (6.230)$$

Inserting (6.230) into (5.40) we find

$$av + 2A_3 v^3 = 0$$

with the solution

$$v = 0 , \quad v = \pm\sqrt{-\frac{a}{2A_3}} . \qquad (6.231)$$

The nontrivial solutions (ii) and (iii) correspond to a statically buckled planar configuration of the system. First, by calculating the absolute value of w according to (5.34) we show that the solutions (ii) and (iii) are identical. Making use of (5.34) it follows for (ii)

$$|w|^2 = \frac{a}{A_3}$$

and for (iii)

$$|w|^2 = (v_1 + iv_2)(v_1 - iv_2) = \frac{a}{2A_3} + \frac{a}{2A_3} = \frac{a}{A_3} .$$

Since any point on the circle $v_1^2 + v_2^2 = a/A_3$ is a solution, the position of the plane which contains the planar buckled tube or the pendulum is not specified. Hence, we obtain a whole orbit of solutions. However, this orbit is not an attractor, since it consists of a family of steady state points. (See the comments in Section 2.3.1 on p. 39)

The stability of the solutions can be determined from a linear stability analysis. Inserting $v_1 = v_{10} + \xi_1$, $v_2 = v_{20} + \xi_2$ into (5.40) and performing the linearization about the equilibrium state v_{10}, v_{20} we obtain

$$\begin{pmatrix} \dot{\xi}_1 \\ \dot{\xi}_2 \end{pmatrix} = \begin{pmatrix} a + A_3(3v_{10}^2 + v_{20}^2) & 2A_3 v_{10} v_{20} \\ 2A_3 v_{10} v_{20} & a + A_3(v_{10}^2 + 3v_{20}^2) \end{pmatrix} \begin{pmatrix} \xi_1 \\ \xi_2 \end{pmatrix} .$$
$$(6.232)$$

Inserting (6.227), (6.229) or (6.231) into (6.232), we find that the trivial solution is stable for $a < 0$ and unstable for $a > 0$. For the nontrivial solutions (6.229) or (6.231) we find from (6.232) that one eigenvalue is always zero. This zero eigenvalue indicates that the plane in which the tube or the double pendulum buckles is not fixed. The stability of the solution in the plane depends only on the coefficient A_3. If $A_3 < 0$ the solution is supercritical and stable and if $A_3 > 0$, it is subcritical and unstable. We recall that the solutions are the amplitudes of the corresponding buckling modes. For the buckled tube with the elastic support acting at half span the mode shape is shown in Fig. 6.49.

For the tube problem A_3 was negative and therefore a supercritical pitchfork bifurcation (Fig. 6.15 plate 0) resulting in a stable planar buckled state is obtained. As mentioned the position of the plane in which the tube is buckled is not determined for the perfect system.

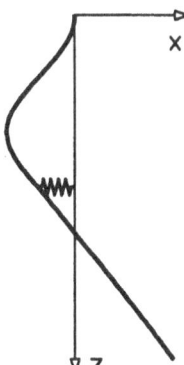

Figure 6.49. Planar eigenmode of the statically buckled tube corresponding to a stable post-bifurcation state

2) Two purely imaginary pairs

This is the case of Hopf bifurcation. Equations (5.40) describing the
case of two zero roots are already in normal form after center manifold
reduction and application of symmetry requirements because only one
nonlinear term remains in (5.38) in complex variables. However, equations (5.47) which describe the symmetric Hopf bifurcation must still be
transformed into normal form. We can do this either by applying the
transformation (6.2) and trying to eliminate all third order terms or by
making use of the averaging method as is explained in Section 6.1.1. Here
we use normal form theory. The calculations are similar to those in Section 6.1.1 and we only indicate which terms cannot be removed. Due to
the application of the transformation (5.46), the linear part in equations
(5.47) is in diagonal form, and hence, the application of normal form
theory is very easy. We use (5.47) with $a = 0$. The resonance condition
(6.11) (see also (6.16) for the ordinary Hopf bifurcation) is

$$\mu_i = i\omega(j - k + \ell - m) \tag{6.233}$$

where for μ_i either $i\omega$ or $-i\omega$ must be inserted. The four entries on
the right hand side follow from the fact that also the complex conjugate
equations to (5.47) must be included (see p. 186). From (6.233), we
obtain Table 6.1 for $(5.47)_1$. The exponents j, k, ℓ, m are to be selected
according to the exponents in $(5.47)_1$. From Table 6.1 follows that the

j	k	ℓ	m	$j - k + \ell - m$	comment
2	1	−	−	1	resonant
2	−	1	−	3	non-resonant
1	1	−	1	−1	non-resonant
1	−	1	1	1	resonant
−	1	−	2	−3	non-resonant
−	−	1	2	−1	non-resonant

Table 6.1. Resonance conditions for the symmetric Hopf bifurcation

terms with coefficients c_1 and c_4 cannot be removed from $(5.47)_1$. A
similar result holds for $(5.47)_2$. This results in the bifurcation equations

$$\dot{z}_1 = (a + i\omega)z_1 + c_1|z_1|^2 z_1 + c_4|z_2|^2 z_1$$
$$\dot{z}_2 = (a + i\omega)z_2 + c_1|z_2|^2 z_2 + c_4|z_1|^2 z_2 , \tag{6.234}$$

where only two complex coefficients c_1 and c_4 are present. Next, we
introduce polar coordinates

$$z_1 = r_1 e^{i\varphi_1} , \qquad z_2 = r_2 e^{i\varphi_2} \tag{6.235}$$

together with $c_1 = A_1 + iB_1$ and $c_4 = A_4 + iB_4$ into (6.234) to obtain

$$\begin{aligned}
\dot{r}_1 &= (a + A_1 r_1^2 + A_4 r_2^2)r_1 \\
\dot{r}_2 &= (a + A_1 r_2^2 + A_4 r_1^2)r_2 \\
\dot{\varphi}_1 &= \omega + B_1 r_1^2 + B_4 r_2^2 \\
\dot{\varphi}_2 &= \omega + B_1 r_2^2 + B_4 r_1^2 .
\end{aligned} \qquad (6.236)$$

We see that the amplitude equations decouple completely from the equations for the phases. Hence, the discussion of the steady state solutions of (6.236) can be performed by first solving the amplitude equations. There are three solutions:

(i)

$$r_1 = r_2 = 0 . \qquad (6.237)$$

This is the trivial state. For example, the downhanging tube.

(ii)

$$r_1 \neq 0, \; r_2 = 0 \quad \text{or} \quad r_1 = 0 \text{ and } r_2 \neq 0.$$

From $(6.236)_1$ the solution

$$r_1 = \sqrt{-\frac{a}{A_1}} \qquad (6.238)$$

follows for $r_2 = 0$. Substituting (6.238) into (6.235) yields

$$z_1 - r_1 e^{i\varphi_1} = r_1 e^{i(\omega t + \varphi_0)}, \; z_2 = 0 ,$$

where, by a suitable time shift we may set $\varphi_0 = 0$. Going one more step back, according to the transformation (5.46) we obtain

$$z_1 = w_1 + iw_2, \; 0 = \overline{w}_1 + i\overline{w}_2 \quad \text{and} \quad \overline{z}_1 = \overline{w}_1 - i\overline{w}_2, \; 0 = w_1 - iw_2.$$

The solution for w_1 and w_2 of this set of equations is

$$\begin{aligned}
w_1 &= \frac{1}{2}z_1 = \frac{1}{2}r_1 e^{i\omega t} = \frac{1}{2}r_1(\cos \omega t + i \sin \omega t)
\end{aligned}$$

$$ (6.239)$$

$$w_2 = \frac{1}{2i}z_1 = -\frac{i}{2}r_1 e^{i\omega t} = \frac{1}{2}r_1(\sin \omega t - i \cos \omega t) .$$

Inserting (6.239) into (5.41) yields the result for the amplitudes v_1, v_2, v_3, v_4 of the critical modes. However, w_1 and w_2 are functions of time which rotate with angular velocity $\omega + O(|r|^2)$ and constant amplitude. In physics, such a phenomenon is called a *travelling* or *rotating wave* (TW). Clearly, this means that the tube forms a three-dimensional space curve which rotates about the vertical axis with constant angular velocity, without changing its shape.

(iii)

$$r_1 = r_2 \,. \tag{6.240}$$

From $(6.236)_1$, we obtain

$$a r_1 + (A_1 + A_4) r_1^3 = 0$$

with the solutions

$$r_1 = 0 \,, \qquad r_1 = \sqrt{-\frac{a}{A_1 + A_4}} \,. \tag{6.241}$$

From $(6.236)_{3,4}$, we obtain with (6.240)

$$\dot{\varphi}_1 = \dot{\varphi}_2 \qquad \text{or} \qquad \varphi_1 - \varphi_2 = 2\varphi_0 = \text{const.} \tag{6.242}$$

Inserting (6.241) and (6.242) into (6.235), we find

$$z_1 = r_1 e^{i(\omega t + \varphi_0)}, \qquad z_2 = r_1 e^{i(\omega t - \varphi_0)}. \tag{6.243}$$

Substituting (6.243) into (5.46) yields

$$w_1 + i w_2 = r_1 e^{i(\omega t + \varphi_0)}, \qquad \overline{w}_1 + i \overline{w}_2 = r_1 e^{i(\omega t - \varphi_0)}$$

or

$$w_1 = r_1 e^{i\varphi_0} \cos \omega t \qquad \text{and} \qquad w_2 = r_1 e^{i\varphi_0} \sin \omega t \,. \tag{6.244}$$

In order to discuss (6.244), we first assume $\varphi_0 = 0$. This assumption in combination with (5.41) leads to

$$v_1 = r_1 \cos \omega t, \quad v_2 = 0, \quad v_3 = r_1 \sin \omega t, \quad v_4 = 0. \tag{6.245}$$

Solution (6.245) is an oscillation in the plane $\varphi_0 = 0$. This plane is spanned by the z-axis and the x-axis. If $\varphi_0 \neq 0$, the plane in which the oscillation occurs does not coincide with the coordinate plane x, z but is rotated by the angle φ_0 about the z-axis. In physics such a solution is called a *standing wave* (SW). As in the case of the statically buckled tube, the angle φ_0 remains unspecified. Hence, a whole orbit of solutions is obtained under variation of φ_0. In Fig. 6.50a and b the eigenmodes of the rotating wave and the standing wave, respectively, are depicted. The dynamics follows from v_1 to v_4 according to (5.41) and (6.239) for the rotating wave and (6.245) for the standing wave. The resulting solution is then given by (5.97).

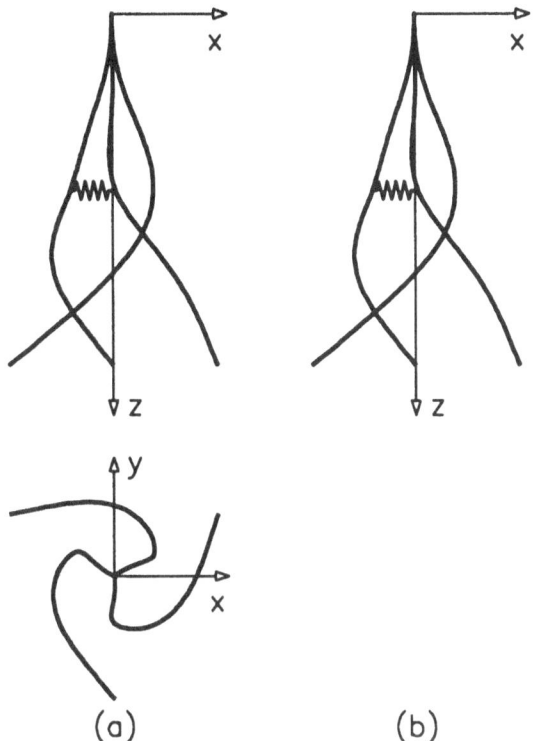

Figure 6.50. Eigenmodes (a) of the travelling or rotating (TW) and (b) of the standing (SW) wave solutions for the symmetric Hopf bifurcation of the fluid-conveying tube

Finally, we treat the question of the stability of these solutions. We proceed as in the case of the divergence bifurcation and obtain from $(6.236)_{1,2}$ equations analogously to (6.232) as follows

$$\begin{pmatrix} \dot{\xi}_1 \\ \dot{\xi}_2 \end{pmatrix} = \begin{pmatrix} a + 3A_1 r_{10}^2 + A_4 r_{20}^2 & 2A_4 r_{10} r_{20} \\ 2A_4 r_{10} r_{20} & a + 3A_1 r_{20}^2 + A_4 r_{10}^2 \end{pmatrix} \begin{pmatrix} \xi_1 \\ \xi_2 \end{pmatrix}. \tag{6.246}$$

The trivial solution $r_{10} = r_{20} = 0$ is stable for $a < 0$. The travelling wave solution (6.238) inserted into (6.246) yields the matrix

$$\begin{pmatrix} -2a & 0 \\ 0 & a\left(1 - \frac{A_4}{A_1}\right) \end{pmatrix}. \tag{6.247}$$

The eigenvalues are $-2a$ and $(1 - A_4/A_1)a$. Therefore, the travelling wave is stable if $a > 0$, $A_1 < 0$ and $A_1 - A_4 > 0$. Similarly, for the standing wave solution (6.241) we obtain from (6.246) the eigenvalues $A_1 + A_4$ and $A_1 - A_4$. Both eigenvalues must be negative for a stable solution.

Following [48], we can draw the bifurcation diagram of Fig. 6.51. From Fig. 6.51, it can be seen that only in two domains stable solutions exist. In [11], the dependence of the coefficients A_i of the nonlinear terms on

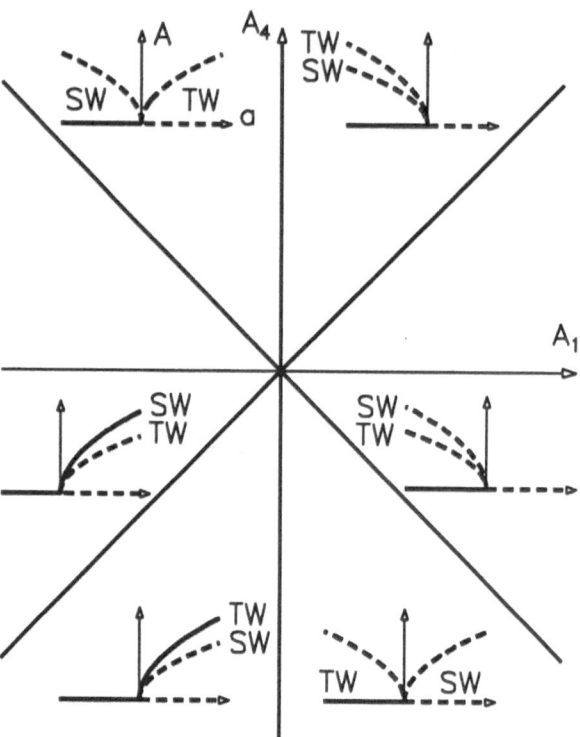

Figure 6.51. Bifurcation diagram for the symmetric Hopf bifurcation in the A_1, A_4 plane which is stratified into six qualitatively different domains. Only in two domains stable solutions (full line) exist

the mass ratio β, according to (5.72) is calculated. It turns out that the size of β determines whether a stable travelling wave or a stable standing wave will be obtained. For the tube the parameter a is given by

$$a = \left.\frac{d\Re\mu(\varrho)}{d\varrho}\right|_{\varrho=\varrho_c} (\varrho - \varrho_c)$$

where $\Re\mu$ is the real part of the critical eigenvalue μ, and ϱ is given by (5.72) and is proportional to the flow rate U.

The travelling and the standing wave solutions are nice examples for the statement on p. 150 that after a *symmetry breaking bifurcation* the bifurcated solutions still have retained some symmetry of the original problem but are less symmetric than the equations. This is easy to understand because the travelling wave is still invariant under a rotation about the z-axis and a corresponding time shift but not under a reflection. The standing wave is invariant only under a reflection about the plane in which it buckles but not under a rotation. If we rotate the plane, we obtain, in general, a different solution (see p. 148). So this is a nice example which shows that for a quite complicated dynamical system after bifurcation not an irregular motion sets in but still a (less) symmetric solution is found.

6.6.5   Symmetric statical systems

In this section, we treat the bifurcation equations (5.67) derived for the buckling problem of the annular plate in Section 5.2. These equations are formally equivalent to (4.173) derived for the rectangular plate. Hence, we are able to make use of all results already obtained in connection with the rectangular plate in Section 6.6.1. The further reduction of the coefficients and the restricted generic unfolding of (5.67) can be given by the transformation (4.187) leading to the form of (6.143)

$$q_0^3 + \mu_1 q_0 q_1^2 - \lambda q_1 + \gamma_0 = 0$$
$$q_1^3 + \mu_2 q_0^2 q_1 - (\lambda + \sigma)q_1 + \gamma_1 = 0 .$$

The modal parameters μ_1 and μ_2 depend on the shape of the plate, the boundary conditions and the type of the loading. The load parameter λ measures the deviation of the thrust p from the critical value p_c and can be analogously derived as in (6.170). The geometry parameter $\sigma \sim b - b_c$ measures the deviation of the diameter b of the hole from the critical value b_c where coincident eigenvalues occur. Finally, γ_0 and γ_1 describe the contribution of a transversal loading to the rotationally symmetric and the lowest order nonsymmetric buckling mode, given by (5.58) for $n = 0$ and $n = 1$, respectively. They can be calculated analogously as in (6.168). Since the evaluation of these parameters has already been explained in detail for the rectangular plate it need not be repeated here.

   In connection with this example we, first, want to explain an important feature of bifurcation problems in the neighborhood of multiple eigenvalues. Suppose the radius b of the annulus to be slightly smaller than the critical value b_c. (We recall that for b_c a three-fold eigenvalue occurs, to which equations (5.67) correspond). In this case, the perfect plate will buckle in the rotationally symmetric mode and for the boundary conditions chosen in Section 5.2, we obtain the bifurcation graph of Fig. 6.52a. This graph shows that for increasing values of λ the plate stays in the symmetric mode with increasing amplitude q_0. Hence, for a perfect plate, even if σ is small, the presence of the second eigenvalue does not have any influence on the buckling behavior.

   Now we consider the presence of an imperfection γ_1 that gives a contribution only to the second buckling mode. If γ_1 is very small the behavior is described by Fig. 6.52b. The figure shows that after a small predeformation in the second mode and after crossing the bifurcation point the behavior is still determined by the first buckling mode. As depicted in Fig. 6.52c, this is still true for increasing the value of γ_1, but now a jump

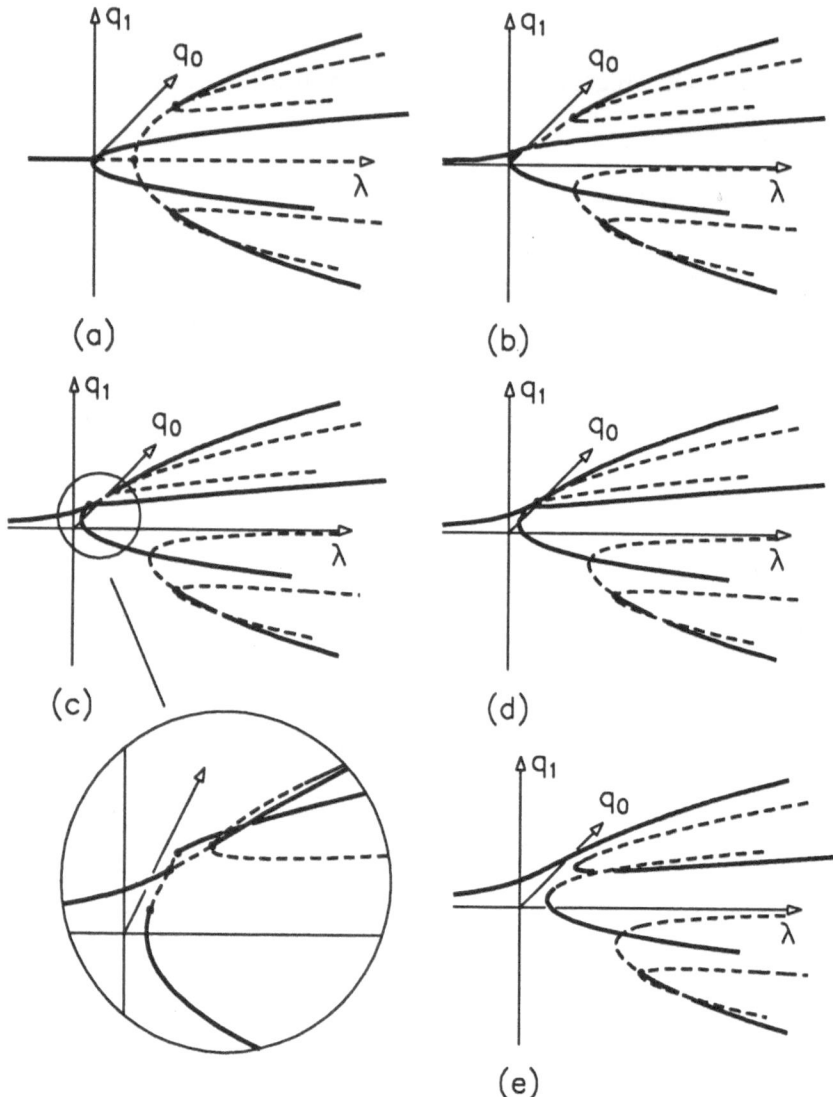

Figure 6.52. Different bifurcation solutions for the annular plate for increasing values of the imperfection γ_1

to the first mode occurs. If the imperfection γ_1 reaches the critical value γ_{1c}, the bifurcation graph of Fig. 6.52d is obtained which shows that the two stable branches of the second mode are joined. This means that, although, the geometry (size of the hole) would require a buckling in the symmetric mode above this value of γ_{1c} the size of the non-symmetric imperfection forces the plate to stay in the non-symmetric mode. This, for example, could never occur for buckling of a simple rod. Finally in

Fig. 6.52e the situation for $\gamma_1 > \gamma_{1c}$ is depicted.

The conclusion to be drawn from this example is that when dealing with buckling problems in the neighborhood of multiple eigenvalues the most complicated situation must always be analysed even if an apparently much simpler cases might apply because the occurrence of unexpected imperfections can change the behavior of the system qualitatively.

Another interesting property of a symmetric system can be explained from Fig. 6.52e ([47]). If we ignore the q_0-coordinate in Fig. 6.52e, we end up with Fig. 6.53a which resembles Fig. 6.15 plate 1 but has different sta-

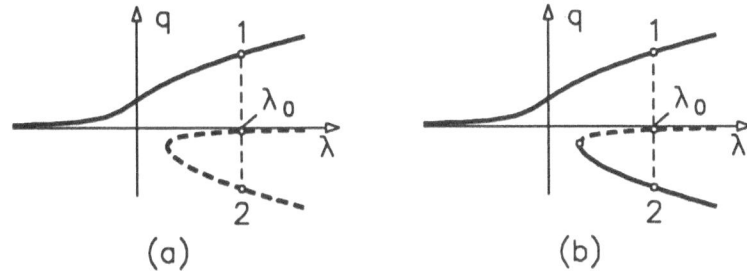

Figure 6.53. The solutions of an imperfect pitchfork bifurcation have different stability properties if the system is rotationally symmetric (a) compared to a nonsymmetric case (b)

bility properties. Let us try to understand this difference by a comparison with the behavior of the rectangular plate which corresponds to Fig. 6.15 plate 1. For convenience, we have depicted it again in Fig. 6.53b. We consider now buckling at a simple (for the rotational symmetric system yet double) eigenvalue and assume that $b > b_c$, that is, the annular plate buckles in the asymmetric mode.

First, we explain the behavior of the rectangular plate depicted in Fig. 4.16 in relation to Fig. 6.53b. Fig. 6.53b shows that if the thrust λ is large enough (for example $\lambda = \lambda_0$) two stable equilibria exist. Of course, in a loading process due to a geometric imperfection with a preference for the second mode, we would end up in state 1. However, if we apply now a load transversal to the plate, a loading path as shown in Fig. 6.7 would be obtained and the plate moves into the stable state q_5 in Fig. 6.7 by a jump. If we remove the transversal load, a state close to q_3 in Fig. 6.7 is reached. This is the stable state 2 in Fig. 6.53 and the deflection due to the plate is opposite to the deflection due to the imperfection.

Now we consider the annular plate with a geometric imperfection that forces the plate into the first non-symmetric mode. Thus for $\lambda = \lambda_0$ we reach state 1 in Fig. 6.53a as it occurred before for the rectangular plate. If we now apply a transversal load the plate will move into a deflected state 2 opposite to the initial deflection of the imperfection just as it was described before. However, if we remove the transversal load, the plate does not remain in a nearby configuration because this state is unstable, since for the annular plate there is the additional possibility to move away from it in the circumferential direction.

To make this important phenomenon better understandable we reconsider the simple buckling problem of a rod (Fig. 2.7), however, now for a rotationally symmetric cross-section. We compare two models. First we allow three-dimensional deformations and second we allow only the usual planar deformation. In the first case we obtain the bifurcation solutions of Fig. 6.53a and in the second case those of Fig. 6.53b. This is easy to understand because in the first case the geometrically imperfect buckled rod will under a transversal loading snap from position 1 into position 2 but after removing the transversal load will move back into position 1 on a three-dimensional path which is not possible for the planar buckling problem.

The explanation for these ways of behavior of the three-dimensional rotationally symmetric rod and the annular plate is that the planar rod and the rectangular plate have in state 1 a unique minimum of the corresponding potential whereas for the perfect three-dimensional symmetric rod and the perfect annular plate any deflected position on the circle is an equilibrium position and only by an imperfection a special isolated stable buckled position is specified.

Appendix A

Linear spaces and linear operators

A.1   Linear spaces

In this section we follow closely [80].

Definition A.1 (Metric space) *By a metric space we mean the pair* (M, ϱ), *consisting of a space* M *and a distance* $\varrho : M \times M \to R$, *which is a single-valued, non-negative, real function defined for all* $\varphi, \psi \in M$ *with the following three properties:*

(1) $\varrho(\varphi, \psi) \geq 0$, $\varrho(\varphi, \psi) = 0$ *if and only if* $\varphi = \psi$.

(2) *Symmetry:* $\varrho(\varphi, \psi) = \varrho(\psi, \varphi)$, *for all* $\varphi, \psi \in M$.   (A.1)

(3) *Triangle inequality:* $\varrho(\varphi, \psi) \leq \varrho(\varphi, \chi) + \varrho(\chi, \psi)$.

Here it is assumed that M is a *linear space*, that is, the eight axioms of a vector space must be satisfied ([80] p. 118).

Definition A.2 (Normed space) *A functional* $p(\psi) = \|\psi\|$ *defined on a linear space* M *is said to be a* norm *if it has the following properties:*

(1) $\|\psi\| \geq 0$ *for all* $\psi \in M$,
       where $\|\psi\| = 0$ *if and only if* $\psi = \mathbf{0}$.

(2) $\|\lambda \psi\| = |\lambda| \, \|\psi\|$ *for all* $\psi \in M$ *and all* $\lambda \in \mathbb{R}$.

(3) *Triangle inequality:* $\|\psi + \varphi\| \leq \|\psi\| + \|\varphi\|$ *for all* $\psi, \varphi \in M$.

Every normed linear space M becomes a metric space if we set

$$\varrho(\varphi, \psi) = \|\varphi - \psi\| \ . \tag{A.2}$$

Examples of normed linear spaces are:

Example 1: The real line R, with $\|x\| = |x|$.

Example 2: The real n-dimensional space with

$$\|x\|_1 = \sum_{i=1}^{n} |x_i|$$

$$\|x\|_2 = \left(\sum_{i=1}^{n} x_i^2\right)^{1/2} \tag{A.3}$$

$$\|x\|_\infty = \max_{1\le i\le n}\{|x_i|\} .$$

In two dimensions the three cases (A.3) can be given a simple geometric meaning (Fig. A.1).

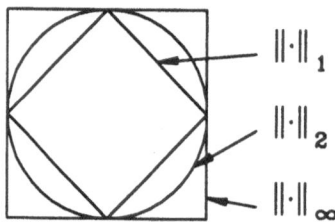

$\|\cdot\|_1$

$\|\cdot\|_2$

$\|\cdot\|_\infty$

Figure A.1. The norms $\|\ \|_1$, $\|\ \|_2$, $\|\ \|_\infty$ of (A.3) in $\mathbb{R}^2$

Example 3: The space $C_\infty[a,b]$ of all functions f continuous on the interval $[a,b]$ with the norm

$$\|f\|_\infty = \max_{a\le x\le b} |f(x)| . \tag{A.4}$$

Definition A.3 (Banach space) *A complete normed linear space with metric (A.2) is called a Banach space B.*

A space B is called *complete* if each *Cauchy sequence* in B converges to an element in B. A sequence $\varphi_i \in B$ is called a Cauchy sequence if for any $\varepsilon > 0$ there exists an N such that $\varrho(\varphi_i, \varphi_j) < \varepsilon$ for $i,j > N$. If for any $\varepsilon > 0$ there exists an N such that $\varrho(\varphi_i, \varphi) < \varepsilon$ for $i > N$, the sequence φ_i converges to the element φ. Each finite-dimensional normed space is complete, and therefore, a Banach space.

Definition A.4 (Scalar product) *By a scalar product in a real linear space we mean a function defined for every pair of elements $\varphi, \psi \in B$. We denote this product by (φ, ψ). It has the following properties:*

$$
\begin{aligned}
&(1) &&(\varphi,\varphi) \ge 0,\ (\varphi,\varphi)=0 \quad \text{if and only if}\quad \varphi = 0\\
&(2) &&(\varphi,\psi) = (\psi,\varphi)\\
&(3) &&(\lambda\varphi,\psi) = \lambda(\varphi,\psi)\\
&(4) &&(\varphi,\psi+\chi) = (\varphi,\psi)+(\varphi,\chi)
\end{aligned}
\tag{A.5}
$$

where λ is a real number.

Definition A.5 (Euclidean space) *A linear space B together with a scalar product is called Euclidean space E.*

Any two elements φ, ψ of an Euclidean space satisfy the Schwarz inequality

$$|(\varphi, \psi)| \leq \|\varphi\|_2 \|\psi\|_2 \tag{A.6}$$

where

$$\|\varphi\|_2 = \sqrt{(\varphi, \varphi)}, \qquad \|\psi\|_2 = \sqrt{(\psi, \psi)} . \tag{A.7}$$

For a simple proof of (A.6) see ([80] p. 142).

An Euclidean space becomes a *normed linear space* when it is equipped with the norm (A.7) ([80] p. 142).

For two elements $\varphi, \psi \in E$, we define

$$\cos \vartheta = \frac{(\varphi, \psi)}{\|\varphi\|_2 \|\psi\|_2} \qquad 0 \leq \vartheta \leq \pi , \tag{A.8}$$

where ϑ denotes the angle between φ and ψ. From (A.6) follows that the right-hand side of (A.8) cannot be greater than one. Hence, (A.8) determines a unique angle. If $(\varphi, \psi) = 0$, (A.8) implies that $\vartheta = \pi/2$, and the two elements are said to be *orthogonal*.

A system of non-zero elements $\{\phi_i\}$ in E is said to be orthogonal if

$$(\phi_i, \phi_j) = 0 , \qquad i \neq j . \tag{A.9}$$

If $\{\phi_i\}$ is an *orthogonal* system, then

$$\{\gamma_i\} = \left\{ \frac{\phi_i}{\|\phi_i\|_2} \right\} \tag{A.10}$$

is an *orthonormal* system.

An orthogonal system of elements $\{\phi_i\}$ is called *complete* if it is impossible to add even one element to it, not identically equal to zero, which is orthogonal to all $\{\phi_i\}$.

Example 4: The real n-space $\mathbb{R}^n$, that is, the set of ordered n-tuples $\boldsymbol{x} = (x_1, \ldots, x_n)^T$, $\boldsymbol{y} = (y_1, \ldots, y_n)^T$ with $\boldsymbol{x} + \boldsymbol{y} = \boldsymbol{z}$, $\boldsymbol{z} = (x_1 + y_1, \ldots, x_n + y_n)^T$ and $\alpha \boldsymbol{x} = (\alpha x_1, \ldots, \alpha x_n)^T$. The scalar product, the norm and the distance are defined by

$$(\boldsymbol{x}, \boldsymbol{y}) = \sum_{i=1}^{n} x_i y_i , \qquad \|\boldsymbol{x}\|_2 = \left(\sum_{i=1}^{n} x_i^2 \right)^{1/2} ,$$

$$\varrho(\boldsymbol{x}, \boldsymbol{y}) = \|\boldsymbol{x} - \boldsymbol{y}\|_2 = \left(\sum_{i=1}^{n} (x_i - y_i)^2 \right)^{1/2} . \tag{A.11}$$

A possible basis consists of the unit vectors

$$\boldsymbol{e}_i = (0 \ldots 0, 1, 0 \ldots 0) , \qquad i = 1, \ldots, n \tag{A.12}$$

with a 1 at the i-th spot.

Example 5: The space $L_2[a, b]$ of all square integrable functions on $[a, b]$. We define the space L_p by

$$L_p(B) := \left\{ f : B \to \mathbb{R} : \int_B |f(x)|^p dx < \infty \right\} .$$

The scalar product and the norm are given by, respectively

$$(f, g) = \int_a^b f(t)g(t)dt , \qquad \|f\|_2 = \left(\int_a^b f^2(x)dx \right)^{1/2} . \qquad (A.13)$$

where

$$\|f\|_p = \left(\int_B |f(x)|^p dx \right)^{1/p} , \qquad p \geq 1 .$$

Among the various orthogonal bases in $L_2[a, b]$ is one of the most important the system of trigonometric functions

$$\frac{1}{\sqrt{2\pi}} , \qquad \frac{1}{\sqrt{\pi}} \cos nx , \qquad \frac{1}{\sqrt{\pi}} \sin nx \qquad (n = 1, 2, \ldots) \qquad (A.14)$$

where $[a, b] = [-\pi, \pi]$. It can be shown that (A.14) is orthonormal and complete.

Under certain mild restrictions it is possible to expand any function $f(x)$ as a series in the system of complete orthonormal functions $\{\gamma_i\}$

$$f(x) = \sum_{i=1}^{\infty} a_i \gamma_i(x) , \qquad (A.15)$$

where the coefficients a_j of this expansion can be calculated as

$$a_j = (f, \gamma_j) = \int_a^b f(x)\gamma_j(x)dx . \qquad (A.16)$$

Substituting (A.15) into (A.13), we find

$$(f, f) = \|f\|_2^2 = \sum_{i=1}^{\infty} a_i^2 . \qquad (A.17)$$

Geometrically, (A.17) means that the square of the norm of the function is equal to the sum of the squares of its projection onto a complete system of mutually orthogonal directions. An orthonormal system $\{\gamma_i\}$ for which (A.17) holds is called *closed.*

Definition A.6 (Hilbert space) *If an Euclidean space E is complete, separable and infinite-dimensional it is called Hilbert space H.*

A metric space is called *separable* if it has a countable everywhere dense subset. For example, the set of rational points is dense on the real line $\mathbb{R}^1$. Also the set of all polynomials with rational coefficients is dense in the space $C[a, b]$, the space of continuous functions on the interval $[a, b]$. (If in addition we assume that the functions are k-times continuously differentiable we denote the space by C^k.) If we take the norm (A.4), the function space C is denoted by C_∞ and if the norm (A.13) is taken, it is denoted by C_2. Another function space with (A.13) but not necessarily continuous functions is $L_2[a, b]$ ([80] p. 383).

However, only the space $L_2[a, b]$ is a Hilbert space. This is, because $C_\infty[a, b]$ with norm (A.4) is complete but (A.4) cannot be derived from an inner product and $C_2[a, b]$ with norm (A.13) is not complete as the following simple example shows:

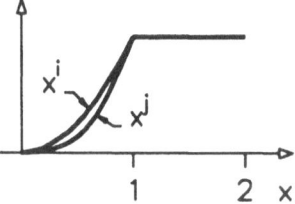

Figure A.2. Sequence of continuous functions on [0, 2] with discontinuous limit

Example 6: We define a sequence of functions on the interval $[0, 2]$ as shown in Fig. A.2 by

$$f_i(x) = \begin{cases} x^i & \text{on} & [0, 1] \\ 1 & \text{on} & (1, 2] \end{cases} .$$

This sequence is certainly a Cauchy sequence, because we have (the integral on $(1, 2]$ is zero)

$$(f_i, f_j) = \|x^i - x^j\|^2 = \int_0^1 (x^i - x^j)^2 dx$$

$$= \frac{1}{2i + 1} - \frac{2}{i + j + 1} + \frac{1}{2j + 1} < \varepsilon^2$$

for any ε, provided i, j are large enough. However, for $i \to \infty$ the limiting function is not continuous, and therefore, not in $C_2[a, b]$, but in $L_2[a, b]$. Hence, using the norm (A.13), the square integrable functions are more appropriate than the continuous functions: $L_2[a, b] \supseteq C[a, b]$.

An important question is, which numbers a_i in (A.15) can be the co-efficients of the expansion of a function in a Hilbert space. Referring to (A.17), at least the necessary condition follows, that the series must converge. It also turns out that this condition is sufficient. That is, a sequence of numbers a_i is the sequence of coefficients of the expansion by an orthogonal system of functions in a Hilbert space if and only if (A.17) converges. This fundamental theorem holds if the Hilbert space is interpreted as the collection of all square integrable functions in the sense of *Lebesque*. If one confines oneself to a space of continuous functions, then the solution of the problem, for which the numbers a_i can be the coefficients of an expansion, would become unnecessarily complicated.

We also remark that it is important to distinguish between the concept of a complete set of functions (A.14) and the concept of a complete space. We recall, for example, that (A.14) is complete in $C_2[a, b]$. However, the space $C_2[a, b]$ is not complete, whereas the space $C_\infty[a, b]$ according to (A.4) is complete but not a Hilbert space.

In conclusion, we can, roughly speaking, state that a Hilbert space H means a vector space with a scalar product. It also can be extended to complex elements. In our applications we assume that H is a set of functions satisfying the boundary conditions and possessing enough smoothness (that is, the functions are in C^k) such that the necessary spatial derivatives of the evolution equations can be performed.

A.2 Linear operators

Let X and Y be Banach spaces. The mapping $\mathbf{A} : X \to Y$ is called linear if it has the property

$$\mathbf{A}(\alpha \boldsymbol{x}_1 + \beta \boldsymbol{x}_2) = \alpha \mathbf{A} \boldsymbol{x}_1 + \beta \mathbf{A} \boldsymbol{x}_2 \qquad (A.18)$$

with $\alpha, \beta \in \mathbb{R}$.

Examples are the matrix (a_{ij}) in an orthonormal basis in $\mathbb{R}^n$ or the differential operator $A : C^2[0, L] \to C[0, L]$ of the buckling problem of the rod of Section 4.3.2 given by

$$A\psi = \frac{\partial^2 \psi}{\partial s^2} + p\psi . \qquad (A.19)$$

Definition A.7 (Norm of an operator) *By the norm of the operator* **A***, we define the number*

$$\|\mathbf{A}\| = \sup_{x \neq 0} \frac{\|\mathbf{A}\boldsymbol{x}\|_Y}{\|\boldsymbol{x}\|_X} \qquad (A.20)$$

where $\|\cdot\|_Y$ and $\|\cdot\|_X$ denote the norms in the corresponding spaces. We suppress these indices in the following. Geometrically, $\|\mathbf{A}\|$ is a measure of the biggest stretching coefficient of the mapping $\mathbf{A}$. Clearly, (A.20) is independent of the length of $\boldsymbol{x}$ because

$$\frac{\|\mathbf{A}\alpha\boldsymbol{x}\|}{\|\alpha\boldsymbol{x}\|} = \frac{|\alpha|\,\|\mathbf{A}\boldsymbol{x}\|}{|\alpha|\,\|\boldsymbol{x}\|} .$$

In the above expression is $\alpha \in \mathbb{R}$. If $\|\mathbf{A}\| < \infty$ the operator $\mathbf{A}$ is called bounded, otherwise it is called unbounded. A bounded operator is *continuous*, because $\|\mathbf{A}\boldsymbol{x}\| \le k\|\boldsymbol{x}\|$.

Examples for the norm of an operator $\mathbf{A}$ in $\mathbb{R}^n$ include the following,

$$\|\mathbf{A}\|_1 = \sum_{i,j} |a_{ij}|$$

$$\|\mathbf{A}\|_2 = \max_i \{\sqrt{|\lambda_i|}\} \quad \text{where } \lambda_i \text{ are the eigenvalues of } \mathbf{A}^T\mathbf{A}$$

$$\|\mathbf{A}\|_\infty = \max_i \sum_{j=1}^n |a_{ij}| . \tag{A.21}$$

We now define two important spaces related to the operator A.

Definition A.8 (Image or range) *The image or range $R(\mathbf{A})$ of an operator $\mathbf{A}$ is given by*

$$R(\mathbf{A}) = \{y \in Y : \quad y = \mathbf{A}\boldsymbol{x}, \boldsymbol{x} \in X\} . \tag{A.22}$$

Definition A.9 (Kernel or nullspace) *The kernel or nullspace $N(\mathbf{A})$ of an operator $\mathbf{A}$ is given by*

$$N(\mathbf{A}) = \{\boldsymbol{x} \in X : \quad \mathbf{A}\boldsymbol{x} = \mathbf{0}\}. \tag{A.23}$$

Since $\mathbf{A}$ is linear, $N(\mathbf{A})$ and $R(\mathbf{A})$ are subspaces of X and Y, respectively. If $\mathbf{A} : \mathbb{R}^n \to \mathbb{R}^m$ then the rank $r(\mathbf{A})$ is given by

$$r(\mathbf{A}) = \dim R(\mathbf{A})$$

whereas the dimension of $N(\mathbf{A})$ is given by

$$\dim N(\mathbf{A}) = n - r.$$

Adjoint operator

There are at least two reasons to introduce the concept of the adjoint operator $\mathbf{A}^\star$. One is to characterize the range $R(\mathbf{A})$ of $\mathbf{A}$ in a simple way. The other one is that for operators $\mathbf{A}$ which are not self-adjoint the eigenvectors and the eigenfunctions of the adjoint operator are needed for projections onto subspaces. This is further elaborated on in Appendix D.

To define $\mathbf{A}^\star$, we consider the spaces $X^\star$ and $Y^\star$, dual to X and Y. The *dual space* $X^\star$ is the space of linear functionals $\mathbf{L}$ on X. A linear functional is a linear mapping $\mathbf{L} : X \to \mathbb{R}^1$. If X is a Hilbert space then the space $X^\star$ is *isomorphic* to X. In other words, there exists an invertible and unique relationship between X and $X^\star$. The mapping $\mathbf{L}$ also satisfies

$$\mathbf{L}(\alpha X_1 + \beta X_2) = \alpha \mathbf{L} X_1 + \beta \mathbf{L} X_2 .$$

Assume that $X = \mathbb{R}^n$ and that we have chosen a basis in X, then an element $\ell \in \mathbf{L}$ is given by the $(1 \times n)$-matrix $(\ell_1, \ldots, \ell_n) = \ell^T$. Thus, we have

$$\mathbf{L}x = \ell^T x = \ell_1 x_1 + \ldots + \ell_n x_n = (\ell, x) .$$

Take an arbitrary x in X and a fixed $y^\star$ in $Y^\star$ and form the scalar quantity

$$(y^\star, \mathbf{A}x) .$$

The above expression is a linear functional in X, because x varies in X. Alternatively, each linear functional in X can be represented by a fixed $x^\star$ in $X^\star$. This is done by forming the inner product $(x^\star, x)$. Thus we have

$$(y^\star, \mathbf{A}x) = (x^\star, x) , \qquad \forall x \in X . \tag{A.24}$$

Equation (A.24) gives a relationship: $y^\star \to x^\star$, which defines the adjoint operator $\mathbf{A}^\star : Y^\star \to X^\star$ by $\mathbf{A}^\star y^\star = x^\star$. Inserting this expression in (A.24) immediately yields

$$(y^\star, \mathbf{A}x) = (\mathbf{A}^\star y^\star, x). \tag{A.25}$$

The relation (A.25) serves to define the adjoint operator in the infinite-dimensional case, provided a suitable inner product has been defined.

For the finite-dimensional case, (A.25) may be expressed as

$$y^{\star T} \mathbf{A} x = (\mathbf{A}^\star y^\star)^T x = y^{\star T} \mathbf{A}^{\star T} x .$$

This relation leads to the fundamental result

$$\mathbf{A}^\star = \mathbf{A}^T , \tag{A.26}$$

that is, the adjoint operator is equal to the transposed operator.

Fredholm operator

Definition A.10 (Fredholm operator) *Let X and Y be Banach spaces. The operator $\mathbf{A} : X \to Y$ is called a Fredholm operator if*

(a) $\dim N(\mathbf{A})$ *is finite*

(b) $R(\mathbf{A})$ *is closed in Y*

(c) $\dim N(\mathbf{A}^\star)$ *is finite.*

We recall that a space X is closed if any Cauchy sequence u_ν possesses a limit u in X.

The *index $i(\mathbf{A})$* of an operator $\mathbf{A}$ is defined as

$$i(\mathbf{A}) = \dim N(\mathbf{A}) - \dim N(\mathbf{A}^\star). \qquad (A.27)$$

In the finite-dimensional case, we have

$$i(\mathbf{A}) = n - r - (m - r) = n - m . \qquad (A.28)$$

Hence, $i(\mathbf{A}) = 0 \Longleftrightarrow n = m$.

Fredholm alternative

By means of the following simple example we give a formulation of the Fredholm alternative and explain the significance of the adjoint operator. Of course, these concepts could be explained more generally. Let $\mathbf{A}$ be a $n \times n$ matrix. We consider the equation

$$\mathbf{A}x = b . \qquad (A.29)$$

The *Fredholm alternative* for (A.29) is:

(i) the homogeneous problem has only the trivial solution and (A.29) has a unique solution, for arbitrary $b \in R(A)$, because the inverse $\mathbf{A}^{-1}$ of $\mathbf{A}$ exists and $x = \mathbf{A}^{-1}b$ or

(ii) the homogeneous problem has a nontrivial solution, but then (A.29) admits a solution if and only if b is orthogonal to each $y \in N(\mathbf{A}^T)$.

This can, alternatively, be expressed in a way that is very important for practical computations. Multiplying (A.29) with $y \in Y$ we find

$$y^T \mathbf{A}x = y^T b$$

or

$$(\mathbf{A}^T y)^T x = y^T b . \qquad (A.30)$$

From (A.30) we can draw the important conclusion that for $\mathbf{A}^T y = 0$, that is, $y \in N(\mathbf{A}^T)$, the right-hand side $y^T b = 0$. This latter relation means that $b \in R(\mathbf{A})$, if it is orthogonal to any $y \in N(\mathbf{A}^T)$. If $N(\mathbf{A}^T)$ is spanned by the k vectors $y_1, \ldots, y_k$, it follows

$$b \in R(\mathbf{A}) \Longleftrightarrow (y_1, b) = 0, \ldots, (y_k, b) = 0. \qquad (A.31)$$

Decomposition of spaces

By means of the adjoint operator it is possible to decompose the spaces X and Y in the form

$$Y = R(\mathbf{A}) \oplus N(\mathbf{A}^\star) \qquad\qquad (A.32)$$

and

$$X = R(\mathbf{A}^\star) \oplus N(\mathbf{A}) . \qquad\qquad (A.33)$$

From (A.32), for example, follows that each y in $\mathbb{R}^m$ has the unique decomposition

$$y = y_1 + y_2 ,$$

where y_1 is in $R(\mathbf{A})$ and y_2 in $N(\mathbf{A}^T)$ and y_1 is complementary to y_2. The general solution of (A.29) is given by

$$x = x_h + x_p \qquad\qquad (A.34)$$

where $x_h \in N(\mathbf{A})$ is arbitrary and x_p is a special solution of $\mathbf{A}x_p = b$. The existence of x_p requires that $b \in R(\mathbf{A})$.

Example 1: We take $\mathbf{A}$ in $\mathbb{R}^3$ given by:

$$\mathbf{A} = \begin{pmatrix} 0 & 0 & 0 \\ 1 & 0 & 0 \\ 0 & 0 & 1 \end{pmatrix} .$$

The adjoint operator follows from (A.26) to

$$\mathbf{A}^T = \begin{pmatrix} 0 & 1 & 0 \\ 0 & 0 & 0 \\ 0 & 0 & 1 \end{pmatrix} .$$

The kernel

$$N(\mathbf{A}) = \text{span}\{(0,1,0)^T\}$$

follows from (A.23). Similarly, we obtain

$$N(\mathbf{A}^T) = \text{span}\{(1,0,0)^T\} .$$

For the range $R(\mathbf{A})$ we find from (A.31)

$$R(\mathbf{A}) = \text{span}\{(0,1,0)^T, (0,0,1)^T\} = N(\mathbf{A}^T)^\perp .$$

Therefore, (A.29) has no solution if $b = (\mu,0,0)^T$, $\mu \neq 0$ but has a solution for $b = (0,\mu,\nu)^T$. For example, if $b = (0,2,3)^T$ we find for (A.34)

$$x = t \begin{pmatrix} 0 \\ 1 \\ 0 \end{pmatrix} + \begin{pmatrix} 2 \\ 0 \\ 3 \end{pmatrix}$$

where t is an arbitrary parameter.

Appendix B

Transformation of matrices to diagonal or Jordan form

In this section we give two examples how to transform matrices to *diagonal or Jordan form* ([59], [159]). This also gives us the opportunity to make some comments on eigenvalues and eigenvectors of real valued matrices.

Example 1:

$$\mathbf{A} = \begin{pmatrix} 1 & 0 & 0 \\ 0 & 2 & -3 \\ 1 & 3 & 2 \end{pmatrix}. \tag{B.1}$$

The characteristic equation is

$$\det(\mathbf{A} - \mu\mathbf{E}) = (1 - \mu)((2 - \mu)^2 + 9) = 0,$$

with the roots: $\mu_1 = 1$, $\mu_{2,3} = 2 \pm 3i$.

The eigenvectors $\boldsymbol{u}, \boldsymbol{v}, \boldsymbol{w} \in \mathbb{R}^3$ follow from the solutions of the corresponding homogeneous systems. For μ_1, we obtain

$$(\mathbf{A} - \mu_1\mathbf{E})\boldsymbol{u} = \boldsymbol{0}$$

$$\begin{pmatrix} 0 & 0 & 0 \\ 0 & 1 & -3 \\ 1 & 3 & 1 \end{pmatrix} \begin{pmatrix} u_1 \\ u_2 \\ u_3 \end{pmatrix} = 0.$$

Setting $u_3 = 1$ results in $u_2 = 3$ und $u_1 = -10$.

Similarly, it follows for μ_2

$$(\mathbf{A} - \mu_2\mathbf{E})\boldsymbol{v} = \boldsymbol{0}$$

$$\begin{pmatrix} -1 - 3i & 0 & 0 \\ 0 & -3i & -3 \\ 1 & 3 & -3i \end{pmatrix} \begin{pmatrix} v_1 \\ v_2 \\ v_3 \end{pmatrix} = 0.$$

A solution is given by $v = (0, i, 1)^T$. A third eigenvector w is given by $w = \bar{v} = (0, -i, 1)^T$, as can be easily verified.

The transformation matrix $\mathbf{B}$ is given by

$$\mathbf{B} = \begin{pmatrix} -10 & 0 & 0 \\ 3 & 1 & 0 \\ 1 & 0 & 1 \end{pmatrix},$$

where the first column is given by the eigenvector u. The second and third column are the imaginary and the real part of the complex eigenvector v. The inverse matrix $\mathbf{B}^{-1}$ is

$$\mathbf{B}^{-1} = \begin{pmatrix} -1/10 & 0 & 0 \\ 3/10 & 1 & 0 \\ 1/10 & 0 & 1 \end{pmatrix}.$$

and thus

$$\mathbf{J} = \mathbf{B}^{-1}\mathbf{A}\mathbf{B} = \begin{pmatrix} 1 & 0 & 0 \\ 0 & 2 & -3 \\ 0 & 3 & 2 \end{pmatrix}.$$

Example 2:

$$\mathbf{A} = \begin{pmatrix} -1 & 1 & -2 \\ 0 & -1 & 4 \\ 0 & 0 & 1 \end{pmatrix}.$$

The characteristic equation is

$$(\mu + 1)^2(\mu - 1) = 0.$$

We see that the eigenvalue $\mu = -1$ has multiplicity $\nu = 2$. In general the number ν_i of repeated values of μ_i is called the *algebraic multiplicity* of μ_i. Furthermore, we have the simple eigenvalue $\mu = 1$. The computation of the eigenvectors for $\mu = -1$ follows from

$$\begin{pmatrix} 0 & 1 & -2 \\ 0 & 0 & 4 \\ 0 & 0 & 2 \end{pmatrix} \begin{pmatrix} u_1 \\ u_2 \\ u_3 \end{pmatrix} = 0.$$

There exists only one linear independent solution $u = (1, 0, 0)^T$. In general the number ϱ_i of the linearly independent eigenvectors belonging to μ_i is called the *geometric multiplicity* of μ_i. If an eigenvalue is repeated then there exist at least one eigenvector and at most ν_i linearly independent eigenvectors, that is, $\varrho_i \leq \nu_i$. If $\varrho_i < \nu_i$, the concept of *principal vectors* must be introduced to obtain ν_i linearly independent vectors in

the *invariant subspace* S_i. The invariant subspace S_i corresponding to
the eigenvalue μ_i is spanned by the eigenvectors and the principal vectors
and is characterized by the property that $AS_i \subset S_i$. In this example one
principal vector $\mathbf{h}$ must be calculated.

Let us consider the concept of principal vectors ([59]). Starting with an
$n \times n$ matrix in *Jordan form* (for example, $n = 4$)

$$\mathbf{J} = \begin{pmatrix} \mu & 1 & 0 & 0 \\ 0 & \mu & 1 & 0 \\ 0 & 0 & \mu & 1 \\ 0 & 0 & 0 & \mu \end{pmatrix}$$

we attempt to calculate the eigenvectors. It can be seen that there exists
only one eigenvector $\mathbf{e}_1 = (1, 0, 0, \ldots, 0)^T$. However, $(n - 1)$ other linear
independent vectors $\mathbf{h}_i$ $(i = 2, \ldots, n)$ can be obtained using the following
scheme

$$\mathbf{J}\mathbf{h}_2 = \mu\mathbf{h}_2 + \mathbf{e}_1$$
$$\mathbf{J}\mathbf{h}_3 = \mu\mathbf{h}_3 + \mathbf{h}_2$$

$$\vdots$$

$$\mathbf{J}\mathbf{h}_n = \mu\mathbf{h}_n + \mathbf{h}_{n-1}\ .$$

The vectors $\mathbf{h}_2, \mathbf{h}_3, \ldots, \mathbf{h}_n$ are called principal vectors of first, second,
$\ldots$, $n - 1$-st order. They form a basis in $\mathbb{R}^n$ in which the corresponding
matrix takes the Jordan form.

Returning to the example, the principal vector $\mathbf{h}$ of first order is calculated from the equation

$$(\mathbf{A} - \mu_1\mathbf{E})\mathbf{h} = \mathbf{u}$$

which reads

$$\begin{pmatrix} 0 & 1 & -2 \\ 0 & 0 & 4 \\ 0 & 0 & 2 \end{pmatrix} \begin{pmatrix} h_1 \\ h_2 \\ h_3 \end{pmatrix} = \begin{pmatrix} 1 \\ 0 \\ 0 \end{pmatrix}\ .$$

A solution is given by $\mathbf{h} = (0, 1, 0)^T$. Finally, the eigenvector $\mathbf{v}$ corresponding to $\mu = 1$ follows to $\mathbf{v} = (0, 2, 1)^T$.

Thus the transformation matrix $\mathbf{B}$ is given by

$$\mathbf{B} = \begin{pmatrix} 1 & 0 & 0 \\ 0 & 1 & 2 \\ 0 & 0 & 1 \end{pmatrix}$$

with $\det \mathbf{B} = 1$ and

$$\mathbf{B}^{-1} = \begin{pmatrix} 1 & 0 & 0 \\ 0 & 1 & -2 \\ 0 & 0 & 1 \end{pmatrix}\ .$$

Calculating

$$\mathbf{B^{-1}AB} = \begin{pmatrix} 1 & 0 & 0 \\ 0 & 1 & -2 \\ 0 & 0 & 1 \end{pmatrix} \begin{pmatrix} -1 & 1 & -2 \\ 0 & -1 & 4 \\ 0 & 0 & 1 \end{pmatrix} \begin{pmatrix} 1 & 0 & 0 \\ 0 & 1 & 2 \\ 0 & 0 & 1 \end{pmatrix}$$

$$= \begin{pmatrix} -1 & 1 & -2 \\ 0 & -1 & 2 \\ 0 & 0 & 1 \end{pmatrix} \begin{pmatrix} 1 & 0 & 0 \\ 0 & 1 & 2 \\ 0 & 0 & 1 \end{pmatrix} = \begin{pmatrix} -1 & 1 & 0 \\ 0 & -1 & 0 \\ 0 & 0 & 1 \end{pmatrix}$$

the Jordan form is obtained.

Appendix C

Adjoint and self-adjoint linear differential operators

C.1 Calculation of the adjoint operator

We treat two examples:

Example 1: We seek to determine the adjoint operator $A^\star$ of the linear differential operator A given by

$$A = D^2 + 2kD + \omega^2 , \qquad D = \frac{d}{dx} . \tag{C.1}$$

A is defined on the domain $\{u \in C^2[0, a] : u(0) = u(a) = 0\}$.

We recall from (A.25) the definition of the *adjoint operator*

$$(Au, v) = (u, A^\star v) . \tag{C.2}$$

The definition of (C.2) requires an appropriately selected inner product as it is given below.

Inserting (C.1) into the left-hand side of (C.2) yields

$$(Au, v) = \int_0^a (u'' + 2ku' + \omega^2 u)v\, dx . \tag{C.3}$$

The right-hand side of (C.2) follows from (C.3) by partial integration. We obtain

$$(u, A^\star v) = \int_0^a u(v'' - 2kv' + \omega^2 v)dx + (u'v - uv' + uv)\Big|_0^a . \tag{C.4}$$

Making use of the boundary conditions (C.1), (C.4) yields the adjoint operator $A^\star$ and the corresponding boundary conditions

$$A^\star = D^2 - 2kD + \omega^2 , \qquad \{v \in C^2[0, a] : v(0) = v(a) = 0\} .$$

Example 2: We seek the adjoint operator A^* of the operator A defined by

$$Au = \ddot{u} + \mu^*\dot{u} + 2\sqrt{\beta}\varrho\dot{u}' + \alpha^*\dot{u}^{IV} + \varrho^2 u'' + u^{IV} + \gamma[(s-1)u']' . \quad (C.5)$$

The boundary and jump conditions for (C.5) are

$$
\begin{array}{lll}
s = 0: & u = 0, & u' = 0 \\
s = \xi: & (u''' + \alpha^*\dot{u}''')|_{\xi_-}^{\xi_+} + cu(\xi) = 0 & \quad\quad (C.6) \\
s = 1: & u'' + \alpha^*\dot{u}'' = 0, & u''' + \alpha^*\dot{u}''' = 0 .
\end{array}
$$

At $s = \xi$ all quantities up to and including the second derivative of s are continuous.

Equations (C.5) and (C.6) describe the linear oscillation problem of the fluid-conveying tube of Section 5.3.

From relation (C.2) and repeated partial integration with respect to s and t follows

$$\int_{t_0}^{t_1}\left\{\int_0^1 [\ddot{u} + \mu^*\dot{u} + 2\sqrt{\beta}\varrho\dot{u}' + \alpha^*\dot{u}^{IV} + \varrho^2 u'' + u^{IV} + \right.$$

$$\left. +\gamma[(s-1)u']' + cu\delta(s-\xi)]vds \right\}dt$$

$$= \int_{t_0}^{t_1}\left\{\int_0^1 u[\ddot{v} - \mu^*\dot{v} + 2\sqrt{\beta}\varrho\dot{v}' - \alpha^*\dot{v}^{IV} + \varrho^2 v'' + v^{IV} + \right. \quad (C.7)$$

$$+\gamma[(s-1)v']' + cv\delta(s-\xi)]ds +$$

$$+[-2\sqrt{\beta}\varrho u\dot{v} + \alpha^*(\dot{u}'''v - \dot{u}''v' + \dot{u}'v'' - \dot{u}v''') +$$

$$+\varrho^2(u'v - uv') + (u'''v - u''v' + u'v'' - uv''')]\Big|_0^1\bigg\}dt .$$

Here we have used a trick by representing the elastic point support, which is responsible for the jump discontinuity in the transversal force, by means of a Dirac δ-function in the integral.

We shortly explain that such a δ-function supplies exactly the jump conditions (C.6)$_2$ at $s = \xi$. If we add to (C.5) the term $c\delta(s-\xi)u(s)$ we obtain

$$\ddot{u} + \mu^*\dot{u}' + 2\sqrt{\beta}\varrho\dot{u}' + \alpha^*\dot{u}^{IV} + \varrho^2 u'' + u^{IV} + \gamma[(s-1)u']' + c\delta(s-\xi)u(s) = 0 .$$

Now we multiply this expression with ds, integrate it from $\xi - \varepsilon$ to $\xi + \varepsilon$ and take the limes $\varepsilon \to 0$. All quantities which are continuous at $s = \xi$ drop out and we finally obtain the jump condition (C.6)$_2$.

Performing the same operation on the right hand side of (C.7) for the adjoint differential operator with respect to the variable s we obtain the jump condition at

$$s = \xi : \qquad (v''' - \alpha^* \dot{v}''')|_{\xi_-}^{\xi_+} + cv(\xi) = 0 .$$

Now we make use of the boundary conditions (C.6) at $s = 0$ and $s = 1$. By partial integration with respect to t we transform the term

$$\alpha^* (\dot{u}'''v - \dot{u}''v' + \dot{u}'v'' - \dot{u}v''')$$

into

$$-\alpha^* (u'''\dot{v} - u''\dot{v}' + u'\dot{v}'' - u\dot{v}''') .$$

The corresponding boundary terms at $t = t_0$ and $t = t_1$ have no influence on the differential operator. The requirement that the boundary terms in the integral must vanish, yields the following results

$$
\begin{aligned}
s = 0 : \quad & u'''(-\alpha^* \dot{v} + v) = 0 \\
& u''(\alpha^* \dot{v}' - v') = 0 \\
s = 1 : \quad & u'[-\alpha^* \dot{v}'' + v'' + \varrho^2 v] = 0 \\
& u[-2\sqrt{\beta}\varrho\dot{v} + \alpha^* \dot{v}''' - v''' - \varrho^2 v'] = 0 .
\end{aligned}
$$

Hence, the operator A^* adjoint to A in (C.5) is

$$A^* v = \ddot{v} - \mu^* \dot{v} + 2\sqrt{\beta}\varrho\dot{v}' - \alpha^* \dot{v}^{IV} + \varrho^2 v'' + v^{IV} + \gamma[(s-1)v']' \qquad \text{(C.8)}$$

with the boundary conditions

$$
\begin{aligned}
s = 0 : \quad & v - \alpha^* \dot{v} = 0 \\
& v' - \alpha^* \dot{v}' = 0 \\
s = \xi : \quad & (v''' - \alpha^* \dot{v}''')|_{\xi_-}^{\xi_+} + cv(\xi) = 0 \qquad \text{(C.9)} \\
s = 1 : \quad & v'' - \alpha^* \dot{v}'' + \varrho^2 v = 0 \\
& v''' - \alpha^* \dot{v}''' + \varrho^2 v' + 2\sqrt{\beta}\varrho\dot{v} = 0 .
\end{aligned}
$$

C.2 Self-adjoint differential operators

Let $y(x) \in C^n[a, b]$ be an *n-times continuously differentiable function* and $-\infty < a < b < \infty$. Let the linear differential operator be given by

$$(Ay)(x) = a_n(x)y^{(n)}(x) + a_{n-1}(x)y^{(n-1)}(x) + \ldots + a_0(x)y(x)$$

where

$$a_n \neq 0 \qquad \text{and} \qquad x \in [a, b].$$

The operator $A^{\star}$ adjoint to A is given by

$$(A^{\star}z)(x) = (-1)^n(\bar{a}_n z)^{(n)} + (-1)^{n-1}(\bar{a}_{n-1}z)^{(n-1)} + \ldots + \bar{a}_0 z$$

where the bars designate conjugate complex quantities. $A^{\star}z = 0$ is the differential equation adjoint to $Ay = 0$. The operator A is called *formally self-adjoint*, that is, $A^{\star} = A$ if and only if

$$Ay = \sum_{k=0}^{m}(-1)^k \left(a_k y^{(k)}\right)^{(k)} \tag{C.10}$$

for m even and a_k real. The significance of the adjoint operator is given by the *Lagrange identity*. For $a_k \in C^k[a, b]$ and $u, v \in C^n[a, b]$, the equation

$$\bar{v}Au - uA^{\star}\bar{v} = [u, v]' \tag{C.11}$$

holds. Here, $[u, v]$ is a bilinear form in $(u, u', \ldots, u^{(n-1)})$ and $(v, v', \ldots, v^{(n-1)})$ given by

$$[u, v] = \sum_{m=1}^{n} \sum_{j+k=m-1} (-1)^j u^{(k)}(a_m \bar{v})^{(j)} , \qquad j \geq 0, k \geq 0 . \tag{C.12}$$

From Lagrange's identity follows *Green's formula* by integration ($a \leq x_1 < x_2 \leq b$)

$$\int_{x_1}^{x_2} (\bar{v}Au - uA^{\star}\bar{v})dx = [u, v](x_2) - [u, v](x_1). \tag{C.13}$$

Consider now the boundary value problem

$$Au = f$$

$$\tag{C.14}$$

$$U_i u = \sum_{j=1}^{n} \left(\mu_{ij} u^{(j-1)}(x_1) + \gamma_{ij} u^{(j-1)}(x_2)\right) = 0 , \qquad i = 1, \ldots, n .$$

If we define an inner product $(u, v) = \int_a^b u\bar{v}dx$, it follows from (C.13) that $(x_1 = a, x_2 = b)$

$$(Au, v) - (u, A^{\star}v) = [u, v](b) - [u, v](a). \tag{C.15}$$

Expression (C.15) corresponds to (A.25) in the finite-dimensional case. It is important to note that on the right-hand side of (C.15) the boundary conditions appear. Hence, the boundary value problem (C.14) is called self-adjoint if $A = A^{\star}$ and the boundary values are such that the right-hand side in (C.15) vanishes. Then the two operators A and $A^{\star}$ are identical. In *linear elasticity* the self-adjointness follows from *Maxwell's principle* ([97]).

Example 1: We consider the following boundary value problem

$$Ay = y'' + \lambda y = 0$$

with the boundary conditions in $x = 0$ and $x = \ell$ (α, β are constants)

$$
\begin{aligned}
U_1 y &= y'(0) - \alpha y(0) &&= 0 \\
U_2 y &= y'(\ell) - \beta y(\ell) &&= 0.
\end{aligned}
$$

For A to be self-adjoint, the following relation must hold for two functions u, v

$$\int_0^\ell \{(u'' + \lambda u)v - u(v'' + \lambda v)\} dx = \int_0^\ell (u''v - uv'') dx = 0 \ .$$

By partial integration

$$\int_0^\ell (u''v - uv'') dx = [u'v - uv']\Big|_0^\ell + \int_0^\ell (u'v' - u'v') dx$$

is obtained. The first term on the right-hand side vanishes due to the boundary conditions, so the boundary value problem is self-adjoint. In general, the self-adjointness of a differential operator can be checked by performing partial integrations ([30] p. 60).

Appendix D

Projection operators

D.1 General considerations

Projection operators are simple mappings which serve to study more complicated mappings. Let X be a linear space and X_1, X_2 be two subspaces of X. Consider the decomposition of X as the direct sum

$$X = X_1 \oplus X_2 \ .$$

Then, for each $x \in X$ there exist one $x_1 \in X_1$ and one $x_2 \in X_2$ such that

$$x = x_1 + x_2.$$

For an x_3 in both X_1 and X_2 follows that $x_3 = 0$. The concept of decomposition is closely related to that of projection.

Definition D.1 (Projection) *Let $P : X \to X$, be a linear operator. P is a projection operator if it is idempotent, that is,*

$$P^2 = P \circ P = P \ , \tag{D.1}$$

where $\circ$ denotes the composition of operators.

The following *theorem* holds:

Theorem D.1 *Each decomposition of X as a direct sum of X_1 and X_2 corresponds exactly to a projection P with $X_1 = R(P)$ and $X_2 = N(P)$ (R and N are defined on p. 293).*

P is called projection on X_1 in the direction of X_2 (Fig. D.1). That is, each element in X_2 is mapped to zero.

There is a second important projection $Q = E - P$ which is complementary to P. E is the identity mapping. Also Q has the property to be idempotent, that is,

$$Q^2 = (E - P)(E - P) = E - 2P + P^2 = E - P = Q.$$

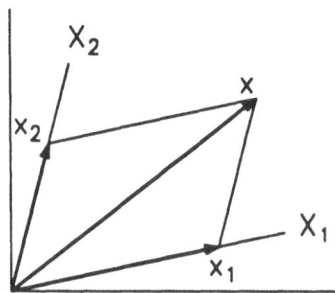

Figure D.1. Decomposition of the space X by the projection operators P and Q

By means of the projection operators P and Q a decomposition of a space can be given. For each element $x \in X$, the following decomposition

$$x = Px + (E - P)x = x_1 + x_2 \qquad (\text{D.2})$$

holds.

Of course, orthogonal projections are important. For them, the complementary space is automatically defined.

Let $\{v_1, \ldots, v_r\}$ be an orthonormal basis of X_1 then

$$Px = \sum_{i=1}^{r}(v_i, x)v_i. \qquad (\text{D.3})$$

In order to show that (D.1) holds for (D.3), it is sufficient to consider a basis vector v_j

$$Pv_j = \sum_{i}(v_i, v_j)v_i = v_j.$$

Expression (D.3) is valid for the infinite-dimensional case, too.

Example 1: Let us return to Example 1 in Appendix A.2. We calculate the projection operators P and $E - P$ onto the range $R(A)$ and the kernel $N(A)$, respectively. We find

$$\mathbf{P} = \begin{pmatrix} 1 & 0 & 0 \\ 0 & 0 & 0 \\ 0 & 0 & 1 \end{pmatrix}, \qquad \mathbf{E} - \mathbf{P} = \begin{pmatrix} 0 & 0 & 0 \\ 0 & 1 & 0 \\ 0 & 0 & 0 \end{pmatrix}.$$

Hence, any vector $x \in \mathbb{R}^3$ is decomposed according to (D.2) into

$$\begin{pmatrix} x_1 \\ x_2 \\ x_3 \end{pmatrix} = \begin{pmatrix} x_1 \\ 0 \\ x_3 \end{pmatrix} + \begin{pmatrix} 0 \\ x_2 \\ 0 \end{pmatrix}.$$

Example 2: We consider the space of functions $L_2[a, b]$, that is, the space of square integrable functions on the interval $[a, b]$ (Appendix A.1). This space includes all continuous functions on $[a, b]$

$$C[a, b] \subset L_2[a, b].$$

Let $\varphi_1, \ldots, \varphi_r$ be an orthonormal basis in X_1. That is,

$$X_1 = \text{span}\{\varphi_1, \ldots, \varphi_r\}.$$

What is the orthogonal projection of functions $x(t) \in C[a, b]$ on X_1? To answer the question we define an inner product by

$$(u, v) = \int_a^b u(t)v(t)dt .$$

For the projection we obtain

$$(Px)(t) = \sum_{i=1}^r (\varphi_i, x)\varphi_i(t) . \tag{D.4}$$

The complementary projection $Q = E - P$ is given by

$$(Qx)(t) = ((E - P)x)(t) = x(t) - \sum_{i=1}^r (\varphi_i, x)\varphi_i(t). \tag{D.5}$$

D.2   Projection for non-self-adjoint operators

In deriving bifurcation equations, for both the finite-dimensional and infinite-dimensional cases certain projections on the spaces spanned by the eigenvectors or eigenfunctions of the kernel of the linear operator at the critical parameter value must be made. In this section we explain why, in general, we need to know the eigenvectors or eigenfunctions of the adjoint operator in order to perform these projections. In order to keep this explanation simple, only the finite-dimensional case is explained ([170] p. 155).

To the eigenvalue problem

$$\mathbf{A}\mathbf{b} = \lambda\mathbf{b} \tag{D.6}$$

one can associate a second one that reads

$$\mathbf{A}^T\mathbf{c} = \mu\mathbf{c} \tag{D.7}$$

which, of course, is the same as

$$\mathbf{c}^T \mathbf{A} = \mu \mathbf{c}^T \ . \tag{D.8}$$

The vectors $\mathbf{b}$ and $\mathbf{c}$ are called *right* and *left* eigenvectors to $\mathbf{A}$, respectively. Usually, instead of *right eigenvector* only the designation *eigenvector* is used. Both problems (D.6) and (D.7) have identical eigenvalues $\lambda_i = \mu_i$ because the determinants of the linear homogeneous systems

$$(\mathbf{A} - \mu \mathbf{E})\mathbf{b} = 0$$

and

$$\mathbf{c}^T(\mathbf{A} - \mu \mathbf{E}) = 0 \quad \text{or} \quad (\mathbf{A} - \mu \mathbf{E})^T \mathbf{c} = 0$$

are the same. However, the corresponding eigenvectors $\mathbf{b}$ and $\mathbf{c}$ are, in general, different. Only in the case that $\mathbf{A} = \mathbf{A}^T$, that is, $\mathbf{A}$ is self-adjoint, are $\mathbf{b}$ and $\mathbf{c}$ identical.

Now consider two different eigenvalues $\lambda_k \neq \lambda_i$. We know that the relations

$$\begin{aligned} \mathbf{A}\mathbf{b}_i &= \lambda_i \mathbf{b}_i \\ \mathbf{c}_k^T \mathbf{A} &= \lambda_k \mathbf{c}_k^T \end{aligned} \tag{D.9}$$

hold. Multiplying (D.9)$_1$ from the left with $\mathbf{c}_k^T$ and (D.9)$_2$ from the right with $\mathbf{b}_i$ and adding these two equations yields

$$0 = (\lambda_i - \lambda_k)\mathbf{c}_k^T \mathbf{b}_i \ . \tag{D.10}$$

Since $\lambda_i \neq \lambda_k$ was assumed, it follows from (D.10) that $\mathbf{c}_k$ is orthogonal to $\mathbf{b}_i$.

For simplicity we assume now that $\mathbf{A}$ is an $n \times n$ matrix which possesses n independent eigenvectors $\mathbf{b}_i$, and hence, also n independent $\mathbf{c}_k$. Then the matrices $\mathbf{B}$ and $\mathbf{C}$ formed by the right and left eigenvectors are non-singular. Their product $\mathbf{N}$ is given by

$$\mathbf{N} = \mathbf{B}^T \mathbf{C} = (\mathbf{b}_i^T \mathbf{c}_j) \tag{D.11}$$

and is non-singular, too. For two different eigenvalues $i \neq j$, the corresponding entry in $\mathbf{N}$ is zero. This implies that the diagonal elements must all be different from zero. By a proper scaling we can always achieve

$$\mathbf{b}_i^T \mathbf{c}_i = 1 \ . \tag{D.12}$$

Hence, for matrices $\mathbf{A}$ that can be diagonalized we get

$$\mathbf{B}^T \mathbf{C} = \mathbf{E} \quad \text{or} \quad \mathbf{C}^T \mathbf{B} = \mathbf{E} \ . \tag{D.13}$$

From (D.13), the important relations follow

$$\mathbf{C}^T = \mathbf{B}^{-1} \quad \text{or} \quad \mathbf{B}^T = \mathbf{C}^{-1} \ . \tag{D.14}$$

Let us now consider the dynamical system (3.4)

$$\dot{x} = Ax + f(x) \,, \tag{D.15}$$

which by $x = By$ (3.7) is transformed into diagonal form

$$\dot{y} = B^{-1}(ABy + f(By)) = C^{T}(ABy + f(By)) \,. \tag{D.16}$$

Hence, in the finite-dimensional case as in (D.16) the eigenvectors of the adjoint operator A^{T} also appear. However, because of (D.14) we did not have to introduce them explicitly in the calculations in Section 3.1.

D.3   Application to the Galerkin reduction

We now explain why, generally, for the application of the Galerkin method the adjoint eigenvectors must be used when performing the projection onto a low-dimensional space of eigenvectors. The dynamical system is given by (D.15). We assume that at loss of stability an eigenvalue $\mu_1 = 0$ occurs with the corresponding eigenvector b_1. In the Galerkin method, an approximation in one variable $y_1(t)$ of the form

$$x = y_1 b_1 \tag{D.17}$$

is substituted into (D.15). This substitution yields

$$\dot{y}_1 b_1 = A y_1 b_1 + f(y_1 b_1) \,. \tag{D.18}$$

The projection of (D.18) clearly must be made with c_1^{T} according to (D.12). Hence, using (D.9) one finds that

$$\dot{y}_i = c_1^{T} A b_1 y_1 + c_1^{T} f(y_1 b_1) = \mu_1 y_1 + c_1^{T} f(y_1 b_1) \,. \tag{D.19}$$

Equation (D.19) gives a bifurcation equation identical to (3.10) obtained in Section 3.1 if the non-critical variables are all set equal to zero.

Appendix E

Spectral decomposition of the inverse of linear differential operators

In this appendix we indicate how to calculate Δ^{-2} the inverse of Δ^2 which is needed in Section 4.3.4 ([159]).

E.1 Derivation of an inversion formula

Let A be a linear self-adjoint differential operator with continuous coefficients (Appendix C.2). We want to solve the boundary value problem

$$\begin{aligned} Aw(x) &= f(x) \\ Mw(a) + Nw(b) &= 0 , \end{aligned} \qquad a \leq x \leq b \qquad \text{(E.1)}$$

where M and N are operators specifying the boundary conditions (Appendix C.2). The function $f(x)$ is given. First, we consider the associated eigenvalue problem

$$(Au)(x) = \lambda u(x), \qquad \text{(E.2)}$$

with the boundary conditions

$$Uu = Mu(a) + Nu(b) = 0 . \qquad \text{(E.3)}$$

Here, $u \neq 0$ is the eigenfunction and λ is the eigenvalue. The following *theorem* holds ([30]):

Theorem E.1 (Eigenvalues and eigenfunctions) *If A is a self-adjoint differential operator then there exists a countable infinite number of real eigenvalues*

$$\lambda_1 \leq \lambda_2 \leq \ldots \leq \lambda_m \leq \ldots$$

311

the multiplicity of which is at most of the order n of the differential operator A. Corresponding to the eigenvalues, there exists a complete orthonormal set of eigenfunctions

$$u_1, u_2, \ldots, u_m, \ldots$$

with

$$(u_i, u_j) = \delta_{ij}, \qquad\qquad (E.4)$$

where

$$(u, v) = \int_a^b uv\,dt$$

denotes the inner product.

If, in addition, the operator A is positive definite, that is

$$(Au_i, u_i) > 0 \qquad \text{for all} \ \ u_i \neq 0\,,$$

then all eigenvalues are positive.

Let us represent w and f by means of the eigenfunctions u_i (Appendices A and D)

$$w = \sum_{i=1}^{\infty} c_i u_i \qquad\qquad (E.5)$$

where the c_i are unknowns to be determined and

$$f = \sum_{i=1}^{\infty} f_i u_i = \sum_{i=1}^{\infty} (f, u_i) u_i. \qquad\qquad (E.6)$$

Introducing (E.5) and (E.6) into (E.1) and using (E.2) results in

$$Aw = \sum_{i=1}^{\infty} c_i Au_i = \sum_{i=1}^{\infty} c_i \lambda_i u_i = \sum_{i=1}^{\infty} (f, u_i) u_i\,.$$

By equating coefficients, we obtain

$$c_i = \frac{1}{\lambda_i}(f, u_i) \qquad\qquad (E.7)$$

provided $\lambda_i \neq 0$. Inserting (E.7) into (E.5) yields the solution of (E.1) which can be written in the form

$$w = A^{-1}f = \sum_{i=1}^{\infty} \frac{(f, u_i)}{\lambda_i} u_i\,. \qquad\qquad (E.8)$$

If $\lambda_1 = 0$ then (E.8) takes the form

$$w = \hat{A}^{-1}f = \sum_{i=2}^{\infty} \frac{(f, u_i)}{\lambda_i} u_i \ . \tag{E.9}$$

If, however, a k-fold zero eigenvalue λ_1 appears, then (E.8) takes the form

$$
\begin{aligned}
w(x) &= \sum_{i=k+1}^{\infty} \frac{(f, u_i)}{\lambda_i} u_i(x) = \sum_{i=k+1}^{\infty} \frac{1}{\lambda_i} \int_a^b f(s) u_i(s) u_i(x) ds \\
&= \int_a^b \left\{ \sum_{i=k+1}^{\infty} \frac{1}{\lambda_i} u_i(s) u_i(x) \right\} f(s) ds = \int_a^b G(x, s) f(s) ds \ .
\end{aligned} \tag{E.10}
$$

For $k = 0$, $G(x, s)$ is the *Green's function* and for $k \neq 0$ $G(x, s)$ is the generalized Green's function. For G the relation

$$G(x, s) = G(s, x)$$

holds. In summary, the solution of (E.1) is given by

$$w(x) = \int_a^b G(x, s) f(s) ds = (A^{-1}f)(x) \ . \tag{E.11}$$

E.2   Three examples

Example 1: In engineering beam theory ([38], eq. 4.45) for a transversely loaded beam with simply supported ends, the following boundary value problem (E.1) arises

$$w^{IV}(x) = f(x), \qquad 0 \leq x \leq \ell \tag{E.12}$$

$$w(0) = w''(0) = w(\ell) = w''(\ell) = 0 \ . \tag{E.13}$$

The associated eigenvalue problem is

$$u^{IV} = \lambda u \tag{E.14}$$

with boundary conditions (E.13). A solution of (E.14) with (E.13) is

$$u_i = \left(\frac{2}{\ell} \right)^{1/2} \sin \frac{i \pi x}{\ell} \tag{E.15}$$

and

$$\lambda_i = \left(\frac{i\pi}{\ell}\right)^4 . \qquad (E.16)$$

If we represent f by means of the eigenfunctions (E.15) in the form

$$f = \left(\frac{2}{\ell}\right)^{1/2} \sum_{i=1}^{\infty} f_i \sin \frac{i\pi x}{\ell} \qquad (E.17)$$

and insert (E.17) into (E.2) we find

$$w = \left(\frac{2}{\ell}\right)^{1/2} \sum_{i=1}^{\infty} \frac{f_i}{(i\pi/\ell)^4} \sin \frac{i\pi x}{\ell} = (A^{-1}f)(x) . \qquad (E.18)$$

Example 2: In the analysis of the nonlinear buckling problem of a rectangular plate occupying the domain Ω: $0 \le x \le \ell$, $0 \le y \le 1$ (Section 4.3.4), one important step is to find the inverse of the operator Δ^2. That is, one has to solve

$$\begin{aligned} \Delta^2 v &= [\varphi, \psi] , & (x,y) \in \Omega & \qquad (E.19) \\ v &= \Delta v = 0 , & (x,y) \in \partial\Omega & \qquad (E.20) \end{aligned}$$

where the operator $[,]$ is given by (J.22) and φ, ψ are sufficiently smooth functions. First, we generalize (E.11) for the case of two independent variables. For this purpose, the right-hand side of (E.19) must be expressed by a series of the eigenfunctions of the operator Δ^2 which are given by (4.156). In order to have an orthonormal set of eigenfunctions, the constants C_{pq} in (4.156) must be selected to be

$$C_{pq} = 2\ell^{-1/2}$$

such that

$$u_{pq} = C_{pq} \sin \frac{p\pi x}{\ell} \sin q\pi y \qquad (E.21)$$

and

$$\lambda_{pq} = \frac{\pi^4(p^2 + \ell^2 q^2)^2}{\ell^4} . \qquad (E.22)$$

Note that the eigenvalues (E.22) are different from (4.157) because (E.2) is different from (4.153). We choose the inner product $(u, v) = \int_{\Omega} uv\,d\Omega$.

Setting

$$f(x,y) = [\varphi, \psi](x,y)$$

we obtain from (E.10)

$$(\Delta^{-2}f)(x,y) = \sum_{\substack{p=1 \\ q=1}}^{\infty} \frac{1}{\lambda_{pq}} \iint_{\Omega} u_{pq}(\xi,\eta)f(\xi,\eta)d\xi d\eta u_{pq}(x,y)$$

$$= \iint_{\Omega} \sum_{\substack{p=1 \\ q=1}}^{\infty} \frac{1}{\lambda_{pq}} u_{pq}(\xi,\eta)u_{pq}(x,y)f(\xi,\eta)d\xi d\eta \qquad (\text{E.23})$$

$$= \iint_{\Omega} G(x,y,\xi,\eta)f(\xi,\eta)d\xi d\eta \ .$$

In (E.23) $G(x,y,\xi,\eta)$ is Green's function for Δ^{-2} and is given by

$$
\begin{aligned}
&G(x,y,\xi,\eta) \\
&= \frac{4\ell^3}{\pi^4} \sum_{\substack{p=1 \\ q=1}}^{\infty} \frac{1}{(p^2+q^2\ell^2)^2} \sin \frac{p\pi x}{\ell} \sin q\pi y \sin \frac{p\pi \xi}{\ell} \sin q\pi\eta. \qquad (\text{E.24})
\end{aligned}
$$

Obviously, the calculation of Δ^{-2} involves the summation of infinite double series.

Example 3: We calculate the *Green's function for the Laplacian* with Dirichlet boundary conditions ($u = 0$ on $\partial\Omega$) which is needed in (4.175). For $\ell = \ell_c$, the orthonormal set of eigenfunctions is

$$u_{pq} = \frac{2}{\sqrt{\ell_c}} \sin \frac{p\pi x}{\ell_c} \sin q\pi y \qquad (\text{E.25})$$

with the corresponding eigenvalues

$$\lambda_{pq} = - \left(\frac{p^2\pi^2}{\ell_c^2} + q^2\pi^2 \right). \qquad (\text{E.26})$$

With (E.25) and (E.26), an expression similar to (E.24) results in

$$G(x,y,\xi,\eta) = \frac{4}{\ell_c} \sum_{\substack{p=1 \\ q=1}}^{\infty} \frac{1}{\lambda_{pq}} \sin \frac{p\pi x}{\ell_c} \sin q\pi y \sin \frac{p\pi \xi}{\ell_c} \sin q\pi\eta \ . \qquad (\text{E.27})$$

Appendix F

Some formulas concerning the shell equations on the complete sphere

In this appendix we provide some formulas useful to understand the shell buckling problem in Section 4.3.5 (b).

F.1 Tensor notations in curvilinear coordinates

In order to treat the shell equations (4.204) for a complete sphere, curvilinear coordinates in three-dimensional Euclidean space must be used ([70]).

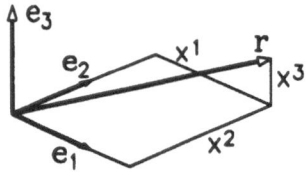

Figure F.1. Rectilinear coordinate system x^1, x^2, x^3

Let r be a vector in three-dimensional space with rectilinear coordinates x^i and the corresponding orthonormal basis e_i (with the usual summation convention implied) (Fig. F.1)

$$r = x^i e_i. \tag{F.1}$$

The total differential reads

$$dr = \frac{\partial r}{\partial x^i} dx^i. \tag{F.2}$$

Alternatively, from (F.1) follows

$$dr = dx^i e_i. \tag{F.3}$$

316

Comparing (F.2) with (F.3) results in

$$e_i = \frac{\partial r}{\partial x^i}. \tag{F.4}$$

Now a curvilinear coordinate system is introduced. Then r is given by the

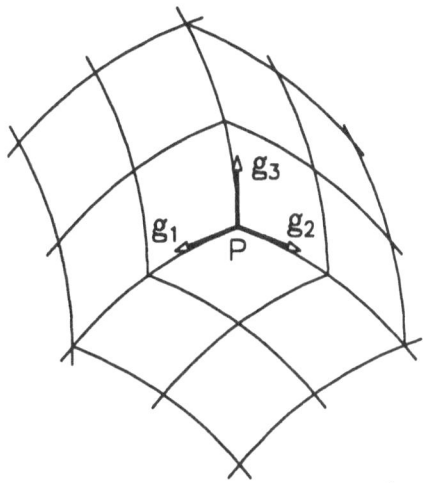

Figure F.2. Curvilinear coordinate system $\vartheta_1, \vartheta_2, \vartheta_3$. The basis vectors g_1, g_2, g_3 point in the directions of the ϑ_1-, ϑ_2- and ϑ_3-curves, respectively

three curvilinear coordinates $\vartheta^1, \vartheta^2, \vartheta^3$ (Fig. F.2). The total differential of r is, analogous to (F.2), given by

$$d r = r_{,i} d\vartheta^i. \tag{F.5}$$

Similar to (F.4), we define the basis

$$g_i = r_{,i} \tag{F.6}$$

where the g_i are dependent on the position P. Now, it is important to know what is the relationship between the basis $g_i(\vartheta^i)$ and the fixed basis e_i? From (F.2) and (F.4) follows

$$\frac{\partial r}{\partial \vartheta^k} = \frac{\partial r}{\partial x^i}\frac{\partial x^i}{\partial \vartheta^k} = e_i \frac{\partial x^i}{\partial \vartheta^k} \ . \tag{F.7}$$

Substituting (F.6) into (F.7) yields

$$g_k = e_i \frac{\partial x^i}{\partial \vartheta^k}. \tag{F.8}$$

Example: We apply (F.8) to spherical coordinates (ϑ, φ, a). The vector r is given by

$$\begin{aligned} x^1 &= a \sin \vartheta \cos \varphi \\ x^2 &= a \sin \vartheta \sin \varphi \\ x^3 &= a \cos \vartheta. \end{aligned} \tag{F.9}$$

From (F.8) and (F.9) follows

$$\mathbf{g}_1 = \mathbf{e}_i \frac{\partial x^i}{\partial \vartheta} = \mathbf{e}_1 a \cos \vartheta \cos \varphi + \mathbf{e}_2 a \cos \vartheta \sin \varphi - \mathbf{e}_3 a \sin \vartheta$$

$$\mathbf{g}_2 = \mathbf{e}_i \frac{\partial x^i}{\partial \varphi} = -\mathbf{e}_1 a \sin \vartheta \sin \varphi + \mathbf{e}_2 a \sin \vartheta \cos \varphi \qquad \text{(F.10)}$$

$$\mathbf{g}_3 = \mathbf{e}_i \frac{\partial x^i}{\partial a} = \mathbf{e}_1 \sin \vartheta \cos \varphi + \mathbf{e}_2 \sin \vartheta \sin \varphi + \mathbf{e}_3 \cos \vartheta.$$

Next, the components of the covariant metric tensor g_{ij} are calculated. The metric tensor appears quite naturally in the quadratic form

$$ds^2 = d\mathbf{r} \cdot d\mathbf{r} = \mathbf{g}_i d\vartheta^i \cdot \mathbf{g}_j d\vartheta^j = \mathbf{g}_i \cdot \mathbf{g}_j d\vartheta^i d\vartheta^j = g_{ij} d\vartheta^i d\vartheta^j. \qquad \text{(F.11)}$$

In the case of a spherical shell is the radius a constant, and hence, only the two variables ϑ, φ appear which, occasionally, we shall call ϑ_1, ϑ_2. Thus $g_{\alpha\beta}$ can be used where α, β take the values $1, 2$. From (F.10) follows

$$g_{\alpha\beta} = \mathbf{g}_\alpha \cdot \mathbf{g}_\beta = \begin{pmatrix} a^2 & 0 \\ 0 & a^2 \sin^2 \vartheta \end{pmatrix} \qquad \text{(F.12)}$$

and

$$ds^2 = g_{\alpha\beta} d\vartheta^\alpha d\vartheta^\beta = a^2 d\vartheta^2 + a^2 \sin^2 \vartheta d\varphi^2. \qquad \text{(F.13)}$$

Inverting (F.12) yields the contravariant metric tensor

$$g^{\alpha\beta} = \begin{pmatrix} \dfrac{1}{a^2} & 0 \\ 0 & \dfrac{1}{a^2 \sin^2 \vartheta} \end{pmatrix}. \qquad \text{(F.14)}$$

Moreover we define the *covariant derivative* "$|$". For a scalar $u(\vartheta_1, \vartheta_2, \vartheta_3)$ the partial derivative is a covariant tensor

$$u\Big|_i = u_{,i}. \qquad \text{(F.15)}$$

If we take the partial derivative of a vector field $\mathbf{A} = A^i \mathbf{g}_i$, we obtain

$$\mathbf{A}_{,j} = A^i_{,j} \mathbf{g}_i + A^i \mathbf{g}_{i,j}. \qquad \text{(F.16)}$$

The term $\mathbf{g}_{i,j}$ in (F.16) follows from (F.8) to

$$\mathbf{g}_{k,l} = \mathbf{e}_i \frac{\partial^2 x^i}{\partial \vartheta^k \partial \vartheta^l}. \qquad \text{(F.17)}$$

Substituting

$$\mathbf{e}_i = \mathbf{g}_j \frac{\partial \vartheta^j}{\partial x^i},$$

obtained from (F.8), into (F.17) yields

$$g_{k,l} = \frac{\partial \vartheta^j}{\partial x^i} \frac{\partial^2 x^i}{\partial \vartheta^k \partial \vartheta^l} g_j = \Gamma^j_{kl} g_j. \tag{F.18}$$

The Γ^j_{kl} are called *Christoffel symbols*. Inserting (F.18) into (F.16) results in

$$\mathbf{A}_{,j} = A^i \Big|_j \mathbf{g}_i \ .$$

The covariant derivative of contravariant vector components is

$$A^i \Big|_j = A^i_{,j} + \Gamma^i_{jk} A^k. \tag{F.19}$$

For covariant vector components instead of (F.19) we obtain

$$A_i \Big|_j = A_{i,j} - \Gamma^k_{ij} A_k \ . \tag{F.20}$$

In tensor analysis ([70]), it is shown that the Christoffel symbols can be calculated from the metric tensor and its derivatives. The following relationship holds

$$\Gamma^k_{lm} = \frac{1}{2} g^{kn} (g_{mn,l} + g_{nl,m} - g_{lm,n}). \tag{F.21}$$

Now we are ready to calculate the Christoffel symbols for the sphere. From (F.12) and (F.14) follows

$$\Gamma^1_{11} = \Gamma^2_{11} = \Gamma^1_{12} = \Gamma^1_{21} = \Gamma^2_{22} = 0. \tag{F.22}$$

Only the following Christoffel symbols are not equal to zero

$$\Gamma^2_{12} = \frac{1}{2} g^{2n} (g_{2n,1} + g_{n1,2} - g_{12,n})$$

$$= \frac{1}{2a^2 \sin^2 \vartheta} (2a^2 \sin \vartheta \cos \vartheta) = \cot \vartheta = \Gamma^2_{21} \tag{F.23}$$

$$\Gamma^1_{22} = -\sin \vartheta \cos \vartheta. \tag{F.24}$$

Given a function $w(\vartheta, \varphi) = w(\vartheta_1, \vartheta_2)$, the first and second covariant derivative $w \Big|_\alpha = w_{,\alpha}$ and $w \Big|_{\alpha\beta}$ can be calculated making use of the formulas (F.15) and (F.20). This results in the following quantities

$$\begin{aligned}
w \Big|_{11} &= w_{,11} - \Gamma^\alpha_{11} w_{,\alpha} &= w_{,11} \\
w \Big|_{12} &= w_{,12} - \Gamma^\alpha_{12} w_{,\alpha} &= w_{,12} - \cot \vartheta_1 w_{,2} \\
w \Big|_{21} &= w_{,21} - \Gamma^\alpha_{21} w_{,\alpha} &= w_{,21} - \cot \vartheta_1 w_{,2} \\
w \Big|_{22} &= w_{,22} - \Gamma^\alpha_{22} w_{,\alpha} &= w_{,22} + \sin \vartheta_1 \cos \vartheta_1 w_{,1}.
\end{aligned} \tag{F.25}$$

Using (F.25), (4.205) can be calculated to be

$$
\begin{aligned}
\Delta w &= g^{\alpha\beta}w\Big|_{\alpha\beta} = \frac{1}{a^2}w_{,11} + \frac{1}{a^2\sin^2\vartheta}(w_{,22} + \sin\vartheta\cos\vartheta w_{,1}) \\
&= \left(\frac{1}{a^2}\frac{\partial^2}{\partial\vartheta^2} + \frac{1}{a^2}\cot\vartheta\frac{\partial}{\partial\vartheta} + \frac{1}{a^2\sin^2\vartheta}\frac{\partial^2}{\partial\varphi^2}\right)w.
\end{aligned}
\tag{F.26}
$$

Δ is the Laplace-Beltrami operator relative to the metric tensor $g_{\alpha\beta}$ (the tilde˜in (4.205) is used because later these expressions are calculated on the unit sphere).

F.2   Spherical harmonics

Here, we indicate how the eigenvalue problem (Δ is now the Laplace-Beltrami operator on the unit sphere)

$$
\Delta Y(\vartheta,\varphi) + \lambda Y(\vartheta,\varphi) = 0
\tag{F.27}
$$

can be reduced to a well-known eigenvalue problem for an ordinary differential equation. We use the method of *separation of variables* in writing $Y(\vartheta,\varphi)$ as product of two functions each depending on only a single variable

$$
Y(\vartheta,\varphi) = \Phi(\varphi)\Theta(\vartheta) .
\tag{F.28}
$$

Substituting (F.28) into (F.27) yields

$$
-\frac{\Phi''}{\Phi} = \frac{\sin\vartheta(\sin\vartheta\Theta')'}{\Theta} + \lambda\sin^2\vartheta .
\tag{F.29}
$$

We argue now in the usual manner using separation of variables. Since the left-hand side is only a function of φ and the right-hand side is only a function of ϑ, both sides must be equal to a constant which we designate by μ. Therefore, (F.27) splits into two ordinary differential equations

$$
\Phi'' + \mu\Phi = 0 \quad\text{and}\quad \sin\vartheta(\sin\vartheta\Theta')' + (\lambda\sin^2\vartheta - \mu)\Theta = 0. \tag{F.30}
$$

Clearly, $\Phi(\varphi)$ must have the period 2π. Therefore, necessarily, $\mu = m^2$ is obtained, where m is an integer. Thus (F.30)$_2$ reads

$$
\sin\vartheta(\sin\vartheta\Theta')' + (\lambda\sin^2\vartheta - m^2)\Theta = 0
\tag{F.31}
$$

where

$$
0 < \vartheta < \pi.
\tag{F.32}
$$

By means of the transformation of variables

$$
\zeta = \cos\vartheta
\tag{F.33}
$$

(F.32) becomes

$$-1 < \zeta < 1. \tag{F.34}$$

Using $d\zeta = -\sin\vartheta\, d\vartheta$, we obtain

$$\frac{d}{d\vartheta} = \frac{d}{d\zeta}\frac{d\zeta}{d\vartheta} = -\sin\vartheta\frac{d}{d\zeta}. \tag{F.35}$$

With (F.35) and $u(\zeta) = \Theta(\vartheta)$ follows from (F.31)

$$-\sin^2\vartheta\frac{\partial}{\partial\zeta}\left(-\sin^2\vartheta\frac{\partial u}{\partial\zeta}\right) + (\lambda\sin^2\vartheta - m^2)u$$

$$= (1-\zeta^2)\frac{d}{d\zeta}\left((1-\zeta^2)\frac{du}{d\zeta}\right) + (\lambda(1-\zeta^2) - m^2)u = 0 .$$

Dividing by $(1-\zeta^2)$ we obtain

$$(1-\zeta^2)u'' - 2\zeta u' + \left(\lambda - \frac{m^2}{1-\zeta^2}\right)u = 0. \tag{F.36}$$

Equation (F.36) is the general form of the *Legendre differential equation* ([65]) for which the eigenvalue λ must be calculated.

First, we study the case $m = 0$. Substituting

$$u = \sum_{n=0}^{\infty} a_n \zeta^n \tag{F.37}$$

into (F.36) results in

$$(1-\zeta^2)\sum_{n=0}^{\infty} n(n-1)a_n\zeta^{n-2} - 2\zeta\sum_{n=0}^{\infty} na_n\zeta^{n-1} + \lambda\sum_{n=0}^{\infty} a_n\zeta^n = 0. \tag{F.38}$$

In order to satisfy (F.38), the coefficients of each power of ζ must be zero. The two lowest powers are ζ^{-2} and ζ^{-1} obtained by setting $n = 0$ and $n = 1$ in the first sum. Both coefficients vanish. Thus (F.38) can be rewritten in the equivalent form

$$\sum_{n=-2}^{\infty} (n+2)(n+1)a_{n+2}\zeta^n + \sum_{n=0}^{\infty} [-n(n-1) - 2n + \lambda]a_n\zeta^n = 0.$$

For $n = -2$ and -1, the coefficients a_0 and a_1 remain undetermined. Equating coefficients of all other powers of ζ yields the following condition for the coefficients

$$(n+2)(n+1)a_{n+2} - n(n-1)a_n - 2na_n + \lambda a_n = 0. \tag{F.39}$$

From (F.39) follows the recursion formula

$$a_{n+2} = \frac{n(n+1) - \lambda}{(n+2)(n+1)} a_n.$$
(F.40)

For $a_0 \neq 0$, $\sum_{k=0}^{\infty} a_{2k}\zeta^{2k}$ is everywhere convergent if it were only a finite sum. This is the case for

$$\lambda = \ell(\ell+1),$$
(F.41)

where $\ell \geq 0$ is an even integer number. A similar condition follows for $a_1 \neq 0$. Here a_0 and a_1 are the values $u(0)$ and $u'(0)$, respectively. This is the solution of the eigenvalue problem (F.36) for $m = 0$. The eigenvalues are given by (F.41). The corresponding eigenfunctions are *Legendre polynomials* $P_\ell(\zeta)$ which are the solutions of the following differential equation

$$(1 - \zeta^2)P_\ell''(\zeta) - 2\zeta P_\ell' + \ell(\ell+1)P_\ell(\zeta) = 0$$
(F.42)

subject to the condition

$$P_\ell(1) = 1.$$
(F.43)

For small ℓ, the Legendre polynomials can be calculated explicitly by means of (F.40) ([65]). For example, from

$$a_{n+2} = \frac{n(n+1) - \ell(\ell+1)}{(n+1)(n+2)} a_n$$

follows that for $\ell = 2$ with $a_0 = 1$: $a_2 = -3$ and satisfying (F.43), we obtain

$$P_2 = \frac{1}{2}(3\zeta^2 - 1).$$

Similarly, for $\ell = 4$ and $a_0 = 1$: $a_2 = -10$, $a_4 = 35/3$, we obtain

$$P_4 = \frac{1}{8}(35\zeta^4 - 30\zeta^2 + 3).$$

By a similar calculation we get $P_1 = \zeta, P_3 = \frac{1}{2}(5\zeta^3 - 3\zeta), \ldots$

Next, the case $m \neq 0$ in (F.36) will be treated. From the general theory of differential equations on the complex plane ([65]) follows that a solution of (F.36) is of the form

$$u = (1 - \zeta^2)^{m/2} f(\zeta)$$
(F.44)

where

$$f(\zeta) = \frac{d^m}{d\zeta^m} P_\ell(\zeta) = P_\ell^{(m)}(\zeta).$$
(F.45)

For (F.44), we may write

$$u = P_\ell^m(\zeta) = (1 - \zeta^2)^{m/2} P_\ell^{(m)}(\zeta)$$
(F.46)

where the P_ℓ^m are for $0 < m \le \ell$ the associated *Legendre polynomials*.

The spherical harmonics $Y_{\ell m}(\vartheta, \varphi)$ of ℓ-th degree are then the $2\ell + 1$ functions

$$P_\ell(\cos \vartheta), \quad P_\ell^m(\cos \vartheta) \cos m\varphi, \quad P_\ell^m(\cos \vartheta) \sin m\varphi; \quad m = 1, \ldots, \ell. \quad \text{(F.47)}$$

Finally, the assertion, made after equation (4.214), that the nonlinear operator (4.210) vanishes for spherical harmonics of order $\ell = 1$ is shown by calculation. With (4.214) and (F.14) follows from (4.210)

$$
\begin{aligned}
\{u, v\} &= 4uv - \left[g^{11} g^{11} u \Big|_{11} v \Big|_{11} + 2 g^{11} g^{22} u \Big|_{12} v \Big|_{12} + g^{22} g^{22} u \Big|_{22} v \Big|_{22} \right] - \\
&\quad - 2uv - 2uv + 2uv \\
&= 2uv - \left[u_{,11} v_{,11} + 2 \frac{1}{\sin^2 \vartheta} (u_{,12} - \cot \vartheta u_{,2})(v_{,12} - \cot \vartheta v_{,2}) + \right. \\
&\quad \left. + \frac{1}{\sin^4 \vartheta} (u_{,22} + \sin \vartheta \cos \vartheta u_{,1})(v_{,22} + \sin \vartheta \cos \vartheta v_{,1}) \right]
\end{aligned}
$$

where (F.25) and (4.205) have been used. The spherical harmonics of degree 1 are the three functions: $\cos \vartheta$, $\sin \vartheta \cos \varphi$, $\sin \vartheta \sin \varphi$. For example, we introduce $u = C_1 \sin \vartheta \cos \varphi$ and $v = C_2 \sin \vartheta \sin \varphi$ in the above relation. This yields $\{u, v\} = 0$.

Appendix G

Some properties of groups

In this appendix we follow the discussions in [55], [120], [142], [31].

G.1 Naive definition of a group

Definition G.1 (Group of transformations) *A group of transformations Γ is an aggregate $\{g_i\}$ of transformations of a given point set which satisfies the following properties:*

1. *It contains the identity transformation.*

2. *For every transformation g_ℓ, it also contains its inverse g_ℓ^{-1}.*

3. *If it includes g_ℓ and g_k, it also includes their composite $g_\ell g_k$. Furthermore the associative law $(g_i g_j)g_k = g_i(g_j g_k)$ must hold.*

Example 1: We consider nonsingular linear transformations in n-dimensional space. With respect to a fixed coordinate system, a linear transformation

$$\boldsymbol{x}' = \mathbf{T}\boldsymbol{x} \tag{G.1}$$

maps the point $\boldsymbol{x}$ with coordinates $(x_1, \ldots, x_n)^T$ into the point $\boldsymbol{x}' = (x_1', \ldots, x_n')^T$. Here $\mathbf{T} = (T_{ij})$, with $i, j = 1, \ldots, n$ is the matrix of the transformation with respect to the coordinate system. It is obvious that the set of nonsingular quadratic matrices forms a group, because all three properties given above are satisfied:

1. The identity transformation is the unit matrix $\mathbf{E}$.

2. If the matrix $\mathbf{T}$ is nonsingular, then there exists the inverse matrix $\mathbf{T}^{-1}$.

3. For two quadratic matrices $\mathbf{T}$ and $\mathbf{R}$ the matrix product $\mathbf{S} = \mathbf{TR}$ is again a quadratic matrix. Furthermore $(\mathbf{TR})\mathbf{S} = \mathbf{T}(\mathbf{RS})$ holds.

G.2 Symmetry groups

Particularly in physics and chemistry, so-called *symmetry transforma-tions* or *symmetry groups* are important. The symmetry of a rigid body, for example, is described by giving the set of all transformations (motions) which preserve the distance between all pairs of points of the body and bring the body into coincidence with itself. Of course, after such a transformation material points of the body have changed position but the body as a whole cannot be distinguished from its former position. Any such tansformation is called a symmetry transformation. It is clear that this set forms a group, the symmetry group of the body. All such distance preserving transformations can be built up from three fundamental types:

(1) Rotation through a definite angle about some axis.

(2) Mirror reflection in a plane.

(3) Parallel displacement (translation).

The last symmetry element, translation, can occur only if the body is infinite in extent (for example, an infinite crystal lattice). For a body of finite extension only the first two types of symmetry are possible. Moreover, all transformations of the symmetry group of a finite body must leave at least one point fixed. In other words, all axes of rotation and all planes of reflection must intersect in (at least) one point. Hence, these groups are called *point groups*.

   First, let us suppose that the body is brought into coincidence with itself if it is rotated about a certain axis through an angle $\psi = 2\pi/n$ (n is an integer). This is the group of rotations of a regular n-gon and will be called $\mathbf{Z}_n$. The trivial case $n = 1$ corresponds to the group containing only the *identity transformation*. The operation of rotation through $2\pi/n$ will be denoted by the symbol C_n. Successive applications of this transformation $C_n^2 = C_n \cdot C_n$, $C_n^3 = C_n \cdot C_n^2$, ... (that is, rotations through $4\pi/n, 6\pi/n, \ldots$) must also bring the body into coincidence with itself. C_n is called the *generator* of the group. It needs not be unique, since a rotation by $-2\pi/n$ gives also a generator if $n > 2$.

   If n is divisible by the integer ℓ, then

$$(C_n)^\ell = C_{n/\ell} \tag{G.2}$$

generates the subgroup $\mathbf{Z}_{n/\ell}$. That is, if $n = 6$, then C_2 and C_3 are also *symmetry rotations*. Hence, the largest n (or the smallest angle ψ) characterizes the symmetry. It is also clear that n successive rotations through $2\pi/n$ about the same axis yield the initial position and produce the identity transformation. Thus

$$(C_n)^n = E, \tag{G.3}$$

where E is the identity transformation.

The second fundamental symmetry operation is *reflection* in a plane which will be denoted by the symbol σ. It does not correspond to a physically possible rigid body motion. Since two successive reflections in the same plane move each point back to its initial position, that is, they yield the identity transformation, we obtain

$$\sigma^2 = E. \tag{G.4}$$

An important symmetry group is the *dihedral group* $\mathbf{D}_n$ consisting of all rotations and reflections in the plane which bring a regular n-gon into coincidence with itself. The composition of a reflection and a rotation yields a reflection about a different axis (see the example $\mathbf{D}_3$ of Fig. G.2). Therefore, the dihedral group $\mathbf{D}_n$ is generated by two elements: (i) a rotation C_n generating the subgroup $\mathbf{Z}_n$ and (ii) a reflection (flip) σ about an axis connecting the center and an edge of the n-gon. In Section G.7 we show that the same properties hold for the continuous group $\mathbf{O}(2)$ which is generated by an infinitesimal rotation and a reflection.

The symmetries studied so far are groups of transformations in three-dimensional space. The group elements themselves give a *representation* of the group in three dimensions. Let us give some examples:

Example 1: The operation $C(\vartheta)$, the rotation about the x_3–axis through an angle ϑ, is given by the transformation

$$\begin{array}{rcl} x_1' &=& x_1 \cos \vartheta + x_2 \sin \vartheta \\ x_2' &=& -x_1 \sin \vartheta + x_2 \cos \vartheta \\ x_3' &=& x_3 \end{array} \tag{G.5}$$

or written in matrix form

$$\begin{pmatrix} x_1' \\ x_2' \\ x_3' \end{pmatrix} = \begin{pmatrix} \cos \vartheta & \sin \vartheta & 0 \\ -\sin \vartheta & \cos \vartheta & 0 \\ 0 & 0 & 1 \end{pmatrix} \begin{pmatrix} x_1 \\ x_2 \\ x_3 \end{pmatrix}.$$

Example 2: An inversion I (Fig. G.1) is the composition of $C_2 \sigma$ according to (G.2) and (G.4), where C_2 and σ have the following matrix representations

$$\mathbf{T}_{C_2} = \begin{pmatrix} -1 & & \\ & -1 & \\ & & 1 \end{pmatrix}, \qquad \mathbf{T}_\sigma = \begin{pmatrix} 1 & & \\ & 1 & \\ & & -1 \end{pmatrix} \tag{G.6}$$

and hence

$$\begin{array}{rcl} x_1'' &=& -x_1 \\ x_2'' &=& -x_2 \\ x_3'' &=& -x_3 \end{array} \quad \text{or} \quad \begin{pmatrix} x_1'' \\ x_2'' \\ x_3'' \end{pmatrix} = \begin{pmatrix} -1 & 0 & 0 \\ 0 & -1 & 0 \\ 0 & 0 & -1 \end{pmatrix} \begin{pmatrix} x_1 \\ x_2 \\ x_3 \end{pmatrix}. \tag{G.7}$$

Thus an inversion moves each point P of the body to the point P'' being its inverse relative to the fixed point O. P and P'' are on opposite sides of O on the straight line POP'' (Fig. G.1). From $I = C_2\sigma$ follow the

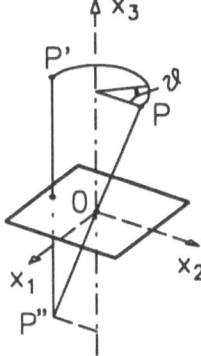

Figure G.1. The inversion I with respect to the fixed point 0 can be composed by a rotation C_2 and a reflection σ

two relations: $C_2 = I\sigma$ and $\sigma = C_2 I$. All three elements I, σ and C_2 commute. If any two of these elements belong to the symmetry group, then so does the third.

Example 3: We consider the symmetries of the equilateral triangle ([142]) to explain that the structure of a finite group can be displayed by its group table. The equilateral triangle (Fig. G.2) allows for six symme-

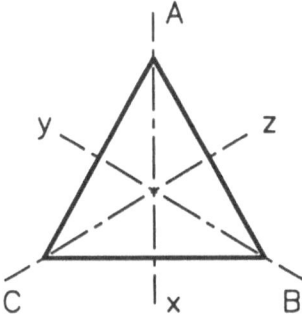

Figure G.2. Equilateral triangle with the axes of reflection x, y, z, which are thought to stay fixed when the triangle moves

try operations. There are the identity E and two counter clockwise rotations C_3 and $(C_3)^2$, which in the notation of (G.2) are rotations through $2\pi/3$ and $4\pi/3$. Further, there are three reflections $\sigma_x, \sigma_y, \sigma_z$ in the lines x, y, z, which are thought as staying fixed when the triangle moves. The set $K = \{E, C_3, (C_3)^2, \sigma_x, \sigma_y, \sigma_z\}$ of symmetries is closed under multiplication. It is called the dihedral group $\mathbf{D}_3$. The following table can be obtained

	E	C_3	$(C_3)^2$	σ_x	σ_y	σ_z
E	E	C_3	$(C_3)^2$	σ_x	σ_y	σ_z
C_3	C_3	$(C_3)^2$	E	σ_y	σ_z	σ_x
$(C_3)^2$	$(C_3)^2$	E	C_3	σ_z	σ_x	σ_y
σ_x	σ_x	σ_z	σ_y	E	$(C_3)^2$	C_3
σ_y	σ_y	σ_x	σ_z	C_3	E	$(C_3)^2$
σ_z	σ_z	σ_y	σ_x	$(C_3)^2$	C_3	E

In order to calculate the element in column σ_x and row C_3 first apply σ_x and then C_3. In other words, $C_3\sigma_x$ means: do σ_x and then do C_3. Hence, the triangle $\underset{C\ \ B}{\overset{A}{\triangle}}$ moves under σ_x to $\underset{B\ \ C}{\overset{A}{\triangle}}$ and then under C_3 to $\underset{A\ \ B}{\overset{C}{\triangle}}$ which gives the same effect as σ_y. (This is an example of a group of finite order for the frequently made statement in Chapter 5 for continuous groups that a reflection about one axis (plane) can be obtained by a rotation and a reflection about another axis (plane).) Let us demonstrate as a second example $\sigma_y\sigma_z$. Under σ_z the triangle $\underset{C\ \ B}{\overset{A}{\triangle}}$ moves to $\underset{C\ \ A}{\overset{B}{\triangle}}$ and under σ_y this goes to $\underset{B\ \ A}{\overset{C}{\triangle}}$ but this is what $(C_3)^2$ does, hence $\sigma_y\sigma_z = (C_3)^2$.

G.3   Representation of groups by matrices

In order to make use of the algebraic properties of groups it is convenient to represent the group elements by matrices, as we saw already in Section G.2. To each element g a matrix $\mathbf{T}_g$ is assigned, such that the group multiplication corresponds to the matrix multiplication

$$\mathbf{T}_{g_1g_2} = \mathbf{T}_{g_1}\mathbf{T}_{g_2}. \tag{G.8}$$

Example 1: The dihedral group $\mathbf{D}_3$ of Section G.2 can be represented by the following set of 3×3 matrices

$$
E = \begin{pmatrix} 1 & 0 & 0 \\ 0 & 1 & 0 \\ 0 & 0 & 1 \end{pmatrix}, \qquad
T_{C_3^1} = \begin{pmatrix} 0 & 1 & 0 \\ 0 & 0 & 1 \\ 1 & 0 & 0 \end{pmatrix},
$$

$$
T_{C_3^2} = \begin{pmatrix} 0 & 0 & 1 \\ 1 & 0 & 0 \\ 0 & 1 & 0 \end{pmatrix}, \qquad
T_{\sigma_x} = \begin{pmatrix} 1 & 0 & 0 \\ 0 & 0 & 1 \\ 0 & 1 & 0 \end{pmatrix}, \qquad \text{(G.9)}
$$

$$
T_{\sigma_y} = \begin{pmatrix} 0 & 0 & 1 \\ 0 & 1 & 0 \\ 1 & 0 & 0 \end{pmatrix}, \qquad
T_{\sigma_z} = \begin{pmatrix} 0 & 1 & 0 \\ 1 & 0 & 0 \\ 0 & 0 & 1 \end{pmatrix}.
$$

The property (G.8) can be checked easily. For example

$$
T_{C_3} T_{\sigma_x} = \begin{pmatrix} 0 & 0 & 1 \\ 0 & 1 & 0 \\ 1 & 0 & 0 \end{pmatrix} = T_{\sigma_y} .
$$

If a different basis is chosen a different representation would be obtained.

All the matrices in (G.9) map the vector $v_1 = (1,1,1)^T$ to itself. Therefore, we can expect to obtain a 2-dimensional representation by chosing two base vectors v_2, v_3 orthogonal to v_1. In this coordinate system a representation of the group D_3 by 2×2 matrices can be given ([143] p. 22).

On a Hilbert space H a representation T_g is called *unitary* provided that

$$
(T_g u, T_g v) = (u, v) \qquad \text{(G.10)}
$$

for all $u, v \in H$ and all $g \in \Gamma$.

G.4 Transformation of functions and operators

Suppose a transformation T represents an element of the group of transformations Γ. The action of the transformation takes x into $x' : x' = Tx$. An important question is: how does a scalar field $f(x)$ (for example, a temperature, density or displacement field) change under the action of the group Γ? To show this, a linear operator T_g is associated with T which acts on the function $f(x)$: for any function $f(x)$, the effect of the operator T_g on f is to change it to a function

$$
f' \equiv T_g f . \qquad \text{(G.11)}
$$

If f' is the scalar function for the same quantity calculated in the new co-ordinate system then the equivalence of the two representations requires

$$f'(\boldsymbol{x}') \equiv T_g f(\boldsymbol{x}') = f(\boldsymbol{x}) \qquad \text{if} \quad \boldsymbol{x}' = \mathbf{T}\boldsymbol{x}. \tag{G.12}$$

In other words, the transformed function $f' \equiv T_g f$ takes the same value at the image point $\boldsymbol{x}'$ that the original function f had at the object point $\boldsymbol{x}$. Or one may say that the point P (with coordinates $\boldsymbol{x}$) moves to its image point P' (with coordinates $\boldsymbol{x}'$) under the transformation $\mathbf{T}$ carrying with it the numerical value f at P. A physical interpretation of (G.12) is that the scalar function f for the same quantity must be the same in the two different representations. Equation (G.12) can be rewritten as

$$T_g f(\mathbf{T}\boldsymbol{x}) = f(\boldsymbol{x}) \tag{G.13}$$

or if $\boldsymbol{x}$ is replaced by $\mathbf{T}^{-1}\boldsymbol{x}$ then

$$T_g f(\boldsymbol{x}) = f(\mathbf{T}^{-1}\boldsymbol{x}). \tag{G.14}$$

This last form is most useful: T_g operating on f replaces $\boldsymbol{x}$ by $\mathbf{T}^{-1}\boldsymbol{x}$.

Example 1: We consider $x' = Tx = x + t$ (one dimension!), then $f'(x)$ is obtained from $f(x)$ by sliding the graph of $f(x)$ t units to the right: $f'(x) = T_g f(x) = f(T^{-1}x) = f(x - t)$ (Fig. G.3), where $T^{-1}x = x - t$.

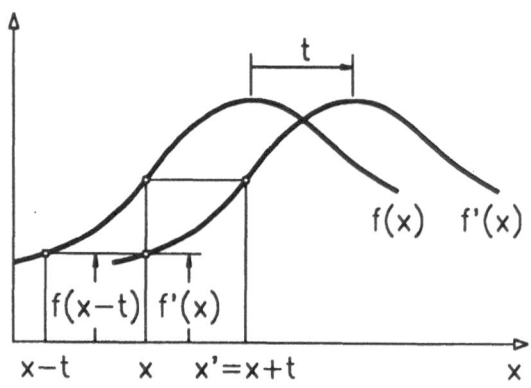

Figure G.3. The operator $Tx = x + t$ results in $f'(x) = T_g f(x) = f(T^{-1}x) = f(x - t)$

It may happen that $T_g f$ is identical with f, that is,

$$T_g f(\boldsymbol{x}) \equiv f(\boldsymbol{x}) \tag{G.15}$$

so that

$$f(\boldsymbol{x}) \equiv f(\mathbf{T}^{-1}\boldsymbol{x}) \qquad \text{or} \qquad f(\mathbf{T}\boldsymbol{x}) \equiv f(\boldsymbol{x}) \tag{G.16}$$

and the function takes on the same value at the image point $\mathbf{T}\boldsymbol{x}$ as at the point $\boldsymbol{x}$. In this case the function f is said to be *invariant* under the operator T_g or under the transformation $\mathbf{T}$. For example, the function $f(\boldsymbol{x}) = x_1^4 + x_2^2$ is invariant under inversion and the function $f(\boldsymbol{x}) = x_1^2 + x_2^2$ is invariant under rotations. To test for invariance of a function, the arguments $\boldsymbol{x}$ are replaced by their images $\mathbf{T}\boldsymbol{x}$ and then one has to check whether the same expression is obtained.

The operator T_g is linear, that is

$$T_g(f(\boldsymbol{x}) + h(\boldsymbol{x})) = T_g f(\boldsymbol{x}) + T_g h(\boldsymbol{x}). \tag{G.17}$$

From (G.14) it is also clear that

$$T_g(f(\boldsymbol{x})h(\boldsymbol{x})) = (f(\mathbf{T}^{-1}\boldsymbol{x})h(\mathbf{T}^{-1}\boldsymbol{x})) = T_g f(\boldsymbol{x})T_g h(\boldsymbol{x}). \tag{G.18}$$

Consider now an operator $H(\boldsymbol{x})$ acting on the function $f(\boldsymbol{x})$, giving the function $e(\boldsymbol{x}) = H(\boldsymbol{x})f(\boldsymbol{x})$. Then

$$T_g(H(\boldsymbol{x})f(\boldsymbol{x})) = T_g e(\boldsymbol{x}) = e(\mathbf{T}^{-1}\boldsymbol{x}) = H(\mathbf{T}^{-1}\boldsymbol{x})f(\mathbf{T}^{-1}\boldsymbol{x}). \tag{G.19}$$

On the other hand, using (G.19), one can write

$$T_g H(\boldsymbol{x})T_g^{-1}T_g f(\boldsymbol{x}) = H(\mathbf{T}^{-1}\boldsymbol{x})f(\mathbf{T}^{-1}\boldsymbol{x}) = H'(\boldsymbol{x})T_g f(\boldsymbol{x}) \tag{G.20}$$

where $H'(\boldsymbol{x}) = H(\mathbf{T}^{-1}\boldsymbol{x}) = T_g H(\boldsymbol{x})T_g^{-1}$ or $H'(\mathbf{T}\boldsymbol{x}) = H(\boldsymbol{x})$. The operators H and H' are in general not the same at a given point $\boldsymbol{x}$. If $H' = H$, then

$$T_g H(\boldsymbol{x})T_g^{-1} = H(\boldsymbol{x}) \tag{G.21}$$

and the operator $H(\boldsymbol{x})$ is said to be invariant under the transformation $\mathbf{T}$.

Let us treat two more examples.

Example 2: We consider the operator $H = a(x)\frac{\partial}{\partial x}$ and the transformation $T : x \to x + t$ as in Example 1. Then we have for a function $f(x)$: $e(x) = H(x)f(x) = a(x)\frac{\partial f}{\partial x}$ and $T_g(H(x)f(x)) = T_g e(x) = e(T^{-1}x) = e(x - t) = a(x - t)\frac{\partial f(x-t)}{\partial x}$. On the other hand we calculate $H(T_g f(x)) = H(f(T^{-1}x)) = H(f(x - t)) = a(x)\frac{\partial f(x-t)}{\partial x}$. Obviously, the operator H is not equivariant under T_g, because $T_g H(f) \neq H(T_g f)$, unless $a(x)$ is a constant or periodic with period t.

Example 3: In $\mathbb{R}^2$ we consider the operator $H = \frac{\partial}{\partial x_1}$, the function $f(x_1, x_2) = x_1^3$ and the transformation $\mathbf{T} = \begin{pmatrix} 0 & 1 \\ -1 & 0 \end{pmatrix}$ which is a rotation by $-\frac{\pi}{2}$. We have $f' = T_g f = f(\mathbf{T}^{-1}\boldsymbol{x}) = f(-x_2, x_1) = -x_2^3$ where $\mathbf{T}^{-1} = \begin{pmatrix} 0 & -1 \\ 1 & 0 \end{pmatrix}$ is used. Further we obtain $e = Hf = \frac{\partial f}{\partial x_1} = 3x_1^2$.

$e' = T_g e = e(\mathbf{T}^{-1}\boldsymbol{x}) = e(-x_2, x_1) = 3x_2^2$. On the other hand $H(T_g f) = \frac{\partial}{\partial x_1}(-x_2^3) = 0$. Obviously, as in Example 2, we have $T_g H f \neq H T_g f$.

Moreover, we show that (G.21) is not satisfied. We obtain

$$T_g H T_g^{-1} f = T_g H f(x_2, -x_1) = T_g \left(\frac{\partial}{\partial x_1} f(x_2, -x_1) \right) = -\frac{\partial}{\partial x_2} f(x_1, x_2)$$

which is different from $\frac{\partial}{\partial x_1} f(x_1, x_2)$.

Finally, we calculate the same example in polar coordinates. We have $x_1 = r \cos\varphi$, $x_2 = r \sin\varphi$, the operator $H = \frac{\partial}{\partial x_1} = \cos\varphi \frac{\partial}{\partial r} - \frac{\sin\varphi}{r} \frac{\partial}{\partial \varphi}$, the transformation $T : (r, \varphi) \to (r, \varphi - \pi/2)$ and the function $f = r^3 \cos^3\varphi$. $f' = T_g f = r^3 \cos^3(\varphi + \pi/2) = -r^3 \sin^3\varphi$, $Hf' = 0$. $e = Hf = 3r^2 \cos^2\varphi$, $e' = T_g e = 3r^2 \sin^2\varphi$. Hence, $H T_g f \neq T_g H f$. Further we have

$$T_g H T_g^{-1} f(r, \varphi) = T_g H f(T(r, \varphi)) = T_g H f \left(r, \varphi - \frac{\pi}{2} \right)$$
$$= T_g \left(\left(\cos\varphi \frac{\partial}{\partial r} - \frac{\sin\varphi}{r} \frac{\partial}{\partial \varphi} \right) f \left(r, \varphi - \frac{\pi}{2} \right) \right)$$
$$= \left(-\sin\varphi \frac{\partial}{\partial r} - \frac{\cos\varphi}{r} \frac{\partial}{\partial \varphi} \right) f(r, \varphi) = -\frac{\partial}{\partial x_2} f(x_1, x_2) .$$

G.5   Examples of invariant functions and operators

Example 1: We show that the function $f(\boldsymbol{x}) = x_1^2 + x_2^2$ is invariant under rotations.

With $\boldsymbol{x}' = \mathbf{T}\boldsymbol{x}$ and $\mathbf{T}$ given by (G.5)

$$\mathbf{T} = \begin{pmatrix} \cos\vartheta & \sin\vartheta \\ -\sin\vartheta & \cos\vartheta \end{pmatrix} \tag{G.22}$$

it must be shown that $x_1'^2 + x_2'^2 = x_1^2 + x_2^2$. This follows immediately by inserting for $\boldsymbol{x}'$.

Example 2: We show that the Laplacian operator $H = \partial^2/\partial x_1^2 + \partial^2/\partial x_2^2$ is invariant under rotations.

It must be shown that

$$\frac{\partial^2}{\partial x_1^2} + \frac{\partial^2}{\partial x_2^2} = \frac{\partial^2}{\partial x_1'^2} + \frac{\partial^2}{\partial x_2'^2}. \tag{G.23}$$

We calculate with (G.22)

$$\frac{\partial}{\partial x_1} = \frac{\partial}{\partial x_1'}\frac{\partial x_1'}{\partial x_1} + \frac{\partial}{\partial x_2'}\frac{\partial x_2'}{\partial x_1} = \cos\vartheta\frac{\partial}{\partial x_1'} - \sin\vartheta\frac{\partial}{\partial x_2'}$$

$$\frac{\partial}{\partial x_2} = \frac{\partial}{\partial x_1'}\frac{\partial x_1'}{\partial x_2} + \frac{\partial}{\partial x_2'}\frac{\partial x_2'}{\partial x_2} = \sin\vartheta\frac{\partial}{\partial x_1'} + \cos\vartheta\frac{\partial}{\partial x_2'}\ .$$

From the above follows

$$\frac{\partial}{\partial x_1'} = \cos\vartheta\frac{\partial}{\partial x_1} + \sin\vartheta\frac{\partial}{\partial x_2},$$

$$\frac{\partial}{\partial x_2'} = -\sin\vartheta\frac{\partial}{\partial x_1} + \cos\vartheta\frac{\partial}{\partial x_2}.$$

The second derivative is given by

$$\frac{\partial^2}{\partial x_1'^2} = \cos^2\vartheta\frac{\partial^2}{\partial x_1^2} + \sin^2\vartheta\frac{\partial^2}{\partial x_2^2} + 2\sin\vartheta\cos\vartheta\frac{\partial^2}{\partial x_1\partial x_2}$$

$$\frac{\partial^2}{\partial x_2'^2} = \sin^2\vartheta\frac{\partial^2}{\partial x_2^2} + \cos^2\vartheta\frac{\partial^2}{\partial x_2^2} - 2\sin\vartheta\cos\vartheta\frac{\partial^2}{\partial x_1\partial x_2}\ .$$

Inserting into the above assertion gives the required result.

Finally we show that (G.21) is satisfied. To simplify the calculation we insert $\vartheta = \pi/2$ in (G.22), and hence, we can use some expressions from Example 3 in Section G.4. We have

$$T_g H T_g^{-1} f(x_1, x_2) = T_g H f(x_2, -x_1)$$

$$= T_g\left(\frac{\partial^2}{\partial x_1^2}f(x_2, -x_1) + \frac{\partial^2}{\partial x_2^2}f(x_2, -x_1)\right)$$

$$= \frac{\partial^2 f(x_1, x_2)}{\partial x_2^2} + \frac{\partial^2 f(x_1, x_2)}{\partial x_1^2} = H f(x_1, x_2)\ .$$

G.6 Abstract definition of a group

Definition G.2 (Group) *We define as a group $\Gamma = (S, M)$ the tou-*
ple (S, M), where S gives the elements of the group and M defines the
composition of the elements, that is $M : S \times S \to S$. The following
requirements must be satisfied $(g, g', g'', \dots \in S)$:

(1) $M(gg') = gg' = g''.$
(2) $(gg')g'' = g(g'g'').$
(3)   There exists a g_1 such that for any g:
$$gg_1 = g_1g = g$$
$g_1 = e$ is the identity element.
(4)   For any g there exists a g^{-1}:
$$gg^{-1} = g^{-1}g = g_1 = e,$$
g^{-1} is the inverse element of g.

Definition G.3 (Order of a group) *The order $|\Gamma|$ of a group Γ is the number of its elements.*

Basically, there are three distinct cases:

$|\Gamma|$:
    (1)   finite
    (2)   countable infinite
    (3)   continuous.

Definition G.4 (Commutative group) *If $gg' = g'g$ for all $g \in \Gamma$, Γ is called a commutative or abelian group.*

In general $gg' \neq g'g$. The symmetry group K of the equilateral triangle in Section G.2 is not commutative. However, the subgroup K' defined below is commutative.

Definition G.5 (Subgroup) *If a subset $S' \subset S$ forms a group under M, the group $\Gamma' = (S', M)$ is called a subgroup.*

To explain what is meant by a subgroup we return to the dihedral group D_3 of the symmetries K of the triangle in Section G.2. If we select the set $K' = \{E, C_3, (C_3)^2\}$ containing only three symmetries, then these form a smaller group in the group K with six elements. These are the symmetries that rotate the triangle but do not overturn it. Not every subset of a group forms a subgroup. The symmetry group K of the equilateral triangle has six subgroups (four of them are nontrivial), which are as follows

$$K, \; K', \; K'' = \{E, \sigma_x\}, \; K''' = \{E, \sigma_y\}, \; K^{IV} = \{E, \sigma_z\}, \; K^{V} = \{E\} \,.$$

This can be seen from the multiplication table in Section G.2, because each of the subgroups satisfies the group requirements defined above. The order of the different subgroups is 6, 3, 2 and 1, which all divide the order 6 of the group K. This follows from the famous *theorem of Lagrange* that the order of the subgroups divides the order of the group ([142]).

By means of a simple example we will explain another important concept, namely the *isomorphism* between two groups. Consider the permutation of three elements a, b, c. At first glance, this seems to be an academic example but in [48] it is shown to be of great importance for the bifurcation analysis of a rectangular solid under uniform tension (Chapter 5). There are six different arrangements of the three elements and they form a group called $S_3 = \{g_1, g_2, g_3, g_4, g_5, g_6\}$, where the g_i are given by the following table:

	g_1	g_2	g_3	g_4	g_5	g_6
a	a	b	c	a	c	b
b	b	c	a	c	b	a
c	c	a	b	b	a	c

For S_3 we obtain exactly the same multiplication table as for the equilateral triangle in Section G.2 if we pair off rigid motions of the triangle given by K with the permutations given by S_3. This allows us to give the following definition ([142] p. 107):

Definition G.6 (Isomorphism of groups) *Two groups G and H are called isomorphic if there is a bijection $f : G \to H$ such that*

$$f(g_1 g_2) = f(g_1) f(g_2)$$

holds for all $g_1, g_2 \in G$.

Isomorphic groups like K and S_3 have the same abstract structure and differ only in their elements. For example, they have the same number and structure of subgroups.

   We now give two examples of groups which occur frequently in physical applications.

G.7   The orthogonal groups **O**(n) and **SO**(n)

The n-dimensional orthogonal group **O**(n) consists of all $n \times n$-matrices **T** satisfying

$$\mathbf{T}^T \mathbf{T} = \mathbf{E} \ . \tag{G.24}$$

Here $\mathbf{T}^T$ is the transpose of **T**.

   The special orthogonal group **SO**(n) consists of all $\mathbf{T} \in \mathbf{O}(n)$ such that $\det \mathbf{T} = 1$.

   The position of a rigid body in $\mathbb{R}^3$ which is fixed at one point can be expressed by an element of **SO**(3) ([5] ch. 6.3). Let $\boldsymbol{x}_1$ and $\boldsymbol{x}_2$ be two arbitrary vectors pointing from the fixed point to two points of the rigid body. After a rotation of the body the vectors $\boldsymbol{x}_i$ are changed to

$x'_i = \mathbf{T}x_i$. However, the inner product of the pairs of vectors has to remain constant. This results in

$$(\pmb{x}_1, \pmb{x}_2) = (\pmb{x}'_1, \pmb{x}'_2) = (\mathbf{T}\pmb{x}_1, \mathbf{T}\pmb{x}_2) \ .$$

Since the vectors are arbitrary, it follows that $\mathbf{T}^T\mathbf{T} = \mathbf{E}$.

If we choose an orthogonal tripod, it is rotated to a tripod of the same orientation. Therefore, the determinant of $\mathbf{T}$ cannot be negative.

$\mathbf{O}(3)$ (and $\mathbf{O}(2)$ in the plane) are special cases of the linear group $\mathbf{GL}(3)$ (and $\mathbf{GL}(2)$) which are given by the linear transformation (G.1) in the appropriate dimension. These transformations depend in general on 9 (respectively 4) parameters. However, only such transformations must be considered which leave the distance $x_1^2 + x_2^2 + x_3^2$ (respectively $x_1^2 + x_2^2$) invariant.

We consider now in detail the planar case, namely, $\mathbf{SO}(2)$ which is the group of rotations in $\mathbb{R}^2$. From (G.1) follows $x'_1 = t_{11}x_1 + t_{12}x_2$, $x'_2 = t_{21}x_1 + t_{22}x_2$ and

$$\begin{aligned} {x'_1}^2 + {x'_2}^2 &= (t_{11}x_1 + t_{12}x_2)^2 + (t_{21}x_1 + t_{22}x_2)^2 \\ &= (t_{11}^2 + t_{21}^2)x_1^2 + (t_{11}t_{12} + t_{21}t_{22})x_1x_2 + (t_{12}^2 + t_{22}^2)x_2^2 = x_1^2 + x_2^2 \ . \end{aligned}$$

For the relation above to hold, the following conditions must be satisfied

$$t_{11}^2 + t_{21}^2 = 1, \quad t_{11}t_{12} + t_{21}t_{22} = 0, \quad t_{12}^2 + t_{22}^2 = 1. \tag{G.25}$$

The four parameters t_{ij} are subjected to three functional relations, and hence, there remains only one free parameter giving a *one-parameter group*. The solution set of (G.25) consists of two disconnected one-parameter families of matrices

$$\begin{aligned} \mathbf{T}_\vartheta &= \begin{pmatrix} \cos\vartheta & \sin\vartheta \\ -\sin\vartheta & \cos\vartheta \end{pmatrix}, \\ \mathbf{T}_s &= \begin{pmatrix} -\cos\vartheta & \sin\vartheta \\ \sin\vartheta & \cos\vartheta \end{pmatrix} = \mathbf{T}_\vartheta \begin{pmatrix} -1 & 0 \\ 0 & 1 \end{pmatrix} . \end{aligned} \tag{G.26}$$

This is easy to understand because from (G.25)$_1$ we conclude $t_{11} = \cos\psi$, $t_{21} = \sin\psi$ and from (G.25)$_3$ that $t_{12} = \sin\vartheta$, $t_{22} = \cos\vartheta$. Finally (G.25)$_2$ yields $\cos\psi\sin\vartheta + \sin\psi\cos\vartheta = \sin(\psi + \vartheta) = 0$. This equation has two solutions $\psi = -\vartheta$ and $\psi = \pi - \vartheta$. If we insert these solutions for ψ into t_{11}, t_{21} we obtain the two matrices $\mathbf{T}_\vartheta$ and $\mathbf{T}_s$.

The transformation $\mathbf{T}_\vartheta$ contains the identity transformation and forms the subgroup $\mathbf{SO}(2)$ of rotations in the plane about the x_3-axis.

This group is abelian and isomorphic to the group

$$S^1 = \{[0, 2\pi], \varphi\vartheta = \varphi + \vartheta \bmod 2\pi\} \ .$$

The angle of the composition of two transformations is the sum of the angles of the individual transformations modulo 2π.

The transformation $\mathbf{T}_s$ can be composed by a reflection about the x_2-axis and a rotation $\mathbf{T}_\vartheta$. $\mathbf{T}_s$ does not form a group since it does not contain the identity transformation and is not closed under multiplication. The product of two elements in $\mathbf{T}_s$ is in $\mathbf{T}_\vartheta$. The square of any member of $\mathbf{T}_s$ is just the identity. The matrix $\mathbf{T}_s$ corresponds to a reflection about the line

$$\begin{pmatrix} x_1 \\ x_2 \end{pmatrix} = t \begin{pmatrix} \sin \vartheta/2 \\ \cos \vartheta/2 \end{pmatrix}. \tag{G.27}$$

This follows from the fact that the eigenvector of $\mathbf{T}_s$ corresponding to the eigenvalue $+1$ is invariant under $\mathbf{T}_s$. Therefore, it gives the direction about which the reflection is performed. The reflection about any other direction can be obtained by a composition of $\mathbf{T}_s$ and $\mathbf{T}_\vartheta$. The union of the two transformations $\mathbf{T}_\vartheta$ and $\mathbf{T}_s$ forms the group **O**(2). It is not abelian, since generally $\mathbf{T}_s\mathbf{T}_\vartheta \neq \mathbf{T}_\vartheta\mathbf{T}_s$.

Let us now turn to the orthogonal group **O**(3). This is the group of linear transformations which leaves $x_1^2+x_2^2+x_3^2$ invariant. Similarly to the case of **SO**(2), one now obtains for **SO**(3) six conditions imposed on the nine parameters in the 3×3 matrix $\mathbf{T}$. Hence, three parameters remain giving a *three-parameter* group. A common choice for these parameters are the *Eulerian angles* $\omega = (\varphi, \vartheta, \psi)$ (for a description of the Eulerian angles see for example [43]). To each element $\omega \in$ **SO**(3) corresponds a unique 3×3 orthogonal real matrix $\mathbf{D}_\omega$ which satisfies the following properties

$$\omega\omega' = \omega'' \to \mathbf{D}_\omega\mathbf{D}_{\omega'} = \mathbf{D}_{\omega\omega'} = \mathbf{D}_{\omega''}.$$

Since the matrices form a compact (bounded and closed) set, the group is called *compact*. Furthermore, the relation

$$\omega^{-1} \to \mathbf{D}_{\omega^{-1}} = \mathbf{D}_\omega^T$$

holds, where $\mathbf{D}_\omega^T$ is the transposed matrix of $\mathbf{D}_\omega$. It is clear that for $x \in \mathbb{R}^3$

$$x' = \mathbf{D}_\omega x$$

with $x' \in \mathbb{R}^3$, as well. The Eulerian angles $(\varphi, \vartheta, \psi)$

$$\begin{array}{ccccc} -\pi & \leq & \varphi & \leq & \pi \\ 0 & < & \vartheta & < & \pi \\ 0 & \leq & \psi & \leq & 2\pi \end{array}$$

fix an element of **SO**(3, $\mathbb{R}$). $\mathbf{D}_\omega$ can be composed by three simple rotations

([43]) as follows

$$\mathbf{D}_\omega = \mathbf{D}(\psi,0,0) \cdot \mathbf{D}(0,\vartheta,0) \cdot \mathbf{D}(0,0,\varphi)$$

$$= \begin{pmatrix} \cos\psi & \sin\psi & 0 \\ -\sin\psi & \cos\psi & 0 \\ 0 & 0 & 1 \end{pmatrix} \begin{pmatrix} 1 & 0 & 0 \\ 0 & \cos\vartheta & \sin\vartheta \\ 0 & -\sin\vartheta & \cos\vartheta \end{pmatrix} \cdot$$

$$\cdot \begin{pmatrix} \cos\varphi & \sin\varphi & 0 \\ -\sin\varphi & \cos\varphi & 0 \\ 0 & 0 & 1 \end{pmatrix} . \tag{G.28}$$

Since $D(0,0,\varphi)$ and $D(\psi,0,0)$ correspond to rotations about the same axis the representation by Eulerian angles is degenerate at $\vartheta = 0$. This can be clearly seen from (G.28) if $\vartheta = 0$ is introduced:

$$\mathbf{D}_\omega = \mathbf{D}(\psi,0,0)\mathbf{D}(0,0,\varphi) = \mathbf{D}(\psi+\varphi,0,0) = \mathbf{D}(0,0,\psi+\varphi) .$$

However, now the inverse transformation is no longer unique. That is, at $\vartheta = 0$ or π, φ and ψ are not unique coordinates.

The identity element $g = e$ is given by Eulerian angles $\omega_e = (0,0,0)$ and the inverse element by $g^{-1} = (-\varphi, -\vartheta, -\psi)$.

The group $\mathbf{O}(3)$ contains also reflections. They do not correspond to a rigid body motion. As already shown for the case $\mathbf{O}(2)$ above, every matrix $\mathbf{T} \in \mathbf{O}(3)$ can be composed by the product of a rotation $\mathbf{D}_\omega \in \mathbf{SO}(3)$ and a matrix $\mathbf{T}_s$ describing a reflection. For example $\mathbf{T}_s = \mathbf{T}_\sigma$ given by $(\text{G.6})_2$ describes a reflection about the x_1, x_2 plane.

The inversion $\mathbf{I} : \boldsymbol{x} \rightarrow -\boldsymbol{x}$, for example, can be composed by a rotation about the x_3-axis by an angle π and a reflection about the (x_1, x_2) plane (see Fig. G.1)

$$\mathbf{I} = \mathbf{T}_\sigma \mathbf{D}(\pi, 0, 0) .$$

G.8 The Euclidean group E(3)

The position of a rigid body in $\mathbb{R}^3$ can be described by the position vector $\mathbf{t}$ of one arbitrary point and the orthogonal matrix $\mathbf{R}$ describing the angular position of a right-handed orthonormal body-fixed frame. This set of positions can be given the structure of a group called $\mathbf{E}(3)$ with elements $g = (\mathbf{R}, \mathbf{t})$ by prescribing the multiplication law

$$(\mathbf{R}_1, t_1)(\mathbf{R}_2, t_2) = (\mathbf{R}_3, t_3) = (\mathbf{R}_1\mathbf{R}_2, \mathbf{R}_1 t_2 + t_1) . \tag{G.29}$$

The identity element is given by $(\mathbf{E}, \mathbf{0})$ and the inverse element to $(\mathbf{R}, \mathbf{t})$ is $(\mathbf{R}, \mathbf{t})^{-1} = (\mathbf{R}^T, -\mathbf{R}^T \mathbf{t})$. A matrix representation is obtained by forming the 4×4 matrix

$$\begin{pmatrix} & \mathbf{R} & & \mathbf{t} \\ 0 & 0 & 0 & 1 \end{pmatrix} \tag{G.30}$$

where the "1" is needed for the matrix multiplication

$$\begin{pmatrix} & \mathbf{R}_1 & & t_1 \\ 0 & 0 & 0 & 1 \end{pmatrix} \begin{pmatrix} & \mathbf{R}_2 & & t_2 \\ 0 & 0 & 0 & 1 \end{pmatrix} = \begin{pmatrix} & \mathbf{R}_3 & & t_3 \\ 0 & 0 & 0 & 1 \end{pmatrix}. \tag{G.31}$$

This group is a *six-parameter* group and it is no longer compact, because the translation parameter t is not confined to a finite domain. $\mathbf{R}_3$ follows from two successive rotations and the translation t_3 from the multiplication in (G.29).

Appendix H

Stability boundaries in parameter space

First, we give a definition of the *stability boundary* in parameter space making use of the well known fact that the asymptotics of solutions, as $t \to \infty$, of a linear system are determined by that eigenvalue of the operator which has the largest real part.

Definition H.1 (Stability boundary) *The stability boundary in parameter space is formed by level surfaces which correspond to those parameter values yielding the largest real part to zero.*

In general, stability boundaries will possess singularities ([7] p. 248). Therefore, it is useful to have some information on the form of stability boundaries in parameter space. Such information allows us to draw certain conclusions concerning the robustness of the system modelling. We assume the dynamical system to be given in the form

$$\dot{x} = \mathbf{A}(\lambda)x \qquad (\text{H.1})$$

where $x \in \mathbb{R}^n, \lambda \in \mathbb{R}^\ell$ and $\mathbf{A}$ is a $n \times n$ matrix. However, the following arguments are not restricted to the finite-dimensional case and (H.1) could also be an infinite-dimensional system. Of particular practical importance is the case $\ell = 2$, that is to consider a stability boundary in a two-dimensional parameter space. As it is explained in detail in Section 3.1, the loss of stability of a stable equilibrium is studied by keeping one of the two main parameters fixed and varying the other (the distinguished parameter) quasi-statically until the stability boundary is reached.

The following fundamental *theorem* describes the qualitative form of the stability boundary in a two-dimensional parameter space ([7] p. 255):

Theorem H.1 (Form of the stability boundary)

1. *The stability boundary of a general two-parameter family of matrices consists of smooth arcs intersecting transversally at their ends.*

2. *At the intersection of two arcs the acute angle of the stability boundary always points into the domain of instability.*

As example, we consider the double pendulum of Fig. 4.3. Setting $c = 0$ and $k = 0$ the classical stability problem of a nonconservatively loaded double pendulum is given. This problem has been investigated for the first time in [168] and later in [58]. In [168], [169] and [58] some surprising results are reported. Of particular interest is the "destabilizing effect" of internal damping. If the critical load P_u is calculated for the undamped system and compared with the critical load P_d for a system with slight viscous internal damping acting at the joints, ($k \neq 0$, but small) it is found that P_d may be smaller than P_u. Therefore, the addition of slight damping to the undamped system may actually lower the critical load. A second remarkable effect has been noted, namely that if one starts with a damped system and goes to the limit of an undamped system the critical load P_d ($k \rightarrow 0$) is different from the critical load P_u which is obtained for the originally undamped system.

A couple of remarks should be made regarding these effects. First of all, the significance of the value P_u as a "critical load" is questionable. The undamped system is unstable if the applied load P is greater than P_u. If P is less than P_u, however, the stability of the system cannot be determined by a linearized analysis. For the damped system, $P > P_d$ implies instability while $P < P_d$ implies asymptotic stability (all eigenvalues have negative real parts). In [110] it is also shown that for slightly damped nonconservative systems it can happen that a stable equilibrium can be made unstable by an increase in the magnitude of damping.

We represent the stability boundary of the system of Fig. 4.3 in a P, c parameter plane. It is especially interesting to see whether there is a qualitative difference between an undamped and a damped system with vanishing damping. From the result shown in Fig. H.1, it can be seen that for the undamped system the two curves which form the stability boundary do not intersect transversally but have a common point of tangency. Hence, according to Theorem H.1 this system is not in general position (structurally stable) but represents a bifurcation point. This is shown in [154] making use of bifurcation diagrams of matrices. On the other hand, it can be seen from Fig. H.1 that for the damped system a transversal intersection of the arcs of the stability boundary is obtained, so this system is in general position and thus insensitive to small perturbations.

From this the important conclusion can be made that if we obtain a stability boundary in parameter space for a mathematical model where

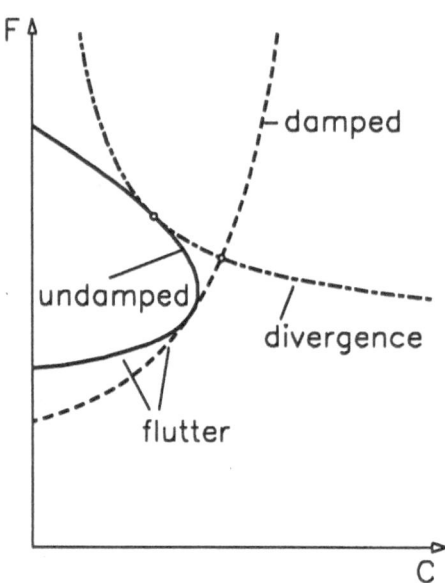

Figure H.1. Stability boundaries of the double pendulum of
Fig. 4.3. For the damped model
the curves intersect and for the
undamped model the curves have
a point of tangency. The former
is structurally stable the latter is
structurally unstable

the arcs do not intersect transversally, this mathematical model is not
a good one. Apparently, it is not a versal deformation of a degenerate
case as is explained in Section 6.4. For the modelling process, one can
say that some relevant parameters have not been included in the model
([154]).

We close this appendix with three remarks:

Remark 1: Theorem H.1 states that at intersections of arcs the acute
angle of the stability boundary points into the domain of instability.
This seems to be quite a general principle in many branches of
science. In [8] it is called "the principle of the fragility of good
things" because it states that good things (for example, stability)
are more fragile than bad things (that is, instability).

Remark 2: From Fig. H.1 follows that the system which sustains the maximum load is the most complicated from the mathematical point
of view. This is not a mere coincidence but a general property of
optimized systems as was mentioned in the Introduction. If we vary
c, this special non-generic case occurs in a structurally stable way
and hence cannot be disregarded as exceptional.

Remark 3: With good arguments in [79] it is claimed that a pure tangential loading (follower force load) as is it used in the model of Fig. 4.1
has not yet been realized technically in a laboratory. Therefore,
the questions treated above may seem to be of more academic than

practical importance. This, however, does not reduce the significance of follower force type of loadings which, for example, occur for a tube carrying fluid or for the force given by a rocket engine. Basically, these loadings lead to the same type of motions of a bar system as a pure tangential load.

Appendix I

Differential equation of an elastic ring

We derive the differential equations of an elastic ring under radial compression ([22], [81]). In Section 4.3.3 we showed that for this problem the Ljapunov-Schmidt reduction up to third order is not trivial.

The main simplifying assumption is that the elongation of the center line may be completely neglected. Thus, the assumption of an inextensional ring is made. It is a good assumption for sufficiently thin rings. The equations are obtained by proceeding in three steps as is explained in detail at the beginning of Appendix J. Here step one is to derive the equilibrium equations and steps two and three describe the bending deformation of the ring.

I.1 Equilibrium equations and bending

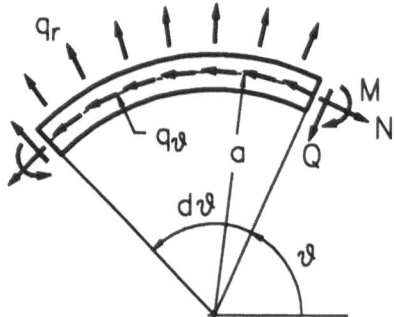

Figure I.1. Ring element with the loadings q_r, q_φ the forces N, Q and the moment M

With the notation shown in Fig. I.1, the moment equation and the two equilibrium equations in tangential and the radial direction, respectively,

344

follow to ([22])

$$M' - Q = 0 \tag{I.1}$$

$$N' + q_\vartheta + \frac{Q}{r} = 0 \tag{I.2}$$

$$Q' + q_r - \frac{N}{r} = 0. \tag{I.3}$$

Here a prime designates differentiation with respect to the arc length s and r is the radius of curvature. From (I.1)–(I.3), N and Q can be eliminated in order to obtain a single equation for the bending moment M as follows

$$(rM'')' + \frac{1}{r}M' + (rq_r)' + q_\vartheta = 0. \tag{I.4}$$

Equation (I.4) defines the equilibrium of a ring in the deformed geometry. Here, q_r and q_ϑ are always taken to remain normal and tangential to the central fiber of the ring, respectively.

Considering the mechanical model of an inextensionable ring, only bending deformations are possible. The original shape of the ring can be completely defined by specifying the initial curvature $1/r_0(s)$ in the undeformed configuration. With the geometrical assumption of a linear relationship for the fiber stretching depending on the distance from the central fiber and with a linear stress-strain relationship (Hooke's law), the following relationship results

$$M = EJ \left(\frac{1}{r} - \frac{1}{r_0} \right). \tag{I.5}$$

Here E and J are Young's modulus and the moment of inertia of the cross-section, respectively.

I.2   Ring equations

Equations (I.4) and (I.5) are the basic equations from which by proper manipulation the desired ring equation will be derived. We set $k(s) := 1/r$ and restrict ourselves to the initially circular ring, that is,

$$k_0(s) = \frac{1}{r_0(s)} = \frac{1}{a} = \text{const.} \tag{I.6}$$

Introducing (I.5) into (I.4), making use of (I.6) and taking $EJ = \text{constant}$, we obtain

$$(rk'')' + \frac{1}{2}(k^2)' + \frac{1}{EJ}[(rq_r)' + q_\vartheta] = 0. \tag{I.7}$$

Now some further simplifications are made. First, we set $q_\vartheta \equiv 0$, that is, only radial pressure acts on the ring. Second, we assume that $q_r = -\tau f(s) EJ$ where τ prescribes the magnitude of the radial load. From (I.7) then follows

$$\left(\frac{1}{k}k''\right)' + \frac{1}{2}(k^2)' - \tau\left(f\frac{1}{k}\right)' = 0. \tag{I.8}$$

Equation (I.8) can be integrated once to obtain

$$\frac{k''}{k} + \frac{1}{2}k^2 - \tau f\frac{1}{k} = \frac{1}{2a^2} - \mu . \tag{I.9}$$

The right-hand side of (I.9) is a suitably selected constant of integration. Finally, we stipulate $f(s) = 1$, that is, a ring under a uniform pressure is studied.

For the circular ring, it is advantageous to introduce the angle $\gamma(s)$ (Fig. I.2). In the undeformed configuration $\gamma_0 \equiv \vartheta$. Further the relations

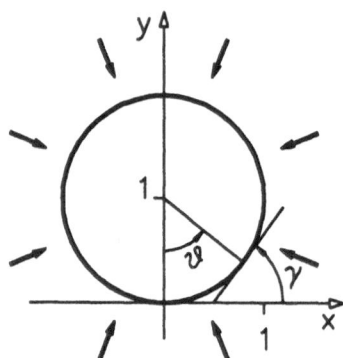

Figure I.2. Geometry of the circular ring

$$k(s) = \gamma'(s), \qquad k_0(s) = \gamma_0'(s) = \frac{1}{a} \tag{I.10}$$

hold. To study the fundamental solution $k_0(s) = \gamma_0'(s)$ the difference between $k(s)$ and $k_0(s)$ is designated by

$$v(s) = k(s) - k_0(s) = \gamma' - \frac{1}{a}. \tag{I.11}$$

Integrating (I.11) between 0 and $2\pi a$, the important relationship

$$\int_0^{2\pi a} v(t)dt = 0 \tag{I.12}$$

follows because the term on the right-hand side of (I.11) is zero after integration. Introducing $k(s) = 1/a + v(s)$ from (I.11) into (I.9) results in

$$\left(\frac{1}{a} + v\right)'' + \frac{1}{2}\left(\frac{1}{a} + v\right)^3 - \tau = \left(\frac{1}{2a^2} - a(\tau + \delta)\right)\left(\frac{1}{a} + v\right)$$

(I.13)

$$v'' + \left(\frac{1}{a^2} + a(\tau + \delta)\right)v = -\delta - \frac{1}{2}v^2\left(v + \frac{3}{a}\right).$$

In (I.13), a new constant δ given by $\mu = a(\tau + \delta)$ has been introduced to replace μ in (I.9). The constant δ in (I.13) must be selected such that the solution defines a closed ring without discontinuities in its slope.

Introducing in (I.13) the dimensionless variables

$$\xi = \frac{s}{a}, \qquad w = av, \qquad (\ )' = \frac{d}{ds} = \frac{1}{a}\frac{d}{d\xi} = \frac{1}{a}(\ )^{\cdot}$$

results in

$$\ddot{w} + (1 + a^3(\tau + \delta))w = -\delta a^3 - \frac{1}{2}w^2(w + 3).$$

(I.14)

Introducing further the parameter p and the constant β defined by

$$p = 1 + a^3\tau + \beta, \qquad \beta = \delta a^3,$$

(I.15)

the boundary value problem

$$w'' + pw = -\beta - \frac{1}{2}w^2(w + 3)$$

(I.16)

$$w(0) = w(2\pi) \qquad \text{and} \qquad w'(0) = w'(2\pi)$$

with the additional condition (I.12)

$$\int_0^{2\pi} w(t)dt = 0$$

(I.17)

can be formulated. Equation (I.16) has the trivial solution $w \equiv 0$ and $\beta = 0$ for all values of the loading parameter p.

Appendix J

Static shallow shell and plate equations for moderately large deformations

In this appendix we give a derivation of the differential equations for both plates and shallow shells for deformations beyond the linear domain. We follow the presentation in [96]. We start with the shallow shell structure. The plate equations then follow from the limiting case with zero initial curvatures.

Instead of using natural coordinates on the curved middle surface of the shallow shell (see Appendix F) cartesian coordinates x, y are introduced to represent the undeformed middle surface of the shell as a

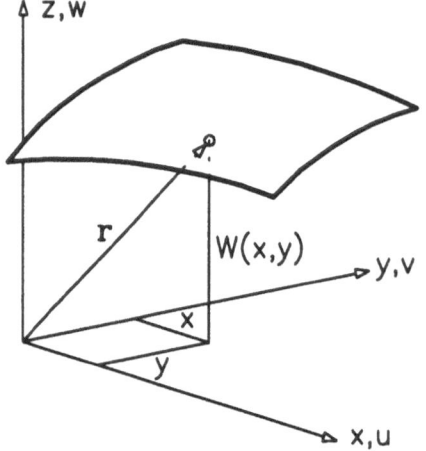

Figure J.1. Representation of the middle surface of the shallow shell in cartesian coordinates

function $z = W(x, y)$ (Fig. J.1). The displacements of the middle surface corresponding to the x, y, z directions are given by u, v, w, respectively.

To perform the derivation of the nonlinear shell equations, we proceed in three steps:

348

(1) The deformation of the structure is analysed in order to obtain a relationship between the displacements and the strain tensor. In this step the usual assumptions made in the linear theory of plates and shells ([152]) are extended by retaining nonlinear terms in the membrane strains while still assuming a linear relationship for the bending strains.

(2) The strain tensor is related to the stress tensor by means of a constitutive law. This law depends on the material of the shell. We use Hooke's law, which represents linear elastic behavior.

(3) The equations of equilibrium are derived which relate the stress tensor to the external loading.

By combining these three steps, we will obtain a relationship between the deformation of the structure and its external loading.

J.1 Deformation of the shell

An infinitesimally small element of the shell with volume $dV = hdxdy$ is considered. The vector r to a point P in the interior of the shell located at a distance ζ from the middle surface is given by (Fig. J.1 and Fig. J.2)

$$r = xe_1 + ye_2 + (W + \zeta)e_3. \qquad (J.1)$$

In order to express the displacement of P by the displacement components u, v, w of the middle surface of the shell in a manner as simple as possible use is made of a quasi *Kirchhoff hypothesis*. This states that sections $x = $ const. and $y = $ const. of the undeformed shell remain plane after deformation, and furthermore, maintain their angle with the de-

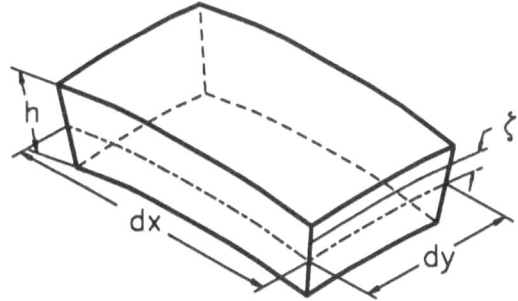

Figure J.2.   Differential shell element showing the coordinate ζ measured from the middle surface

formed middle surface. This angle is in general an oblique one. Thus, designating partial derivatives by $\partial/\partial x(\ ) = (\ )_{,x}, \quad \partial/\partial y(\ ) = (\ )_{,y}$ the relations

$$u(\zeta) = u - \zeta w_{,x}, \qquad v(\zeta) = v - \zeta w_{,y}, \qquad w(\zeta) = w \qquad (J.2)$$

are obtained. The vector $\hat{\boldsymbol{r}}$ to the point $\hat{P}$ in the interior of the deformed shell is given by

$$\hat{\boldsymbol{r}} = (x + u - \zeta w_{,x})\boldsymbol{e}_1 + (y + v - \zeta w_{,y})\boldsymbol{e}_2 + (W + \zeta + w)\boldsymbol{e}_3. \qquad (\text{J}.3)$$

To obtain the components of the strain tensor, the length of the differential line elements $(\mathbf{d}\boldsymbol{r})^2$ and $(\mathbf{d}\hat{\boldsymbol{r}})^2$ must be calculated. From (J.1) follows

$$\mathbf{d}\boldsymbol{r} = \boldsymbol{r}_{,x}dx + \boldsymbol{r}_{,y}dy = (\boldsymbol{e}_1 + (W+\zeta)_{,x}\boldsymbol{e}_3)dx + (\boldsymbol{e}_2 + (W+\zeta)_{,y}\boldsymbol{e}_3)dy \; . \quad (\text{J}.4)$$

Therefore, we obtain

$$(\mathbf{d}\boldsymbol{r})^2 = [1 + (W+\zeta)_{,x}^2]dx^2 + \qquad\qquad\qquad (\text{J}.5)$$
$$+ 2[(W+\zeta)_{,x}(W+\zeta)_{,y}]dxdy + [1 + (W+\zeta)_{,y}^2]dy^2.$$

The relation for the deformed shell follows from (J.3) to

$$\begin{aligned}
(\mathbf{d}\hat{\boldsymbol{r}})^2 &= [(1 + u_{,x} - (\zeta w_{,x})_{,x})^2 + (v_{,x} - (\zeta w_{,y})_{,x})^2 + \\
&\quad + (W + \zeta + w)_{,x}^2]dx^2 + \\
&\quad + 2[(1 + u_{,x} - (\zeta w_{,x})_{,x})(u_{,y} - (\zeta w_{,x})_{,y}) + \\
&\quad + (v_{,x} - (\zeta w_{,y})_{,x})(1 + v_{,y} - (\zeta w_{,y})_{,y}) + \\
&\quad + (W + \zeta + w)_{,x}(W + \zeta + w)_{,y}]dxdy + \\
&\quad + [(u_{,y} - (\zeta w_{,x})_{,y})^2 + (1 + v_{,y} - (\zeta w_{,y})_{,y})^2 + \\
&\quad + (W + \zeta + w)_{,y}^2]dy^2 \; .
\end{aligned} \qquad (\text{J}.6)$$

In (J.4), (J.5) and (J.6) ζ is still considered to be a function of x, y because thus far shells with variable thickness have not been excluded.

The strain tensor components $\varepsilon_{xx}, \varepsilon_{yy}, \varepsilon_{xy}$ are defined as

$$\frac{1}{2}((\mathbf{d}\hat{\boldsymbol{r}})^2 - (\mathbf{d}\boldsymbol{r})^2) = \varepsilon_{xx}dx^2 + 2\varepsilon_{xy}dxdy + \varepsilon_{yy}dy^2. \qquad (\text{J}.7)$$

Inserting (J.5) and (J.6) into (J.7) and collecting corresponding terms yields

$$\begin{aligned}
\varepsilon_{xx} &= u_{,x} + \frac{1}{2}(u_{,x}^2 + v_{,x}^2 + w_{,x}^2) + W_{,x}w_{,x} + \\
&\quad + \zeta_{,x}w_{,x} - (\zeta w_{,x})_{,x} + \ldots \\
&= u_{,x} + \frac{1}{2}(u_{,x}^2 + v_{,x}^2 + w_{,x}^2) + W_{,x}w_{,x} - \zeta w_{,xx} + \ldots \qquad (\text{J}.8) \\
\varepsilon_{yy} &= v_{,y} + \frac{1}{2}(u_{,y}^2 + v_{,y}^2 + w_{,y}^2) + W_{,y}w_{,y} - \zeta w_{,yy} + \ldots \\
2\varepsilon_{xy} &= u_{,y} + v_{,x} + u_{,x}v_{,y} + w_{,x}w_{,y} + \\
&\quad + W_{,x}w_{,y} + W_{,y}w_{,x} - 2\zeta w_{,xy} + \ldots \; .
\end{aligned}$$

The points in (J.8) indicate that terms of the form $u_{,x}(\zeta w_{,x})_{,x}$ and $((\zeta w_{,x})_{,x}^2$ have been neglected. This is because they are at least of third order. Furthermore it can be seen that the terms involving $\zeta_{,x}$, which are retained with respect to shells of variable thickness drop out. Therefore, equation (J.8) is still valid for shells of variable thickness.

Taking into consideration that for stability problems of thin-walled structures the displacement w orthogonal to the middle surface is much larger than the displacements u, v in the middle surface the quadratic terms in u, v are neglected compared to those in w in (J.8). This results in

$$
\begin{aligned}
\varepsilon_{xx} &= u_{,x} + \frac{1}{2}w_{,x}^2 + W_{,x}w_{,x} - \zeta w_{,xx} \\
\varepsilon_{yy} &= v_{,y} + \frac{1}{2}w_{,y}^2 + W_{,y}w_{,y} - \zeta w_{,yy} \\
2\varepsilon_{xy} &= u_{,y} + v_{,x} + w_{,x}w_{,y} + W_{,x}w_{,y} + W_{,y}w_{,x} - 2\zeta w_{xy}.
\end{aligned} \tag{J.9}
$$

J.2   Constitutive law

Hooke's law is used, as constitutive law describing the behavior of the material of the shell. It gives a linear relationship between the stress tensor and strain tensor. This, of course, implies that the strains must be small. However, the restriction to small strains does not necessarily require the deformation or the deformation gradients to be small ([167]). Furthermore, the usual assumption for thin-walled structures is made that the stress component $\sigma_{zz} \cong 0$. The components σ_{xz} and σ_{yz} which will appear in the equations of equilibrium do not contribute to the constitutive relationship because the corresponding strains are zero due to the generalized Kirchhoff hypothesis. Hence, a problem of plane stress over the thickness of the shell is obtained and the constitutive equations are

$$
\begin{aligned}
hE\left(u_{,x} + \frac{1}{2}w_{,x}^2 + W_{,x}w_{,x}\right) &= N_x - \nu N_y \\
hE\left(v_{,y} + \frac{1}{2}w_{,y}^2 + W_{,y}w_{,y}\right) &= N_y - \nu N_x \\
Gh(u_{,y} + v_{,x} + w_{,x}w_{,y} + W_{,x}w_{,y} + W_{,y}w_{,x}) &= N_{xy} .
\end{aligned} \tag{J.10}
$$

In (J.10) $G = E/(2(1 + \nu))$ is the shear modulus, E is Young's modulus, ν is Poisson's number and

$$
N_x = \int_{-h/2}^{+h/2} \sigma_{xx}dz, \qquad N_y = \int_{-h/2}^{+h/2} \sigma_{yy}dz, \qquad N_{xy} = \int_{-h/2}^{+h/2} \sigma_{xy}dz
$$

are the averaged stresses over the small shell thickness h. The bending moment M_x follows from (J.9) and (J.10) as

$$
\begin{aligned}
M_x &= \int_{-h/2}^{+h/2} \xi \sigma_{xx} d\xi \\
&= \int_{-h/2}^{+h/2} \xi \frac{E}{1-\nu^2}(\varepsilon_{xx} + \nu\varepsilon_{yy}) d\xi = -K(w_{,xx} + \nu w_{,yy}) .
\end{aligned}
\tag{J.11}
$$

In (I.11), the plate stiffness K is introduced by

$$
K = \frac{Eh^3}{12(1-\nu^2)}.
\tag{J.12}
$$

Similar expressions are obtained for M_y and M_{xy}, as follows

$$
\begin{aligned}
M_y &= -K(w_{,yy} + \nu w_{,xx}) \\
M_{xy} &= -(1-\nu)Kw_{,xy}.
\end{aligned}
\tag{J.13}
$$

J.3   Equations of equilibrium

Since we do not intend to study dynamic effects the equations of motion reduce to the *equations of equilibrium*. In a nonlinear theory, the equations of equilibrium for the stresses and loads must be calculated in the deformed geometry. However, in the nonlinear theory of shallow shells the approximation is made that the undeformed geometry can still be used for the calculation of the stresses and loads. This is an assumption which restricts the validity of the equations obtained to moderately large deformations. Thus, the equations of equilibrium of linear shell and plate theory ([152] cha. 12) can be used. Furthermore, we assume the shell thickness h to be constant. From Fig. J.3 and Fig. J.4 follow three force equilibrium equations

$$
\begin{aligned}
N_{x,x} + N_{xy,y} &= 0 \\
N_{xy,x} + N_{y,y} &= 0
\end{aligned}
\tag{J.14}
$$

$$
Q_{xz,y} + Q_{yz,x} + N_x(W+w)_{,xx} + N_y(W,w)_{,yy} + 2N_{xy}(W+w)_{,xy} + t = 0 \tag{J.15}
$$

where $t(x,y)$ is a distributed loading and two moment equilibrium equations

$$
\begin{aligned}
M_{xy,z} - M_{y,y} + Q_{yz} &= 0 \\
M_{yx,z} + M_{x,x} + Q_{xz} &= 0 .
\end{aligned}
\tag{J.16}
$$

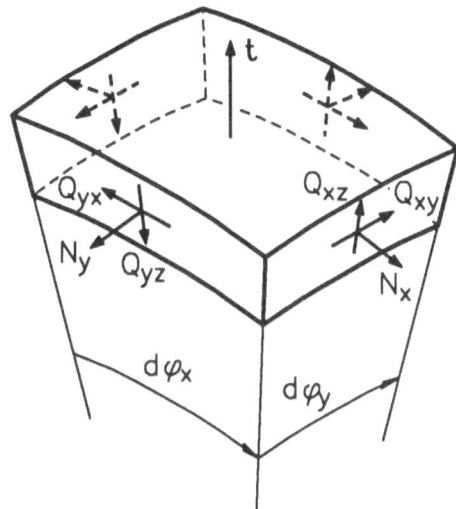

Figure J.3. Stress resultants and distributed loading $t(x, y)$ acting on a differential shell element

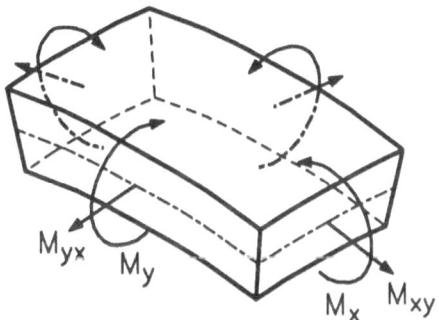

Figure J.4. Moments acting on a differential shell element

The moment equilibrium equation about the z-axis is identically satisfied. Differentiating the first equation in (J.16) with respect to x and the second equation with respect to y and inserting these expressions into (J.15) the shear forces can be eliminated and we obtain

$$M_{x,xx} - 2M_{xy,xy} + M_{y,yy} +$$
$$+ N_x(W + w)_{,xx} + 2N_{xy}(W + w)_{,xy} + N_y(W + w)_{,yy} + t = 0 . \tag{J.17}$$

The 12 equations (J.9), (J.10), (J.11), (J.13), (J.14), (J.17) determine the problem because they allow to calculate the 12 unknowns: $u, v, w, \varepsilon_{xx},$ $\varepsilon_{yy}, \varepsilon_{xy}, N_x, N_y, N_{xy}, M_x, M_y, M_{xy}$.

For the further treatment of the problem a reduction of this set of equations to two nonlinear differential equations in two variables can be made by introducing the stress function $F(x, y)$ defined by

$$N_x = F_{,yy}, \qquad N_y = F_{,xx}, \qquad N_{xy} = -F_{,xy}. \tag{J.18}$$

Substituting (J.18) into (J.14) satisfies these equations identically. Substituting (J.11), (J.13) and (J.18) into (J.17) results in a nonlinear partial differential equation

$$K\Delta^2 w = F_{,yy}(W+w)_{,xx}+F_{,xx}(W+w)_{,yy}-2F_{,xy}(W+w)_{,xy}+t(x,y) \quad (J.19)$$

in the two unknowns $w(x,y)$ and $F(x,y)$. A second equation follows by forming the identity

$$(u_{,x})_{,yy} + (v_{,y})_{,xx} - (u_{,y} + v_{,x})_{,xy} = 0. \qquad (J.20)$$

Inserting from (J.10) into (J.20) and making use of (J.18) yields

$$\Delta^2 F = Eh(w_{,xy}^2 - w_{,xx}w_{,yy} + 2W_{,xy}w_{,xy} - W_{,xx}w_{,yy} - W_{,yy}w_{,xx}). \quad (J.21)$$

Equations (J.19) and (J.21) are two nonlinear elliptic partial differential equations describing the deformation of the shallow shell. To give them a more compact form, a nonlinear operator is defined by

$$[a, b] = a_{,yy}b_{,xx} - 2a_{,xy}b_{,xy} + a_{,xx}b_{,yy}. \qquad (J.22)$$

The introduction of [,] shows a striking symmetry in equations (J.19) and (J.21) which are now specified for three special geometries, namely the plate, the sphere and the cylinder.

J.4   Special cases

J.4.1   Plate

In this case $W \equiv 0$. Equations (J.19) and (J.21) become

$$\begin{aligned} \frac{1}{Eh}\Delta^2 F &= -\frac{1}{2}[w, w] \\ K\Delta^2 w &= [F, w] + t \ . \end{aligned} \qquad (J.23)$$

These are the *von Karman* plate equations.

J.4.2   Sphere

Here $W_{,xx} = -1/a$, $W_{,yy} = -1/a$, $W_{,xy} = 0$ and a is the radius of the middle surface of the sphere. The equations are

$$\begin{aligned} \frac{1}{Eh}\Delta^2 F &= -\frac{1}{2}[w, w] + \frac{1}{a}(w_{,xx} + w_{,yy}) \\ K\Delta^2 w &= [F, w] + t - \frac{1}{a}(F_{,xx} + F_{,yy}) \ . \end{aligned} \qquad (J.24)$$

J.4.3 Cylinder

The axis of the cylinder is assumed to be in x-direction. Then $W_{,xx} = W_{,xy} = 0$ and the equations take the form

$$
\begin{aligned}
\frac{1}{Eh}\Delta^2 F &= -\frac{1}{2}[w,w] - W_{,yy}w_{,xx} \\
K\Delta^2 w &= [F,w] + t + F_{,xx}W_{,yy} \ .
\end{aligned}
\tag{J.25}
$$

Appendix K

Shell equations for axisymmetric deformations

The aim of this appendix is the derivation of nonlinear shell equations for finite symmetrical deflections of thin elastic shells of revolution. The main references for this appendix are [116], [115].

Similar to Appendix F, the analysis proceeds in three steps. First, the geometry is studied deriving a relationship between the displacement field and the strains. Second, since only static buckling is considered in Section 4.3.5, the equilibrium equations giving the relationship between the stress resultants and the couples on the one hand and the loading on the other hand are derived. Finally, in the third step the stress resultants and the couples are related to the strains by means of a constitutive relationship, namely Hooke's law.

K.1 Geometrical relations

The equation of the middle surface of the shell (Fig. K.1) is defined in parametric form by

$$r = r(\vartheta), \qquad z = z(\vartheta) . \tag{K.1}$$

The angle ϑ together with the polar angle φ in the x, y plane are the coordinates on the middle surface. The sloping angle γ of the tangent to a meridian curve is given by

$$\tan \gamma = \frac{dz}{dr} . \tag{K.2}$$

From (K.2) follows that (Fig. K.2)

$$\cos \gamma \;=\; \frac{dr}{ds} \;=\; \frac{dr}{\sqrt{dr^2 + dz^2}} \;=\; \frac{r'}{\sqrt{r'^2 + z'^2}} \;=\; \frac{r'}{\alpha}$$

$$(K.3)$$

$$\sin \gamma \; = \; \frac{z'}{\alpha} \qquad \text{with} \qquad \alpha = \sqrt{r'^2 + z'^2} \; .$$

The prime denotes differentiation with respect to ϑ. Let us now define the radial and circumferential unit vectors $\boldsymbol{j}_r$ and $\boldsymbol{j}_\varphi$ in terms of the unit vectors $\boldsymbol{i}, \boldsymbol{j}, \boldsymbol{k}$ in x, y and z direction, respectively (Fig. K.1 and Fig. K.2)

$$\boldsymbol{j}_r = \cos \varphi \boldsymbol{i} + \sin \varphi \boldsymbol{j}, \qquad \boldsymbol{j}_\varphi = -\sin \varphi \boldsymbol{i} + \cos \varphi \boldsymbol{j}. \qquad (K.4)$$

For the tangential and normal unit vectors $\boldsymbol{j}_\vartheta$ and $\boldsymbol{n}$ follows

$$\boldsymbol{j}_\vartheta = \cos \gamma \boldsymbol{j}_r + \sin \gamma \boldsymbol{k}, \qquad \boldsymbol{n} = -\sin \gamma \boldsymbol{j}_r + \cos \gamma \boldsymbol{k}. \qquad (K.5)$$

The radius vector $\mathbf{R}$ to a point of the shell may now be written in the

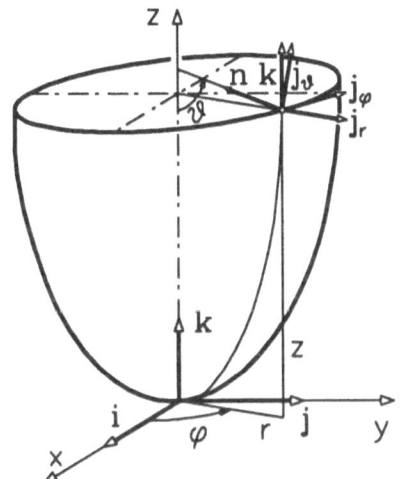

Figure K.1. Description of the middle surface of an axisymmetrical shell by $r(\vartheta)$ and $z(\vartheta)$

following form

$$\mathbf{R} = r\boldsymbol{j}_r + z\boldsymbol{k} + \zeta \boldsymbol{n} \; . \qquad (K.6)$$

In (K.6), ζ represents the distance of the point from the middle surface.

The radius vector $\mathbf{R}(\vartheta, \varphi, \zeta)$ depends on $\vartheta, \varphi, \zeta$ which define a system of orthogonal curvilinear coordinates in space. Inserting $\boldsymbol{j}_r$ and $\boldsymbol{n}$ from (K.4) and (K.5) into (K.6), $\mathbf{R}$ can be expressed in terms of $\boldsymbol{i}, \boldsymbol{j}, \boldsymbol{k}$ as

$$\mathbf{R} = (r - \zeta \sin \gamma) \cos \varphi \boldsymbol{i} + (r - \zeta \sin \gamma) \sin \varphi \boldsymbol{j} + (z + \zeta \cos \gamma) \boldsymbol{k} \; . \qquad (K.7)$$

The linear element $\mathbf{dR}$, follows to

$$\mathbf{dR} = \frac{\partial \mathbf{R}}{\partial \vartheta} d\vartheta + \frac{\partial \mathbf{R}}{\partial \varphi} d\varphi + \frac{\partial \mathbf{R}}{\partial \zeta} d\zeta \; . \qquad (K.8)$$

Carrying out the differentiations in (K.8), we obtain

$$dS^2 = \mathbf{dR} \cdot \mathbf{dR} = \alpha^2 \left(1 - \frac{\zeta \gamma'}{\alpha}\right)^2 d\vartheta^2 + r^2 \left(1 - \frac{\zeta}{r}\sin\gamma\right)^2 d\varphi^2 + d\zeta^2 .$$

(K.9)

From $R_\vartheta d\gamma = ds$ and (K.3) follows

$$\frac{1}{R_\vartheta} = \frac{\gamma'}{\alpha} \qquad \text{and} \qquad \frac{1}{R_\varphi} = \frac{\sin\gamma}{r} .$$ (K.10)

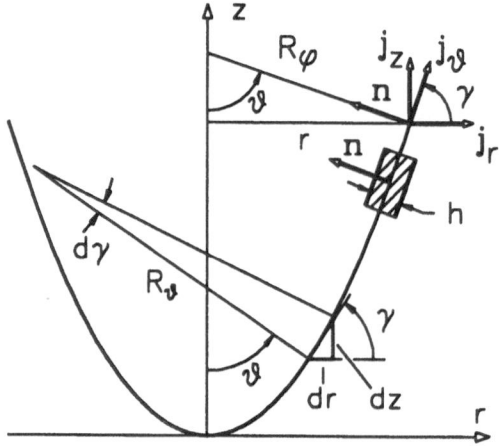

Figure K.2. Section of the middle surface of the axisymmetrical shell

The ratios in (K.10) are the principal radii of curvature of the middle surface of the shell.

We designate the quantities referring to the undeformed middle surface by a subscript 0. Then the location of the deformed middle surface is written in the form

$$r = r_0 + u, \qquad z = z_0 + w .$$ (K.11)

Here, u and w are the components of the displacement in the radial and axial directions, respectively. Moreover, with

$$\gamma = \gamma_0 - \beta$$ (K.12)

a new variable β, the angle enclosed by the tangents to the deformed and undeformed meridian, is defined (Fig. K.3 and Fig. K.4).

For the bending deformation, the usual assumption is made that the normal to the undeformed middle surface is deformed without extension into the normal to the deformed middle surface. This is an assumption similar to those of Bernoulli and Kirchhoff for rods and plates, respectively. It means that deformations due to transverse shear stresses $\tau_{\vartheta\zeta}$

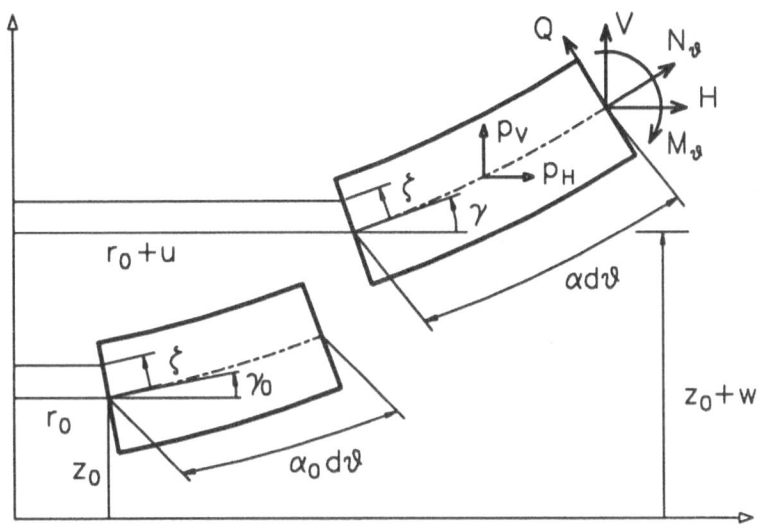

Figure K.3. Shell element in the undeformed and deformed state

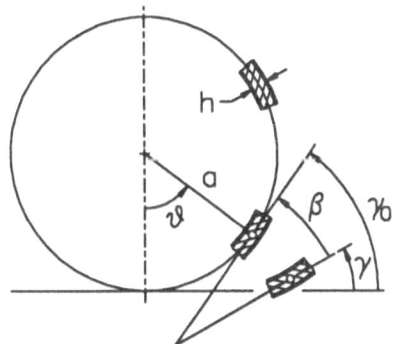

Figure K.4. Deformation of the spherical shell expressed by the angle β

and transverse normal stresses $\tau_{\zeta\zeta}$ are neglected in comparison with the deformations resulting from the remaining stresses.

The expression for the undeformed shell similar to equation (K.9) reads

$$dS_0^2 = a_0^2 \left(1 - \frac{\zeta \gamma_0'}{a_0}\right)^2 d\vartheta^2 + r_0^2 \left(1 - \frac{\zeta}{r_0} \sin \gamma_0\right)^2 d\varphi^2 + d\zeta^2 . \qquad (\mathrm{K}.13)$$

Since $\vartheta, \varphi, \zeta$ are orthogonal curvilinear coordinates, the strain components ε_ϑ and ε_φ follow from (K.9) and (K.13) to

$$\varepsilon_\vartheta = \frac{\alpha\left(1 - \frac{\zeta\gamma'}{\alpha}\right) - \alpha_0\left(1 - \frac{\zeta\gamma_0'}{\alpha_0}\right)}{\alpha_0\left(1 - \frac{\zeta\gamma_0'}{\alpha_0}\right)} = \frac{\alpha - \alpha_0 - \zeta(\gamma' - \gamma_0')}{\alpha_0\left(1 - \frac{\zeta\gamma_0'}{\alpha_0}\right)}$$

$$\varepsilon_\varphi = \frac{r\left(1 - \frac{\zeta}{r}\sin\gamma\right) - r_0\left(1 - \frac{\zeta}{r_0}\sin\gamma_0\right)}{r_0\left(1 - \frac{\zeta}{r_0}\sin\gamma_0\right)} \tag{K.14}$$

$$= \frac{r - r_0 - \zeta(\sin\gamma - \sin\gamma_0)}{r_0\left(1 - \frac{\zeta}{r_0}\sin\gamma_0\right)}.$$

We restrict the consideration to thin shells in the sense that the thickness h is small compared with the magnitudes of the radii of curvature R_ϑ and R_φ, defined in (K.10). Thus the termes with ζ in the denominator of (K.14) can be neglected and we replace (K.14) with

$$\varepsilon_\vartheta = \varepsilon_{\vartheta m} + \zeta\kappa_\vartheta, \qquad \varepsilon_\varphi = \varepsilon_{\varphi m} + \zeta\kappa_\varphi. \tag{K.15}$$

In (K.15), $\varepsilon_{\vartheta m}$ and $\varepsilon_{\varphi m}$ are the strains along the middle surface of the shell, whereas the terms $\zeta\kappa_\vartheta$ and $\zeta\kappa_\varphi$ represent the contribution from the bending deformation. Making use of (K.3) and (K.11), we obtain

$$\varepsilon_{\vartheta m} = \frac{\alpha - \alpha_0}{\alpha_0} = \frac{ds - ds_0}{ds_0}$$

$$= \frac{\cos\gamma_0}{\cos\gamma}\frac{dr}{dr_0} - 1 = \frac{\cos\gamma_0}{\cos\gamma}\left(1 + \frac{u'}{r_0'}\right) - 1$$

$$\varepsilon_{\varphi m} = \frac{r - r_0}{r_0} = \frac{u}{r_0} \tag{K.16}$$

$$\kappa_\vartheta = -\frac{\gamma' - \gamma_0'}{\alpha_0} = \frac{\beta'}{\alpha_0}$$

$$\kappa_\varphi = -\frac{\sin\gamma - \sin\gamma_0}{r_0}.$$

A relevant compatibility condition between $\varepsilon_{\vartheta m}$ and $\varepsilon_{\varphi m}$ is obtained by eliminating u from (K.16)$_1$ by inserting from (K.16)$_2$ to

$$\cos\gamma_0(r_0\varepsilon_{\varphi m})' - \cos\gamma(r_0'\varepsilon_{\vartheta m}) = r_0'(\cos\gamma - \cos\gamma_0). \tag{K.17}$$

Finally, an expression to calculate the displacements u follows immediately from (K.16)$_2$. We obtain

$$u = r_0\varepsilon_{\varphi m}. \tag{K.18}$$

Although the displacement w does not appear explicitly in (K.16), inserting from (K.3) $z' = \alpha\sin\gamma$ into (K.11)$_2$, we find

$$w' = \alpha\sin\gamma - z_0' = \alpha_0(1 + \varepsilon_{\vartheta m})\sin(\gamma_0 - \beta) - z_0'. \tag{K.19}$$

K.2 Stress resultants, couples and equilibrium equations

Due to the rotational symmetry and the assumptions made in Section K.1 the only non-vanishing stress components are the three components σ_ϑ, σ_φ, $\tau_{\vartheta\zeta}$. The stress resultants and couples are defined as follows:

$$N_\vartheta = \int_{-h/2}^{h/2} \sigma_\vartheta d\zeta \;, \qquad N_\varphi = \int_{-h/2}^{h/2} \sigma_\varphi d\zeta \;, \qquad Q = \int_{-h/2}^{h/2} \tau_{\vartheta\zeta} d\zeta \;,$$

$$M_\vartheta = \int_{-h/2}^{h/2} \zeta\sigma_\vartheta d\zeta \;, \qquad M_\varphi = \int_{-h/2}^{h/2} \zeta\sigma_\varphi d\zeta \;. \tag{K.20}$$

In the integrals in (K.20) the expression $(1+\zeta/r)d\zeta$ has been replaced by $d\zeta$, which is an admissible simplification for thin shells ([167]). Resultants

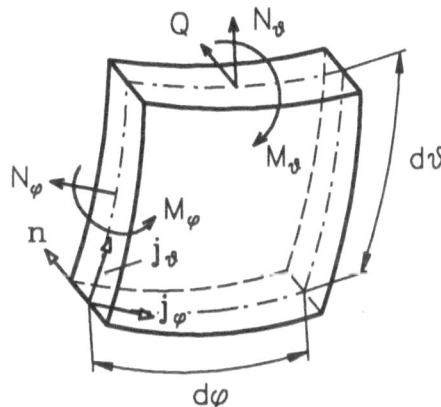

Figure K.5. Stress resultants and couples acting on an axisymmetric shell element

and couples may be combined to form resultant and couple vectors as follows (Fig. K.5)

$$\begin{aligned}
\mathbf{N}_\vartheta &= N_\vartheta \mathbf{j}_\vartheta & \mathbf{Q} &= Q\mathbf{n} & \mathbf{N}_\varphi &= N_\varphi \mathbf{j}_\varphi \\
\mathbf{M}_\vartheta &= M_\vartheta \mathbf{j}_\varphi & \mathbf{M}_\varphi &= M_\varphi \mathbf{j}_\vartheta .
\end{aligned} \tag{K.21}$$

In addition to these quantities, a load intensity vector $\mathbf{p}$ is introduced in the form

$$\mathbf{p} = p_\vartheta \mathbf{j}_\vartheta + p_n \mathbf{n}. \tag{K.22}$$

For what follows, it is convenient to write $\mathbf{N}_\vartheta$ and $\mathbf{p}$ in the alternative form which is (Fig. K.3)

$$\mathbf{N}_\vartheta = H\mathbf{j}_r + V\mathbf{k}, \qquad \mathbf{p} = p_H\mathbf{j}_r + p_V\mathbf{k}. \tag{K.23}$$

The radial (horizontal) stress resultant H and the axial (vertical) stress resultant V are related to N_ϑ and Q as follows

$$N_\vartheta = H \cos \gamma + V \sin \gamma, \qquad Q = -H \sin \gamma + V \cos \gamma. \qquad (K.24)$$

The differential equations of equilibrium for an element of the shell are for the forces

$$\frac{\partial}{\partial \vartheta}(r\mathbf{N}_\vartheta) + \frac{\partial}{\partial \varphi}(\alpha \mathbf{N}_\varphi) + \mathbf{p}r\alpha = 0 \qquad (K.25)$$

and the moments

$$\frac{\partial}{\partial \vartheta}(r\mathbf{M}_\vartheta) + \frac{\partial}{\partial \varphi}(\alpha \mathbf{M}_\varphi) + r\alpha[\mathbf{j}_\vartheta \times (\mathbf{Q} + \mathbf{N}_\vartheta) + \mathbf{j}_\varphi \times \mathbf{N}_\varphi] = 0. \qquad (K.26)$$

Inserting from (K.21) and (K.23) into (K.25) and (K.26) and making use of (K.4) and (K.5) results in

$$\begin{aligned} (rV)' + r\alpha p_V &= 0 \\ (rH)' - \alpha N_\varphi + r\alpha p_H &= 0 \\ (rM_\vartheta)' - \alpha \cos \gamma M_\varphi - r\alpha Q &= 0. \end{aligned} \qquad (K.27)$$

The first two equations in (K.27) are the force equilibrium equations in the axial and radial directions, respectively, while the third equation is the only nontrivial moment equilibrium equation.

K.3   Stress strain relations

We recall that the effect of transverse shear and transverse normal stresses on the deformations is neglected. Then, we obtain for an isotropic, homogeneous material with linear elastic behavior from Hooke's law the following relations

$$\varepsilon_\vartheta = \frac{1}{E}(\sigma_\vartheta - \nu\sigma_\varphi), \qquad \varepsilon_\varphi = \frac{1}{E}(\sigma_\varphi - \nu\sigma_\vartheta). \qquad (K.28)$$

Inserting from (K.16) into (K.28) and making use of (K.20) gives

$$\begin{aligned} N_\vartheta &= \frac{Eh}{1-\nu^2}(\varepsilon_{\vartheta m} + \nu\varepsilon_{\varphi m}) , \qquad & N_\varphi &= \frac{Eh}{1-\nu^2}(\varepsilon_{\varphi m} + \nu\varepsilon_{\vartheta m}) \\ M_\vartheta &= K(\kappa_\vartheta + \nu\kappa_\varphi) , \qquad & M_\varphi &= K(\kappa_\varphi + \nu\kappa_\vartheta) \end{aligned} \qquad (K.29)$$

where K is given by (J.12).

Equations (K.16), (K.27) and (K.29) are 11 equations for the 11 unknowns: u, β, $\varepsilon_{\vartheta m}$, $\varepsilon_{\varphi m}$, κ_ϑ, κ_φ, V, H, N_φ, M_ϑ and M_φ.

K.4 Spherical shell

For a spherical shell follows from Fig. K.4 that $r_0 = a \sin \vartheta$ and $z_0 = a(1 - \cos \vartheta)$ and from (K.3) $\alpha_0 = a$. From (K.16) follow with $\gamma_0 = \vartheta$ the relations

$$
\begin{aligned}
\varepsilon_{\vartheta m} &= \frac{\cos \vartheta}{\cos(\vartheta - \beta)} \left(1 + \frac{u'}{r_0'}\right) - 1 \\[2mm]
\varepsilon_{\varphi m} &= \frac{u}{r_0} \\[2mm]
\kappa_\vartheta &= \frac{\beta'}{a} \\[2mm]
\kappa_\varphi &= -\frac{\sin(\vartheta - \beta) - \sin \vartheta}{r_0} .
\end{aligned}
\tag{K.30}
$$

The compatibility condition (K.17) now becomes

$$
(r_0 \varepsilon_{\varphi m})' \cos \vartheta - r_0' \varepsilon_{\vartheta m} \cos(\vartheta - \beta) = r_0'[\cos(\vartheta - \beta) - \cos \vartheta] .
\tag{K.31}
$$

Equations (K.29) do not change. For the equilibrium equations, we consider $(K.27)_1$. With $(K.16)_1$ and $(K.16)_2$, we find

$$
[r_0(1 + \varepsilon_{\varphi m})V]' + (1 + \varepsilon_{\varphi m})(1 + \varepsilon_{\vartheta m})r_0 \alpha_0 p_V = 0 .
$$

Since we use Hooke's law, which is applicable only if the strains are small, it is justified to neglect $\varepsilon_{\varphi m}$ and $\varepsilon_{\vartheta m}$ as compared to unity. However, it has been pointed out in [102] that, for example, $\varepsilon_{\varphi m} \ll 1$ does not necessarily imply $c'_{\varphi m}$ to be small, and hence, these terms might be missing for large finite deflections in the following set of equations

$$
\begin{aligned}
(r_0 V)' + r_0 a p_V &= 0 \\
(r_0 H)' - a N_\varphi + r_0 a p_H &= 0 \\
(r_0 M_\vartheta)' - a M_\varphi \cos(\vartheta - \beta) - \\
- r_0 a[V \cos(\vartheta - \beta) - H \sin(\vartheta - \beta)] &= 0
\end{aligned}
\tag{K.32}
$$

given in [115]. Hence, according to [102], (K.32) might be valid for moderate rotations or small finite deflections, which is sufficient to describe the initial nonlinear post-buckling behavior.

We assume two different types of pressure loading. First, we consider dead loading, that is, p keeps always the direction pointing towards the origin of the sphere. In this case

$$
p_V = p \cos \vartheta, \qquad p_H = -p \sin \vartheta.
\tag{K.33}
$$

Second, a follower force type of loading is considered, that is, the pressure is always orthogonal to the shell surface. In this case

$$
p_V = p \cos(\vartheta - \beta), \qquad p_H = -p \sin(\vartheta - \beta).
\tag{K.34}
$$

From (K.32)$_1$ follows

$$r_0 V = - \int_0^\vartheta pa^2 \cos \tilde\gamma \sin \eta \, d\eta \qquad \text{(K.35)}$$

where

$$
\begin{aligned}
\tilde\gamma(\vartheta) &= \vartheta & \text{for dead loading} \\
\tilde\gamma(\vartheta) &= \vartheta - \beta & \text{for follower force loading.}
\end{aligned}
\qquad \text{(K.36)}
$$

Similar to the situation in Appendices F and J and for equations (4.128) and (4.204), it is possible to reduce the given set of equations to two nonlinear differential equations in two variables. One variable β describes the deformation and the second $\hat\psi$ is a stress variable defined by

$$\hat\psi = r_0 H \ . \qquad \text{(K.37)}$$

Here H is the horizontal stress resultant (Fig. K.3). The first of the two coupled equations for β and $\hat\psi$ is obtained by inserting M_ϑ and M_φ from (K.29), considering (K.30), (K.35) and (K.37), into (K.32)$_3$. The second equation by introducing $\varepsilon_{\vartheta m}$ and $\varepsilon_{\varphi m}$ from (K.29) expressed in terms of β and $\hat\psi$ by means of (K.24)$_1$, (K.27)$_2$, (K.35) and (K.37) into the compatibility equation (K.31).

Under the assumption that h and p do not depend on ϑ, these two equations can be written in dimensionless variables ([125]) in the following form

$$
\begin{aligned}
\delta \Bigg[\beta'' &+ \beta' \cot \vartheta + \cot^2 \vartheta \frac{\cos(\vartheta - \beta)}{\cos \vartheta} \frac{\sin(\vartheta - \beta) - \sin \vartheta}{\cos \vartheta} - \\
&- \nu \frac{\cos(\vartheta - \beta) - \cos \vartheta}{\sin \vartheta} \Bigg] \\
&= - \psi^* \frac{\sin(\vartheta - \beta)}{\sin \vartheta} - 4\lambda \frac{\cos(\vartheta - \beta)}{\sin \vartheta} \int_0^\vartheta \cos \tilde\gamma \sin \eta \, d\eta \ ,
\end{aligned}
\qquad \text{(K.38)}
$$

$$\delta \left[\psi^{\star\prime\prime} + \psi^{\star\prime} \cot \vartheta - \psi^{\star} \left(\cot^2 \vartheta \frac{\cos^2(\vartheta - \beta)}{\cos^2 \vartheta} - \right.\right.$$
$$\left.\left. - \nu(1 - \beta') \frac{\sin(\vartheta - \beta)}{\sin \vartheta} \right) \right]$$
$$= \frac{\cos(\vartheta - \beta) - \cos \vartheta}{\sin \vartheta} + \delta \left[-4\lambda \cot \vartheta \int_0^{\vartheta} \cos \tilde{\gamma} \sin \eta \, d\eta \cdot \right. \qquad \text{(K.39)}$$
$$\cdot \left(\frac{\sin 2(\vartheta - \beta)}{\sin 2\vartheta} + \nu(1 - \beta') \frac{\cos(\vartheta - \beta)}{\cos \vartheta} \right) +$$
$$\left. + 4\lambda\nu \sin(\tilde{\gamma} - \vartheta + \beta) + 4\lambda \frac{(\sin^2 \vartheta \sin \tilde{\gamma})'}{\sin \vartheta} \right].$$

Here, $\tilde{\gamma}$ is given by (K.36), $\lambda = p/p_c$ is the loading parameter with p_c given by (4.203), δ by (4.200) and

$$\psi^{\star} = \frac{4\hat{\psi}}{a^2 p_c} \qquad \text{(K.40)}$$

is the dimensionless stress function.

The boundary conditions for (K.38, K.39) are

$$\begin{aligned} \psi^{\star}(0) &= \psi^{\star}(\pi) &= 0 \\ \beta(0) &= \beta(\pi) &= 0. \end{aligned} \qquad \text{(K.41)}$$

$(\text{K.41})_1$ follows from (K.37) since $r_0(0) = r_0(\pi) = 0$ while $(\text{K.41})_2$ is a consequence of the rotational symmetry.

Appendix L

Equations of motion of a fluid conveying tube

In this appendix we consider a downhanging flexible tube clamped at the upper end and free at the lower end with an intermediate elastic support conveying an incompressible fluid flow (Fig. 5.7). In the derivation of the equations of motion, we adopt the assumptions made in [89] and

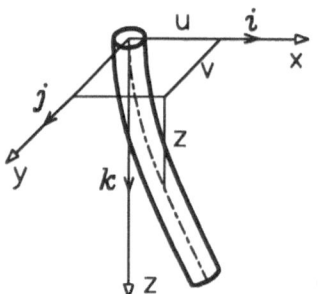

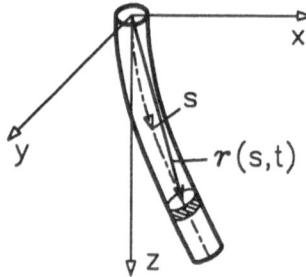

Figure L.1. Coordinates u, v, z describing the center line of the deformed tube

Figure L.2. Radius vector $r(s, t)$ to a fluid element at position s on the center line

follow the derivation given there. The three essential differences between our analysis and that of [89] are that we include damping and gravitational effects which are neglected in [89], and furthermore, a rotationally symmetric elastic support is assumed to act at the tube.

The following simplifying assumptions are made in setting up the mechanical models of the tube and the fluid flow:

(1) The tube has a uniform annular cross-section, even when it is bent.

(2) The tube is slender compared to its diameter.

366

(3) The effects of rotatory inertia and shear deformation are negligible.

(4) The center line of the tube is inextensible.

(5) The tube is made of a visco-elastic material and is initially straight.

(6) Plane sections before deformation remain plane after deformation (Bernoulli-Euler beam theory).

(7) The flow in the tube has constant velocity U relative to the tube and is in the direction of the tangent of the center line of the tube.

(8) The pressure variation in the fluid across a cross-section due to lateral accelerations is negligible.

We proceed similarly as in Appendices I, J and K. First, we describe the geometry of deformation giving the strains. Then the stresses and the strains are related by a material law. Finally, and here we are more general than in the Appendices I, J, K where only equations for a static system behavior were derived, the external loads are related to the stresses by means of the principles of linear and angular momentum, to obtain tube equations describing the dynamics of the tube.

L.1   Geometry of tube deformation

In the undeformed state the tube's axis is assumed to coincide with the z-axis of the introduced Cartesian co-ordinate system. Since the center line of the tube is assumed to be inextensible the arc length s along the deformed tube coincides with the z-axis in the undeformed state.

The location of a material point on the tube axis is given by Fig. L.1 and Fig. L.2

$$\boldsymbol{r} = u(s,t)\boldsymbol{i} + v(s,t)\boldsymbol{j} + z(s,t)\boldsymbol{k} \ . \tag{L.1}$$

Here $\boldsymbol{i}, \boldsymbol{j}, \boldsymbol{k}$ are fixed orthogonal unit vectors with $\boldsymbol{k}$ along the direction of the undeformed tube axis. From (L.1) follows the unit tangent vector $\boldsymbol{t}$ as

$$\boldsymbol{t} = \boldsymbol{r}' = u'\boldsymbol{i} + v'\boldsymbol{j} + z'\boldsymbol{k}. \tag{L.2}$$

Here and below the following notation for partial derivatives is used: $\partial/\partial s = (\ )'$, $\partial/\partial t = (\ )^{\cdot}$. The normal vector $\boldsymbol{n}$ is given by

$$\boldsymbol{t}' = \kappa \boldsymbol{n} \tag{L.3}$$

where κ designates the curvature of the space curve described by the center line. Hence, the tangent vector $\boldsymbol{t}$, the normal vector $\boldsymbol{n}$ and the binormal vector $\boldsymbol{b} = \boldsymbol{t} \times \boldsymbol{n}$ are determined by u, v and z. To describe points of the tube which are not located on the axis, a second coordinate

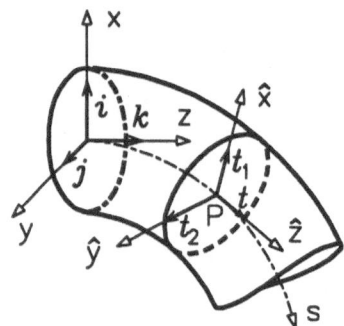

Figure L.3. Coordinate system $\hat{x}, \hat{y}, \hat{z}$ describing the orientation of a section of the deformed tube

system $(\hat{x}, \hat{y}, \hat{z})$ with its origin at the point P on the axis of the deformed tube is introduced (Fig. L.3). The axis $\hat{z}$ points into the direction of the tangent vector at P. The plane $\hat{x}, \hat{z}$ contains the linear material element, which was in the x-direction in the undeformed tube and $\hat{y}$ is perpendicular to this plane. Let $t(s,t)$, $t_1(s,t)$ and $t_2(s,t)$ be the unit vectors in the directions $\hat{z}, \hat{x}$ and $\hat{y}$, respectively, as is shown in Fig. L.3.

From $t_1 \cdot t_1 = 1$ follows

$$t_1' \cdot t_1 = 0 . \qquad (L.4)$$

Therefore t_1' lies in the plane spanned by t and t_2. Its component in the direction t_2 is the twist τ

$$t_1' \cdot t_2 = \tau . \qquad (L.5)$$

The component in the direction t follows from $(t_1 \cdot t)' = 0$ to

$$t_1' \cdot t = -t_1 \cdot t'. \qquad (L.6)$$

Since the projections of the vector t_1' onto three mutually perpendicular directions are known it may be expressed as

$$t_1' = -(t_1 \cdot t')t + \tau t_2 = -(t_1 \cdot t')t + \tau t \times t_1. \qquad (L.7)$$

For a deformed center line, $t(s,t)$ is known and, if in addition, we specify the twist τ then t_1 can be calculated from (L.7). Similarly an equation for t_2 is obtained as

$$t_2' = -(t_2 \cdot t')t + \tau t \times t_2. \qquad (L.8)$$

The coordinates u, v and the twist τ are the three dependent variables describing the geometry of the problem.

Two further important geometric relations can be derived. One is an explicit expression for the curvature κ in terms of the coordinates u, v, z of a point on the deformed axis. From (L.3) follows

$$\kappa^2 = |(t \times t')|^2 = t' \cdot t' = (u'')^2 + (v'')^2 + (z'')^2. \qquad (L.9)$$

The second kinematic relation

$$(u')^2 + (v')^2 + (z')^2 = 1 \tag{L.10}$$

follows from assumption (d) and is the inextensibility constraint.

L.2 Stress-strain relationship

The material of the tube is considered to behave linearly viscoelastic. That is, in the one-dimensional case, as it occurs here a linear stress (σ)-strain (ε) relationship of the form $\sigma = E\varepsilon + \alpha\dot{\varepsilon}$ is stipulated. In this expression, E is Young's modulus and α is a coefficient describing the internal dissipation.

We only indicate how an accurate analysis would have to be performed. However, for reasons given below we shall use a simpler approximative approach. From [88] p. 393 and Fig. L.3 we obtain for ε and $\dot{\varepsilon}$

$$
\begin{aligned}
\varepsilon &= \kappa_1\hat{y} - \kappa_2\hat{x} \\
\dot{\varepsilon} &= \dot{\kappa}_1\hat{y} - \dot{\kappa}_2\hat{x}
\end{aligned}
$$

where κ_1 and κ_2 are the components of curvature of the strained central line. The connection to κ given by (L.9) is

$$\kappa^2 = \kappa_1^2 + \kappa_2^2 \ .$$

As usual, instead of the stresses the resultant forces and couples at the cross-section are calculated. In our case, due to the inextensibility constraint, only the moment $\mathbf{M}$ is relevant. Its components follow from the above expressions to

$$
\begin{aligned}
M_{\hat{x}} &= J_B(E\kappa_1 + \alpha\dot{\kappa}_1) \\
M_{\hat{y}} &= -J_B(E\kappa_2 + \alpha\dot{\kappa}_2) \\
M_{\hat{z}} &= J_T(G\tau + \beta\dot{\tau}) \ .
\end{aligned}
$$

J_B and J_T are the axial and polar area moments of inertia of the cross-section of the tube, respectively. G is the shear modulus and β a material coefficient describing the dissipation due to twisting.

If we express the curvatures κ_1, κ_2 and the twist τ appearing in the components of the moment vector $\mathbf{M}$ by the expressions derived in Section L.1 we obtain very complicated expressions which are not easy to use for the further calculations because the vectors t_1 and t_2 appear in them. Therefore, we make a simpler ansatz for $\mathbf{M}$ in the following form

$$\mathbf{M} = J_B(E t \times t' + \alpha t \times \dot{t}') + J_T(G\tau t + \beta\dot{\tau} t) \ . \tag{L.11}$$

The argument leading to (L.11) is the following. If $\alpha = 0$, then the remaining terms in (L.11) are exactly those following from the equations given above. The damping terms in (L.11) introduced in an engineering manner may be considered as a good approximation to the damping terms obtained from the derivation above. In fact this approximation provides the correct linear damping terms.

L.3   Linear and angular momentum

The loading acting on the tube is due to its acceleration, the flowing fluid and the gravitational force. Let m_T and m_F denote the masses per unit length of the tube and the fluid, respectively. The inertial force per unit length of the tube due to its acceleration is $-m_T \ddot{r}(s,t)$.

To calculate the acceleration of a fluid element, we must notice that for a fluid element $s = s(t)$ and $ds/dt = U$ according to assumption (7) and Fig. L.2. Hence, taking the time derivatives of the position vector $r(s(t), t)$ of a fluid element we obtain for the velocity and the acceleration of the fluid element

$$v_F = \frac{dr}{dt} = \frac{\partial r}{\partial t} + \frac{\partial r}{\partial s}\frac{ds}{dt} = \left(\frac{\partial}{\partial t} + U\frac{\partial}{\partial s}\right) r(s,t)$$

and

$$\mathbf{a}_F = \left(\frac{\partial^2}{\partial t^2} + 2U\frac{\partial^2}{\partial t \partial s} + U^2\frac{\partial^2}{\partial s^2}\right) r(s,t) = \ddot{r} + 2U\dot{r}' + U^2 r''.$$

Under the usual assumptions made in continuum mechanics which are already used in (L.11), it is possible to obtain resultants of the stresses over the cross-section in form of an internal force $\mathbf{Q}(s,t)$ and a moment $\mathbf{M}(s,t)$. Let the change of $\mathbf{Q}$ and $\mathbf{M}$ due to shifting the cross-section from

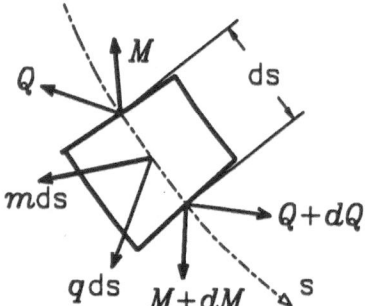

Figure L.4. Resultant force $\mathbf{Q}$, couple $\mathbf{M}$, distributed force $\mathbf{q}$ and distributed moment $\mathbf{m}$ acting on a tube element

s to $s + ds$ be expanded in a Taylor series. Then the force and moment equilibrium equations for an infinitesimally small differential tube and fluid element can be formulated. Taking the limit as $ds \to 0$ the two vector equations become

$$\mathbf{Q}' + \mathbf{q} = \mathbf{0}$$
$$\mathbf{M}' + \mathbf{t} \times \mathbf{Q} + \mathbf{m} = \mathbf{0}. \tag{L.12}$$

Here $\mathbf{q}$ and $\mathbf{m}$ are distributed forces and moments. From d'Alembert's principle ([167]) and the acceleration calculated above immediately follows

$$\mathbf{q} = -m_T\ddot{\mathbf{r}} - m_F(\ddot{\mathbf{r}} + 2U\dot{\mathbf{r}}' + U^2\mathbf{r}'') - \mu\dot{\mathbf{r}} + (m_T + m_F)g\mathbf{k}$$
$$\mathbf{m} \equiv \mathbf{0}. \tag{L.13}$$

Here is μ the coefficient of an assumed velocity proportional external damping and g is the gravitational acceleration. $(L.13)_2$ is a consequence of assumption (3).

Substituting (L.13) into (L.12), we find using (L.2) that

$$\mathbf{Q}' = (m_T + m_F)(\ddot{\mathbf{r}} - g\mathbf{k}) + m_F(2U\dot{\mathbf{t}} + U^2\mathbf{t}') + \mu\dot{\mathbf{r}} \tag{L.14}$$

$$\mathbf{M}' + \mathbf{t} \times \mathbf{Q} = \mathbf{0}. \tag{L.15}$$

It should be noted that friction between fluid and tube, which is an internal force, does not appear in (L.14).

The geometric relations (L.2), (L.3), (L.9), and (L.10), the stress-strain relation (L.11), and the equations of linear and angular momentum (L.14) and (L.15) are the basic equations describing the problem. What remains to be done is to properly rearrange these relationships.

Before we do this, we simplify (L.11) by showing that for the problem under investigation the twist τ can be neglected without influencing the results considerably. Taking the scalar product of (L.15) with $\mathbf{t}$ and combining it with (L.11), we obtain

$$\alpha J_B \mathbf{t} \cdot (\mathbf{t}' \times \dot{\mathbf{t}}') + J_T[G\tau' + \beta\dot{\tau}'] = 0. \tag{L.16}$$

From (L.16) follows for $\alpha = \beta = 0$, that is, for no internal damping, that $\tau' = 0$. This means, that the twist is constant along the axis of the tube. If no axial torque is applied at either end, then $\tau(s,t) \equiv 0$. However, due to (L.16) there is a nonlinear coupling between twist τ and the displacements u, v and z. In [148], numerical values of the material constants appearing in (L.16) are given. Even for silicon rubber the material damping constants are about 1 to 2 orders of magnitude smaller than the Young's modulus or the shear modulus G that appears in (L.16). Hence, it seems to be justified to neglect the nonlinear coupling between displacement and twist in the equations of motion, and to preceed with $\tau \equiv 0$ in the further calculations since no axial torque will be assumed to act on the tube.

L.4   Tube equations and boundary conditions

To write the equations (L.14) in the desired form, some further rearrangements are necessary. To obtain $\mathbf{Q}$, we take the cross product of t and (L.15), that is

$$t \times \mathbf{M}' + t \times (t \times \mathbf{Q}) = \mathbf{0}. \tag{L.17}$$

Applying the vector identity $\mathbf{a} \times (\mathbf{b} \times \mathbf{c}) = (\mathbf{a} \cdot \mathbf{c})\mathbf{b} - (\mathbf{a} \cdot \mathbf{b})\mathbf{c}$ to (L.17) yields

$$\mathbf{Q} = (N - pA)t + t \times \mathbf{M}' . \tag{L.18}$$

Here

$$N - pA = t \cdot \mathbf{Q} \tag{L.19}$$

is introduced. N denotes the axial force in the tube, A is the area of the cross-section of the fluid and p the pressure in the fluid. Before taking the derivative of (L.18) with respect to s and inserting this expression into (L.14) to obtain the equations of motion, the terms on the right-hand side in (L.18) must be calculated. From (L.11) follows (with $\tau \cong 0$)

$$t \times \mathbf{M}' = J_B\{E[(t \cdot t'')t - t''] + \alpha[(t \cdot \dot{t}')t' + (t \cdot \dot{t}'')t - \dot{t}'']\}. \tag{L.20}$$

Inserting (L.20) into (L.18), taking the derivative with respect to s and finally the inner product with t we find

$$
\begin{aligned}
t \cdot \mathbf{Q}' &= (N - pA)' + J_B\{E[(t \cdot t'')' - t \cdot t'''] + \\
&\quad + \alpha[(t \cdot \dot{t}')(t \cdot t'') + (t \cdot \dot{t}'') - (t \cdot \dot{t}''')]\} \\
&= (N - pA)' + J_B\{E(t' \cdot t'') + \alpha[-\kappa^2(t \cdot \dot{t}') + (t' \cdot \dot{t}'')]\}.
\end{aligned} \tag{L.21}
$$

To obtain (L.21), we differentiated (L.3) to obtain $t'' = \kappa'n + \kappa n'$. From Frenet's equations follows $n' = -\kappa t + \tau b$. Inserting this latter expression into the equation for t'' and taking the inner product with t yields $t \cdot t'' = -\kappa^2$, which is used in (L.21).

From (L.9) follows by differentiation that $t' \cdot t'' = \kappa\kappa' = 1/2(\kappa^2)'$. Inserting this expression into (L.21) and rearranging (L.21), we find

$$(N - pA + \frac{1}{2}EJ_B\kappa^2)' = t \cdot \mathbf{Q}' - \alpha J_B[-\kappa^2(t \cdot \dot{t}') + (t' \cdot \dot{t}'')].$$

This equation can be integrated in s. However, we want to have the action of an elastic intermediate or end support acting on the tube included in the equations of motion. Hence, we have to bear in mind that several quantities are discontinuous at the point of attachment of the elastic support. Therefore, the interval of integration must be split into two parts:

$$0 \le s \le \xi_- : \quad \left. \left(N - pA + \frac{1}{2}EJ_B\kappa^2\right)\right|_s^{\xi_-}$$

$$= \int_s^{\xi_-} \{t \cdot Q' - \alpha J_B[-\kappa^2(t \cdot t') + (t' \cdot t'')]\} ds$$

(L.22)

$$\xi_+ \le s \le L : \quad \left. \left(N - pA + \frac{1}{2}EJ_B\kappa^2\right)\right|_s^{L}$$

$$= \int_s^{L} \{t \cdot Q' - \alpha J_B[-\kappa^2(t \cdot t') + (t' \cdot t'')]\} ds$$

where $t \cdot Q'$ follows from the combination of (L.2) and (L.14) to

$$
\begin{aligned}
t \cdot Q' &= t \cdot [(m_T + m_F)(\ddot{r} - g k) + m_F(2U\dot{t} + U^2 t') + \mu \dot{r}] \\
&= (m_T + m_F)(u'\ddot{u} + v'\ddot{v} + z'\ddot{z} - gz') + \mu(u'\dot{u} + v'\dot{v} + z'\dot{z})
\end{aligned}
$$

(L.23)

To write down the equations of motion in the x- and y-direction, we return to (L.18) where for the second term on the right-hand side (L.20) must be inserted. To project Q in the x-direction we form the inner product with i

$$
\begin{aligned}
Q \cdot i &= (N - pA)t \cdot i + J_B\{E[(t \cdot t'')t \cdot i - t'' \cdot i] + \\
&\quad + \alpha[(t \cdot t')t' \cdot i + (t \cdot t'')t \cdot i - t'' \cdot i]\} \\
&= (N - pA)u' + J_B\{-E[\kappa^2 u' + u'''] + \\
&\quad + \alpha[(t \cdot t')u'' + (t \cdot t'')u' - u''']\}.
\end{aligned}
$$

(L.24)

For $Q \cdot j$, we obtain a similar expression if we replace u by v in (L.24). Taking the derivative of (L.24) with respect to s, the left-hand side of (L.14) for the x component is obtained. The equations of motion (L.14) may then be expressed in the form

$$
\begin{aligned}
G_1(u, v, U) &= (m_T + m_F)\ddot{u} + m_F(2U\dot{u}' + U^2 u'') + \\
&\quad + \mu \dot{u} + J_B(Eu^{IV} + \alpha \dot{u}^{IV}) - \\
&\quad - [(N - pA - EJ_B\kappa^2)u' + \\
&\quad + \alpha J_B\{(t \cdot t')u'' + (t \cdot t'')u'\}]' = 0 \\
G_2(u, v, U) &= (m_T + m_F)\ddot{v} + m_F(2U\dot{v}' + U^2 v'') + \\
&\quad + \mu \dot{v} + J_B(Ev^{IV} + \alpha \dot{v}^{IV}) - \\
&\quad - [(N - pA - EJ_B\kappa^2)v' + \\
&\quad + \alpha J_B\{(t \cdot t')v'' + (t \cdot t'')v'\}]' = 0.
\end{aligned}
$$

(L.25)

Equations (L.25) together with (L.10) are the desired equations of motion giving the relationship between z and u and v. The expression $(N - pA)$ is given by (L.22) for the corresponding domain of the tube.

Boundary and intermediate conditions

At $s = 0$, the tube is clamped. At $s = \xi$ there is the elastic support, and at $s = L$, the tube is free. Since we only need the boundary conditions for the solution of the linear eigenvalue problem in Section 5.3 we present below only the linearized expressions. We obtain

for the fixed end at $s = 0$:

$$u = u' = 0, \quad v = v' = 0 . \tag{L.26}$$

For the point of support at $s = \xi$: The pressure p in the fluid and the curvature of the tube are continuous, however, the resultant section force $\mathbf{Q}$ is discontinuous. Hence, the following jump conditions

$$
\begin{aligned}
p(\xi_-) - p(\xi_+) &= 0 \\
\kappa(\xi_-) - \kappa(\xi_+) &= 0 \\
\mathbf{Q}(\xi_-) - \mathbf{Q}(\xi_+) &= \mathbf{S}
\end{aligned}
\tag{L.27}
$$

must hold. Here is $\mathbf{S}$ the force of the linear elastic support with stiffness c_s which points into the direction to the z-axis. It is given by

$$\mathbf{S} = -c_s(u\mathbf{i} + v\mathbf{j}) . \tag{L.28}$$

The linearized intermediate conditions are

$$
\begin{aligned}
u'''(\xi_+) + \frac{\alpha}{E}\dot{u}'''(\xi_+) - u'''(\xi_-) - \frac{\alpha}{E}\dot{u}'''(\xi_-) &= -c_s u(\xi) \\
v'''(\xi_+) + \frac{\alpha}{E}\dot{v}'''(\xi_+) - v'''(\xi_-) - \frac{\alpha}{E}\dot{v}'''(\xi_-) &= -c_s v(\xi)
\end{aligned}
$$

and in addition

$$u, \ u', \ u'' + \frac{\alpha}{E}\dot{u}'' \quad \text{and} \quad v, \ v', \ v'' + \frac{\alpha}{E}\dot{v}''$$

are continuous.

For the free end at $s = L$: The moment $\mathbf{M}$ according to (L.11) and the resultant force $\mathbf{Q}$ according to (L.18) must vanish. The linearized boundary contitions are

$$
\begin{aligned}
u'' + \frac{\alpha}{E}\dot{u}'' &= 0 \\
v'' + \frac{\alpha}{E}\dot{v}'' &= 0
\end{aligned}
\tag{L.29}
$$

Since $(N - pA) = 0$ for $s = L$ we obtain from (L.18) $t \times \mathbf{M}' = \mathbf{0}$, which is exactly (L.20). Writing this equation in components and retaining only the linear terms we obtain

$$
\begin{aligned}
u''' + \frac{\alpha}{E}\dot{u}''' &= 0 \\
v''' + \frac{\alpha}{E}\dot{v}''' &= 0 \,.
\end{aligned}
\qquad\text{(L.30)}
$$

Appendix M

Various concepts of equivalences

In this appendix we follow [48], [109].

The concept of equivalence is fundamental in the treatment of bifurcation problems. To motivate its significance we return to the buckling problem of the double pendulum in Fig. 2.5a. For simplicity we include now only one pendulum. In addition to the load F we apply also a

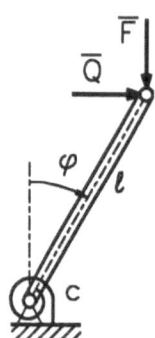

Figure M.1. Simple conservatively loaded pendulum with one degree of freedom

transversal load Q (Fig. M.1). In this case the potential (4.60) is

$$V = \frac{\overline{V}}{c} = \frac{\varphi_1^2}{2} - \frac{\overline{F}}{c}\ell(1 - \cos\varphi_1) - \frac{\overline{Q}}{c}\ell\sin\varphi_1 . \qquad (M.1)$$

Dropping the index and using new dimensionless parameters F, Q we obtain the equilibrium equation

$$f(\varphi, F, Q) = f(\varphi, \overline{\lambda}) = \varphi - F\sin\varphi - Q\cos\varphi = 0 . \qquad (M.2)$$

The question to be asked now is: can we find an expression simpler than (M.2) which describes the bifurcation problem in the neighborhood of the bifurcation point qualitatively correctly? Or in other words: does there exist an equivalence relation between (M.2) and a simpler problem $g(x, \overline{\lambda}) = 0$?

Before we proceed we remark that more precisely the term *"germ"* instead of "function" should be used in the following. However, as all our considerations are local we may neglect this difference and we refer to [48] p. 54 for a more detailed treatment of this point.

There are at least four different cases of *equivalence* which are important in connection with bifurcation problems.

M.1 Right-equivalence

Definition M.1 (right-equivalence) *Two functions* $V, W : \mathbb{R}^n \to \mathbb{R}$ *are said to be right-equivalent around* $\mathbf{0}$ *if there exists a local diffeomorphism (that is a smooth transformation with a smooth inverse)* $\mathbf{X} : \mathbb{R}^n \to \mathbb{R}^n$, *with* $\mathbf{X}(\mathbf{0}) = \mathbf{0}$ *and* $\det\left(\frac{\partial \mathbf{X}}{\partial \mathbf{x}}\right)\big|_0 \neq 0$, *such that*

$$W(\mathbf{x}) = V(\mathbf{X}(\mathbf{x})) + \text{const.} \tag{M.3}$$

The constant serves to adjust the value of the function at $\mathbf{0}$.

We note that the critical points are not changed by the smooth reversible change of coordinates $\mathbf{X}(\mathbf{x})$, because

$$0 = \frac{\partial W}{\partial x_i} = \sum_j \frac{\partial V}{\partial x_j}(\mathbf{X}(\mathbf{x}))\frac{\partial X_j}{\partial x_i}(\mathbf{x}) \ . \tag{M.4}$$

Hence, if $V(\mathbf{x})$ has a critical point at $\mathbf{0}$ so does W. However, by an appropriate choice of $\mathbf{X}(\mathbf{x})$, $W = V(\mathbf{X}(\mathbf{x}))$ could be much easier to analyse than $V(\mathbf{x})$. *Right-equivalence* leads to the normal forms of elementary catastrophe theory.

For the simple buckling problem (Fig. M.1) we know that for the critical parameter value $\overline{\lambda} = \overline{\lambda}_c$, where $\overline{\lambda}_c = (1,0)^T$,

$$V(x, \overline{\lambda}_c) = \frac{1}{4}x^4 \tag{M.5}$$

is right-equivalent to (M.1). The relation between the new and the old variable and the parameters can be obtained following the calculations in Section 6.5.3.

However, one nontrivial point needs some further explanation. So far in the transformation from (M.1) to (M.5) we have only cared about the terms up to fourth order and their coefficients. Since in (M.1) also terms of higher than fourth order are present and in (M.5) only fourth order terms are considered we must say something about what has happened to these higher order terms in the transformation. In fact they have been eliminated by the nonlinear change of coordinates defined above. The approach is similar to that used in Section 6.1. In order to explain the basic idea we treat the potential $V(\varphi)$ of the perfect system $(Q = 0)$ of Fig. M.1

$$V(\varphi) = \frac{\varphi^2}{2} - F(1 - \cos\varphi) . \qquad (M.6)$$

We are going to show by the following calculations that by means of the change of coordinates $\varphi = X(x)$ the potential $W(x) = V(X(x))$ takes the form

$$W(x) = \frac{1}{4}x^4 - \frac{1}{2}\lambda x^2 + \text{const.} \qquad (M.7)$$

The transformation must be composed by several steps. First, we notice that the critical parameter value for F is given by $F = 1$. We introduce a new parameter μ defined by

$$F = 1 + \mu . \qquad (M.8)$$

Inserting (M.8) into (M.6) results in

$$V = -\frac{\mu}{2}\varphi^2 + \frac{1}{24}\varphi^4 - \frac{1}{6!}\varphi^6 - \mu\frac{\varphi^4}{24} + \cdots . \qquad (M.9)$$

Now we make the following change of variables and parameters

$$\varphi = \sqrt[4]{6}y , \qquad \mu = \frac{\lambda}{\sqrt{6}} . \qquad (M.10)$$

Inserting (M.10) into (M.9) yields

$$V = -\frac{\lambda}{2}y^2 + \frac{1}{4}y^4 - \frac{\sqrt{6}}{5!}y^6 - \frac{\lambda}{4\sqrt{6}}y^4 + \cdots . \qquad (M.11)$$

The final steps are to eliminate the higher order terms. We start with the term of sixth order. For its elimination we set (for $\lambda = 0$)

$$y = x + h(x) = x + h_1 x^3 , \qquad (M.12)$$

where we have implicitly made use of the reflectional symmetry of the problem (Appendix G). Inserting (M.12) into (M.11) we obtain

$$h_1 = \frac{\sqrt{6}}{5!}$$

in order to annihilate the term of sixth order. Proceeding in this way all higher order terms can be eliminated. However, we have to note that each transformation changes the remaining other higher order terms.

M.2   Contact equivalence

We assume $f, g : \mathbb{R}^n \to \mathbb{R}^n$, $X : \mathbb{R}^n \to \mathbb{R}^n$ is a diffeomorphism and $\dot{x} = f(x)$.

Definition M.2 (contact equivalence) *Two vector fields f and g are called contact equivalent if*

$$g(x) = S(x)f(X(x)) , \qquad (M.13)$$

where $S(x)$ is a $n \times n$ matrix, $\det(S(0)) \neq 0$, $X(0) = 0$ and $\det(\frac{\partial X}{\partial x})|_0 \neq 0$.

Contact equivalence preserves zeros of vector fields.

M.3   Vector field equivalence

We consider a dynamical system given by

$$\dot{x} = f(x) \qquad (M.14)$$

and we suppose that $Y : \mathbb{R}^n \to \mathbb{R}^n$ is a *diffeomorphic* change of coordinates with $Y(0) = 0$ in the form

$$x = Y(y) . \qquad (M.15)$$

Substituting (M.15) into (M.14) yields with $\dot{x} = \left(\frac{\partial Y}{\partial y}\right) \dot{y}$

$$\left(\frac{\partial Y}{\partial y}\right) \dot{y} = f(Y(y))$$

or

$$\dot{y} = \left(\frac{\partial Y}{\partial y}\right)^{-1} f(Y(y)) = g(y) . \qquad (M.16)$$

Definition M.3 (vector field equivalence) *Two vector fields $f(x)$ and $g(y)$ are vector field equivalent if equation (M.16) is satisfied with some diffeomorphism Y.*

The *vector field equivalence* preserves the whole dynamics. Vector field equivalence is a special case of the contact equivalence because the matrix S in (M.13) is replaced by $\left(\frac{\partial Y}{\partial y}\right)^{-1}$ in (M.16).

Example: We consider the transformation to normal form (6.112). By the way we are also able to prove equations (6.115). We recall that the normal form (6.112) corresponds to a double zero eigenvalue ((4.12) case (4)). In order to keep the formulas as simple as possible we write (M.14) with (4.12)$_4$ in the following form ($\boldsymbol{x}^T = (r, s)$, $\boldsymbol{f}^T = (R, S)$)

$$
\begin{aligned}
\dot{r} &= R(r, s) &&= s + a_{130}r^3 + a_{121}r^2 s + a_{112}rs^2 + \\
&&& \quad + a_{103}s^3 + O(|r|^5 + |s|^5) \\
\dot{s} &= S(r, s) &&= a_{230}r^3 + a_{221}r^2 s + a_{212}rs^2 + \\
&&& \quad + a_{203}s^3 + O(|r|^5 + |s|^5) \ .
\end{aligned}
\tag{M.17}
$$

For the transformation (M.15) we use the notation ($\boldsymbol{y}^T = (u, v)$, $\boldsymbol{Y}^T = (U, V)$)

$$
\begin{aligned}
r &= U(u, v) &&= u + \alpha_{130}u^3 + \alpha_{121}u^2 v + \alpha_{112}uv^2 + \alpha_{103}v^3 \\
s &= V(u, v) &&= v + \alpha_{230}u^3 + \alpha_{221}u^2 v + \alpha_{212}uv^2 + \alpha_{203}v^3 \ .
\end{aligned}
\tag{M.18}
$$

Inserting (M.18) into (M.16) we obtain

$$
\begin{pmatrix} \dot{u} \\ \dot{v} \end{pmatrix} = \begin{pmatrix} U_{,u} & U_{,v} \\ V_{,u} & V_{,v} \end{pmatrix}^{-1} \begin{pmatrix} R(U(u,v),\, V(u,v)) \\ S(U(u,v),\, V(u,v)) \end{pmatrix} = \mathbf{K}^{-1} \begin{pmatrix} R^\star \\ S^\star \end{pmatrix} .
\tag{M.19}
$$

The right-hand side of (M.19) must be calculated up to third order. Hence, we calculate all expressions only up to an order such that we obtain third order terms in (M.19). We get

$$
\begin{aligned}
U_{,u} &= 1 + 3\alpha_{130}u^2 + 2\alpha_{121}uv + \alpha_{112}v^2 &&= 1 + Q_1 \\
U_{,v} &= \alpha_{121}u^2 + 2\alpha_{112}uv + 3\alpha_{103}v^2 &&= Q_2 \\
V_{,u} &= 3\alpha_{230}u^2 + 2\alpha_{221}uv + \alpha_{212}v^2 &&= Q_3 \\
V_{,v} &= 1 + \alpha_{221}u^2 + 2\alpha_{212}uv + 3\alpha_{203}v^2 &&= 1 + Q_4
\end{aligned}
\tag{M.20}
$$

$$
\begin{aligned}
R^\star(u, v) &= R(U(u, v), V(u, v)) = v + (a_{130} + \alpha_{230})u^3 + \\
&\quad + (a_{121} + \alpha_{221})u^2 v + (a_{112} + \alpha_{212})uv^2 + \\
&\quad + (a_{103} + \alpha_{203})v^3 + O(|u|^5 + |v|^5) \\
S^\star(u, v) &= S(U(u, v), V(u, v)) = a_{230}u^3 + a_{221}u^2 v + \\
&\quad + a_{212}uv^2 + a_{203}v^3 + O(|u|^5 + |v|^5) \ .
\end{aligned}
\tag{M.21}
$$

Next we calculate the inverse of the *Jacobian* $\mathbf{K}$ making use of the notation in (M.20). If a matrix $\mathbf{A}$ is given by

$$
\mathbf{A} = \begin{pmatrix} a & b \\ c & d \end{pmatrix} ,
$$

then its inverse $\mathbf{A}^{-1}$ is given by

$$\mathbf{A}^{-1} = \begin{pmatrix} d & -b \\ -c & a \end{pmatrix} \frac{1}{D} \tag{M.22}$$

where $D = ad - bc \neq 0$ is the determinant of $\mathbf{A}$. Inserting from (M.20) into (M.22) we obtain

$$\mathbf{K}^{-1} = \begin{pmatrix} 1+Q_1 & Q_2 \\ Q_3 & 1+Q_4 \end{pmatrix}^{-1} = \begin{pmatrix} 1+Q_4 & -Q_2 \\ -Q_3 & 1+Q_1 \end{pmatrix} \frac{1}{D}$$

$$= \begin{pmatrix} 1-Q_1 & -Q_2 \\ -Q_3 & 1-Q_4 \end{pmatrix} + O(4) \tag{M.23}$$

where $D = 1 + Q_1 + Q_4 + Q_1 Q_4 - Q_2 Q_3$. Since we consider only terms up to third order, we must retain in $\mathbf{K}^{-1}$ only quadratic terms because multiplying $\mathbf{K}^{-1}$ with $(R^\star, S^\star)^T$ we obtain at least third order terms. Therefore, we can use $1/D = 1 - Q_1 - Q_4$. Multiplication gives the result in (M.23). Returning to equation (M.19) we obtain up to third order

$$\begin{pmatrix} \dot{u} \\ \dot{v} \end{pmatrix} = \mathbf{K}^{-1} \begin{pmatrix} R^\star \\ S^\star \end{pmatrix}$$

$$= \begin{pmatrix} v + (a_{130} + \alpha_{230})u^3 + (a_{121} + \alpha_{221} - 3\alpha_{130})u^2 v + \\ a_{230}u^3 + (a_{221} - 3\alpha_{230})u^2 v + \end{pmatrix} \tag{M.24}$$

$$\begin{pmatrix} + (a_{112} + \alpha_{212} - 2\alpha_{121})uv^2 + (a_{103} + \alpha_{203} - \alpha_{112})v^3 \\ + (a_{212} - 2\alpha_{221})uv^2 + (a_{203} - \alpha_{212})v^3 \end{pmatrix}$$

By a proper choice of the coefficients α_{ijk} in (M.18) we are able to simplify (M.24) considerably. Since the linear part in (M.17) corresponds to a degenerate single second order differential equation it might be meaningful to simplify (M.24) in such a way that we obtain a single second order nonlinear differential equation. In this case we must annihilate all cubic terms in the first equation. This results in the following system of equations

$$\begin{aligned} \alpha_{230} &= -a_{130}, \\ \alpha_{221} - 3\alpha_{130} &= -a_{121}, \\ \alpha_{212} - 2\alpha_{121} &= -a_{112}, \\ \alpha_{203} - \alpha_{112} &= -a_{103}. \end{aligned} \tag{M.25}$$

From (M.25) follows $\alpha_{230} = -a_{130}$, which must be inserted into (M.24)$_2$. Hence, the first two terms remain in (M.24)$_2$. The remaining coefficients vanish if we take $\alpha_{221} = \frac{1}{2}a_{212}$ and $\alpha_{212} = a_{203}$. The final result is the Bogdanov-Takens normal form

$$
\begin{aligned}
\dot{u} &= v & +O(|u|^4 + |v|^4) \\
\dot{v} &= A_{30}u^3 + A_{21}u^2v + O(|u|^4 + |v|^4)
\end{aligned}
\tag{M.26}
$$

where $A_{30} = a_{230}$, $A_{21} = 3a_{130} + a_{221}$. Equations (M.26) are identical with the equations (6.112) and the above coefficients are the same as (6.115).

M.4 Bifurcation equivalence

As indicated in Section 6.5.3 the use of the equivalence relation leading to the normal forms of catastrophe theory is not always the best way to treat a bifurcation problem in engineering, because in general a mixing of the distinguished parameter and the imperfection parameters occurs. An improvement is given by the concept of *bifurcation equivalence*. Here one starts with the equilibrium equations

$$
f(\boldsymbol{x}, \overline{\lambda}) = 0 , \qquad f : \mathbb{R}^n \times \mathbb{R}^\ell \to \mathbb{R}^n ,
\tag{M.27}
$$

where $\overline{\lambda}$ is the parameter vector.

Definition M.4 (bifurcation equivalence) *Two vector fields f and g are equivalent if*

$$
\mathbf{S}(\boldsymbol{x}, \overline{\lambda})f(\mathbf{X}(\boldsymbol{x}, \overline{\lambda}), \Lambda(\overline{\lambda})) = g(\boldsymbol{x}, \overline{\lambda})
\tag{M.28}
$$

is satisfied for some orientation preserving maps $\mathbf{X}(\boldsymbol{x}, \overline{\lambda})$ and $\Lambda(\overline{\lambda})$ with $\mathbf{X}(0,0) = 0$ and $\Lambda(0) = 0$.

If $\Lambda(\overline{\lambda}) = \overline{\lambda}$ the vector fields are **strongly** *equivalent.*

The relation (M.28) is a generalization of (M.13).

We consider again the buckling problem described by (M.2). This problem is bifurcation equivalent to

$$
x^3 - \lambda x + \alpha + \beta x^2 = 0 .
\tag{M.29}
$$

The coefficients λ, α, β can, for example, be calculated similarly to those in (6.130) in Section 6.5.3.

The higher order terms in (M.29) have been eliminated by a transformation similarly to (M.12).

M.5 Recognition problem

If one has derived bifurcation equations $f(x, \lambda) = 0$ for a given physical problem then often the question arises to which classified case $g(x, \lambda) = 0$ can this problem be reduced. This is the so-called recognition problem.

We explain it with the scalar example

$$g(x, \lambda) = x^3 - \lambda x = 0 . \qquad (M.30)$$

We state the recognition problem for (M.30) in the form of a theorem:

Theorem M.1 (Recognition problem) *Any $f(x, \lambda) = 0$ is equivalent to $g(x, \lambda) = x^3 - \lambda x = 0$ if at $(x, \lambda) = (0, 0)$*

$$f = \frac{\partial f}{\partial x} = \frac{\partial^2 f}{\partial x^2} = \frac{\partial f}{\partial \lambda} = 0$$

$$\frac{\partial^3 f}{\partial x^3} > 0 , \qquad \frac{\partial^2 f}{\partial x \partial \lambda} < 0 . \qquad (M.31)$$

That is, we only have to perform the operations listed in (M.31) to check whether a bifurcation equation $f(x, \lambda) = 0$ can be reduced to (M.30).

The condition (M.31) includes the problem of *finite determinacy* because only a finite number of conditions on the derivatives of a given function $f(x, \lambda)$ guarantees that $f(x, \lambda)$ is equivalent to a simpler function $g(x, \lambda)$.

Appendix N

Slowly varying parameter

In this book always the assumption is made that the parameter variation is quasi-static (see p. 48). This is a good assumption if we study the stability behavior of a system in the neighborhood of the stability boundary under small quasi-static perturbations of the parameter values (imperfections). However, there exist problems where the assumption of a quasi-static variation of the parameters is not completely accurate.

For example, in vehicle dynamics, if a car or a train accelerates to its final speed, the velocity V, which we have chosen for such problems as distinguished parameter, changes slowly and continuously its values. Now we ask the question: are there any qualitative changes both in the bifurcation analysis and in the results if the parameter varies slowly but continuously compared to the quasi-static parameter variation?

First of all, instead of (3.1) we obtain now a system in the form

$$\begin{aligned} \dot{x} &= F(x, \lambda) \\ \dot{\lambda} &= \varepsilon \end{aligned} \tag{N.1}$$

where it is assumed that $\lambda \in \mathbb{R}^1$ and ε is small.

Before we discuss the question raised above from a more theoretical point of view we look at a numerical experiment. In Fig. N.1 results for the oscillator of Section 4.1.3 are shown. For a quasi-static variation of the distinguished parameter λ a Hopf bifurcation with the bifurcating amplitude curve I occurs at loss of stability at $\lambda = 0$. Curve II indicates that for small ε an influence on the bifurcation behavior in the neighborhood of the quasi-static bifurcation exist. Basically, it is a retardation concerning the onset of instability. This retardation depends on several influence factors: (i) the absolute value of ε (ii) the initial condition P and (iii) whether small perturbations act on the system or not ([104], [105], [40]). After the loss of stability has taken place for the dynamic parameter variation the two amplitude curves are almost identical.

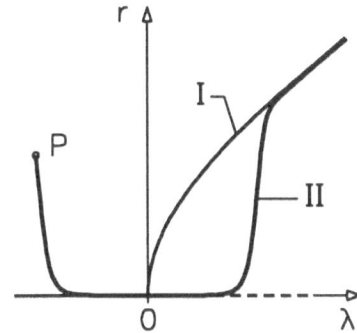

Figure N.1. Comparison of the onset of instability for the double pendulum of Fig. 4.6 for a slowly increasing parameter λ (curve II) and a quasistatic variation of λ (curve I)

In the following we make some comments how to calculate the size of this retardation.

We assume that for (N.1) x_0 is an equilibrium solution and that for a quasi-static parameter variation the bifurcation occurs at $\lambda = 0$.

Under certain conditions the center manifold szenario as it is developed in Chapter 3 still holds. That is, if ε is small enough, the eigenvalues of the linearized operator will also vary slowly with time and in the neighborhood of the critical parameter value $\lambda = 0$ of the quasi-static variation we will have some "critical" eigenvalues with real part close to zero and the other eigenvalues lying in the left half of the complex plane away from the imaginary axis ([73]). Now it would be nice if one could assume that again a separation of the variables in active and passive ones as in equation (3.31) is possible and that the quasistatic center manifold is a first approximation for the reduction of (N.1) to a set of bifurcation equations. This would result in one equation for a zero root and in two equations for a pure imaginary pair. The condition that such an szenario holds is that the other eigenvalues of the linear operator are still in the left half plane sufficiently far away from the imaginary axis even if the parameter values are such that the critical eigenvalues have a positive real part. The result of this reduction process will then be bifurcation equations on a slowly varying center manifold. That is, the parameters in the normal form of these "bifurcation equations" will be slowly varying with time. The analysis of these nonautonomous bifurcation equations can be performed by perturbation methods ([40]).

Appendix O

Transformation of dynamical systems into standard form

In applications, quite frequently, the equations of motion, after they are transformed into a first order system, have the form (4.2)

$$\mathbf{R}(\boldsymbol{x})\dot{\boldsymbol{x}} = \mathbf{N}(\boldsymbol{x}) \tag{O.1}$$

where $\mathbf{R}(\boldsymbol{x}) \in \mathbf{L}(\mathbb{R}^n, \mathbb{R}^n)$ depends on the variables and $\boldsymbol{x}, \mathbf{N} \in \mathbb{R}^n$. The task is to bring (O.1) into the standard form (3.1), that is

$$\dot{\boldsymbol{x}} = \mathbf{F}(\boldsymbol{x}) \ . \tag{O.2}$$

We explain two different procedures to perform this task.

O.1 Power series expansion

We expand $\mathbf{R}(\boldsymbol{x})$ into its power series

$$
\begin{aligned}
\mathbf{R}(\boldsymbol{x}) &= \mathbf{R}_0 + \mathbf{R}_1(\boldsymbol{x}) + \mathbf{R}_2(\boldsymbol{x}^2) + \ldots \\
&= \mathbf{R}_0(\mathbf{E} + \mathbf{R}_0^{-1}\mathbf{R}_1 + \mathbf{R}_0^{-1}\mathbf{R}_2 + \ldots + \mathbf{R}_0^{-1}\mathbf{R}_k + \ldots) \\
&= \mathbf{R}_0(\mathbf{E} + \mathbf{Q}_1(\boldsymbol{x}) + \mathbf{Q}_2(\boldsymbol{x}^2) + \ldots) \ .
\end{aligned}
\tag{O.3}
$$

Formally we can write

$$\mathbf{R}^{-1}(\boldsymbol{x}) = (\mathbf{E} + \mathbf{Q}_1 + \mathbf{Q}_2 + \ldots)^{-1}\mathbf{R}_0^{-1} \ . \tag{O.4}$$

In order to calculate the right hand side of (O.4) we have to calculate the inverse of a matrix which is close to the unit matrix $\mathbf{E}$. That is,

Hence, we obtain

$$\mathbf{R}^{-1}(x) = (\mathbf{E} - \mathbf{Q} + \mathbf{Q}^2 - \mathbf{Q}^3 + \ldots)\mathbf{R}_0^{-1} \tag{O.5}$$

where

$$\mathbf{Q} = \mathbf{Q}_1(x) + \mathbf{Q}_2(x^2) + \ldots + \mathbf{Q}^k(x^k) + \ldots$$

as it is defined by (O.3).

For example, the third order terms of $\mathbf{R}^{-1}(x)$ follow to

$$[-\mathbf{Q}_3(x^3) + (\mathbf{Q}_1(x)\mathbf{Q}_2(x^2) + \mathbf{Q}_2(x^2)\mathbf{Q}_1(x)) - \mathbf{Q}_1^3(x)]\mathbf{R}_0^{-1} \;.$$

The final step is to multiply $\mathbf{R}^{-1}(x)$ according to (O.5) with $\mathbf{N}(x)$ which also is expanded into a power series

$$\mathbf{N}(x) = \mathbf{N}_1(x) + \mathbf{N}_2(x^2) + \mathbf{N}_3(x^3) + \ldots \;. \tag{O.6}$$

Another possibility to calculate $\mathbf{R}^{-1}(x)$, which is used in Section 4.1.1, is by making use of the relation

$$\mathbf{R}(x)\mathbf{R}^{-1}(x) = \mathbf{E} \;. \tag{O.7}$$

Inserting

$$\begin{aligned} \mathbf{R}(x) &= \mathbf{R}_0 + \mathbf{R}_1(x) + \mathbf{R}_2(x^2) + \ldots \\ \mathbf{R}^{-1}(x) &= \mathbf{T}_0 + \mathbf{T}_1(x) + \mathbf{T}_2(x^2) + \ldots \end{aligned} \tag{O.8}$$

into (O.7) we obtain

$$(\mathbf{R}_0 + \mathbf{R}_1(x) + \mathbf{R}_2(x^2) + \ldots)(\mathbf{T}_0 + \mathbf{T}_1(x) + \mathbf{T}_2(x^2) + \ldots) = \mathbf{E} \;.$$

From this equation the recursive formulas for the coefficients of the terms of required order k can be obtained. We write them down up to $k = 3$.

$$\begin{aligned} k = 0: \quad & \mathbf{R}_0\mathbf{T}_0 = \mathbf{E} \\ & \mathbf{T}_0 = \mathbf{R}_0^{-1} \\[6pt] k = 1: \quad & \mathbf{R}_0\mathbf{T}_1 + \mathbf{R}_1\mathbf{T}_0 = 0 \\ & \mathbf{T}_1 = -\mathbf{T}_0\mathbf{R}_1\mathbf{T}_0 \\[6pt] k = 2: \quad & \mathbf{R}_0\mathbf{T}_2 + \mathbf{R}_1\mathbf{T}_1 + \mathbf{R}_2\mathbf{T}_0 = 0 \\ & \mathbf{T}_2 = -\mathbf{T}_0(\mathbf{R}_1\mathbf{T}_1 + \mathbf{R}_2\mathbf{T}_0) \\[6pt] k = 3: \quad & \mathbf{R}_0\mathbf{T}_3 + \mathbf{R}_1\mathbf{T}_2 + \mathbf{R}_2\mathbf{T}_1 + \mathbf{R}_3\mathbf{T}_0 = 0 \\ & \mathbf{T}_3 = -\mathbf{T}_0(\mathbf{R}_1\mathbf{T}_2 + \mathbf{R}_2\mathbf{T}_1 + \mathbf{R}_3\mathbf{T}_0) \;. \end{aligned}$$

O.2   Recursive calculation

Inserting (O.8) into (O.1) we can write the resulting equation in the form

$$\mathbf{R}_0\dot{\boldsymbol{x}} = \mathbf{N}(\boldsymbol{x}) - \mathbf{R}_1(\boldsymbol{x})\dot{\boldsymbol{x}} - \mathbf{R}_2(\boldsymbol{x}^2)\dot{\boldsymbol{x}} - \dots . \tag{O.9}$$

From (O.9) we may calculate in a recursive way the dynamical system in standard form (3.1) up to the required order as it is done in Section 4.2.1.
We obtain for:

(a) linear terms:

$$\mathbf{R}_0\dot{\boldsymbol{x}} = \mathbf{N}_1(\boldsymbol{x}) \quad \text{or} \quad \dot{\boldsymbol{x}} = \mathbf{R}_0^{-1}\mathbf{N}_1(\boldsymbol{x}) = \mathbf{F}_1(\boldsymbol{x})$$

(b) quadratic terms:

$$\begin{aligned}
\mathbf{R}_0\dot{\boldsymbol{x}} &= \mathbf{N}_2(\boldsymbol{x}^2) - \mathbf{R}_1(\boldsymbol{x})\dot{\boldsymbol{x}} = \mathbf{N}_2(\boldsymbol{x}^2) - \mathbf{R}_1(\boldsymbol{x})\mathbf{F}_1(\boldsymbol{x}) \\
\mathbf{F}_2(\boldsymbol{x}) &= \mathbf{R}_0^{-1}(\mathbf{N}_2 - \mathbf{R}_1\mathbf{F}_1)
\end{aligned}$$

(c) cubic terms:

$$\begin{aligned}
\mathbf{R}_0\dot{\boldsymbol{x}} &= \mathbf{N}_3(\boldsymbol{x}^3) - \mathbf{R}_1(\boldsymbol{x})\mathbf{F}_2(\boldsymbol{x}^2) - \mathbf{R}_2(\boldsymbol{x}^2)\mathbf{F}_1(\boldsymbol{x}) \\
\mathbf{F}_3(\boldsymbol{x}) &= \mathbf{R}_0^{-1}(\mathbf{N}_3 - \mathbf{R}_1\mathbf{F}_2 - \mathbf{R}_2\mathbf{F}_1) .
\end{aligned}$$

This procedure can be continued up to the required order and supplies the standard form

$$\dot{\boldsymbol{x}} = \mathbf{F}(\boldsymbol{x}) = \mathbf{F}_1(\boldsymbol{x}) + \mathbf{F}_2(\boldsymbol{x}^2) + \mathbf{F}_3(\boldsymbol{x}^3) + \dots .$$

Bibliography

[1] A. Aceves, H. Adachihara, C. Jones, J. C. Lerman, D. W. Mc Laughlin, J. V. Moloney, and A. C. Newell. Chaos and Coherent Structures in Partial Differential Equations. *Physica*, 18D:85–112, 1986.

[2] V. I. Arnold. Lectures on Bifurcations in Versal Families. *Russ. Math. Surv.*, 27:54–123, 1972.

[3] V. I. Arnold. Normal Forms for Functions Near Degenerate Critical Points, The Weyl Groups of A_k, D_k, E_k and Lagrangian Singularities. *Funct. Anal. Appl.*, 6:254–272, 1972.

[4] V. I. Arnold. Critical Points of Smooth Functions and Their Normal Forms. *Russ. Math. Surveys*, 30:1–75, 1975.

[5] V. I. Arnold. *Mathematical Methods of Classical Mechanics.* Springer-Verlag, New York – Heidelberg – Berlin, 1978.

[6] V. I. Arnold. *Ordinary Differential Equations.* MIT Press, Boston, 1978.

[7] V. I. Arnold. *Geometrical Methods in the Theory of Ordinary Differential Equations.* Springer-Verlag, New York – Heidelberg – Berlin, 1983.

[8] V. I. Arnold. *Catastrophe Theory.* Springer-Verlag, Berlin – Heidelberg – New York – Tokyo, 1984.

[9] B. Aulbach. *Continuous and Discrete Dynamics Near Manifolds of Equilibria*, vol. 1058 of *Lect. Notes in Math.* Springer-Verlag, Berlin – Heidelberg – New York – Tokyo, 1984.

[10] A. K. Bajaj and P. R. Sethna. Bifurcations in Three-Dimensional Motions of Articulated Tubes; Part 1: Linear Systems and Symmetry; Part 2: Nonlinear Analysis. *J. Appl. Mech.*, 49:606–618, 1982.

[11] A. K. Bajaj and P. R. Sethna. Flow Induced Bifurcations to Three-Dimensional Oscillatory Motions in Continuous Tubes. *SIAM J. Appl. Math.*, 44:270–286, 1984.

[12] L. Bauer and E. L. Reiss. Nonlinear Buckling of Rectangular Plates. *J. Soc. Indust. Appl. Math.*, 13:603–626, 1965.

[13] M. S. Berger. On von Karman's Equations and the Buckling of a Thin Elastic Plate; I: The Clamped Plate. *Comm. Pure Appl. Math.*, 20:687–719, 1967.

[14] M. S. Berger. *Nonlinearity and Functional Analysis*. Academic Press, New York, 1977.

[15] L. Berke and R. L. Carlson. Experimental Studies of the Post-Buckling Behavior of Complete Spherical Shells. *Experimental Mech.*, 8:548–553, 1968.

[16] R. D. Blevins. *Flow-induced Vibration.* van Nostrand Reinhold Comp., New York, 1977.

[17] H. Brauchli. *On the Norm-Dependence of the Concept of Stability*, vol. 503 of *Lecture Notes in Mathematics*, p. 235–238. Springer-Verlag, Berlin – Heidelberg – New York, 1976.

[18] L. Brun and M. Potier-Ferry. Constitutive Inequalities and Dynamic Stability in the Linear Theories of Elasticity and Thermoelasticity. *J. of Thermal Stresses*, 7:35–49, 1984.

[19] B. Budiansky. *Theory of Buckling and Post-Buckling Behavior of Elastic Structures*, vol. 14 of *Advances in Applied Mechanics*, p. 1–65. Academic Press, 1984.

[20] F. H. Busse. Patterns of Convection in Spherical Shells. *J. Fluid Mech.*, 72:67–85, 1975.

[21] J. Carr. *Applications of Centre Manifold Theory*, vol. 35 of *Applied Math. Sciences*. Springer-Verlag, New York – Heidelberg – Berlin, 1981.

[22] G. F. Carrier. On the Buckling of Elastic Rings. *J. of Math. and Phys.*, 26:94–103, 1947.

[23] J. Casti. Topological Methods for Social and Behavioral Systems. *Int. J. General Systems*, 8:187–210, 1982.

[24] S. Chandrasekhar. *Hydrodynamic and Hydromagnetic Stability.* Clarendon Press, London, 1961.

[25] P. Chossat. Bifurcation and Stability of Convective Flows in a Rotating or Not Rotating Spherical Shell. *SIAM J. Appl. Math.*, 37:624–647, 1979.

[26] S. N. Chow and J. Hale. *Methods of Bifurcation Theory*, vol. 251 of *Grundlehren der math. Wiss.* Springer-Verlag, New York – Heidelberg, 1982.

[27] S. N. Chow, J. K. Hale, and J. Mallet-Paret. Applications of Generic Bifurcation I. *Arch. Rat. Mech. Anal.*, 59:159–188, 1976.

[28] S. N. Chow, J. K. Hale, and J. Mallet-Paret. Applications of Generic Bifurcation II. *Arch. Rat. Mech. Anal.*, 62:209–236, 1976.

[29] S. N. Chow and J. Mallet-Paret. Integral Averaging and Bifurcation. *J. Differential Equ.*, 26:112–159, 1977.

[30] L. Collatz. *Eigenwertprobleme und ihre numerische Behandlung.* Chelsea Publishing, New York, 1948.

[31] J. F. Cornwell. *Group Theory in Physics*, vol. 1 and 2 of *Techniques in Physics 7.* Academic Press, London – San Diego, 1984.

[32] J. D. Crawford and E. Knobloch. On Degenerate Hopf Bifurcation with Broken $O(2)$ Symmetry. *Nonlinearity*, 1:617–652, 1988.

[33] G. Dangelmayr and E. Knobloch. The Takens-Bogdanov-Bifurcation with $O(2)$ Symmetry. *Phil. Trans. R. Soc. London*, A 322:243–279, 1987.

[34] K. Desoyer, P. Kopacek, and I. Troch. *Industrieroboter und Handhabungsgeräte.* R. Oldenbourg Verlag, München – Wien, 1985.

[35] R. L. Devaney. *An Introduction to Chaotic Dynamical Systems.* Addison-Wesley, Reading, 1987.

[36] E. Doedel. *AUTO: Software for Continuation and Bifurcation Problems in Ordinary Differential Equations.* CALTECH, Pasadena, 1986.

[37] P. G. Drazin. *Solitons*, vol. 85 of *London Math. Soc. Lecture Note Series.* Cambridge University Press, London, 1984.

[38] C. L. Dym and I. Shames. *Solid Mechanics, A Variational Approach.* McGraw-Hill, New York, 1973.

[39] J. A. Ellison, A. W. Sáenz, and H. S. Dumas. Improved N-th Order Averaging Theory for Periodic Systems. *J. of Differential Equations*, 84:383–403, 1990.

[40] T. Erneux and P. Mandel. Imperfect Bifurcation with Slowly Varying Control Parameter. *SIAM J. Appl. Math.*, 46:1–16, 1986.

[41] Z. Gaspar. Critical Imperfection Territory. *J. Struct. Mech.*, 11:297–325, 1983.

[42] R. Gilmore. *Catastrophe Theory for Scientists and Engineers.* J. Wiley and Sons, New York, 1981.

[43] H. Goldstein. *Classical Mechanics.* Addison-Wesley, Reading, 2nd edition, 1980.

[44] M. Golubitsky and D. Schaeffer. A Theory for Imperfect Bifurcations Via Singularity Theory. *Comm. Pure Appl. Math.*, 32:21–98, 1979.

[45] M. Golubitsky and D. Schaeffer. Imperfect Bifurcation in the Presence of Symmetry. *Comm. Math. Phys.*, 67:205–232, 1979.

[46] M. Golubitsky and D. Schaeffer. Bifurcations with $O(2)$ Symmetry Including Applications to the Benard Problem. *Comm. Pure Appl. Math.*, 25:81–111, 1982.

[47] M. Golubitsky and D. Schaeffer. A Discussion of Symmetry and Symmetry Breaking. *Proceedings of Symposium in Mathematics*, 40:499–515, 1983.

[48] M. Golubitsky, I. Stewart, and D. Schaeffer. *Singularities and Groups in Bifurcation Theory*, vol. 1 and 2 of *Applied Math. Sciences.* Springer-Verlag, New York – Heidelberg – Berlin, 1985.

[49] M. Graeff, R. Scheidl, H. Troger, and E. Weinmüller. An Investigation of the Complete Post-Buckling Behavior of Axisymmetric Spherical Shells. *ZAMP*, 36:803–821, 1985.

[50] J. Guckenheimer and P. Holmes. *Nonlinear Oscillations, Dynamical Systems, and Bifurcations of Vector Fields*, vol. 42 of *Applied Math. Sciences.* Springer-Verlag, Berlin – Heidelberg – New York, 1983.

[51] K. Hackl. *Eine Untersuchung des Plattenbeulens im Rahmen der Verzweigungstheorie.* PhD thesis, RWTH-Aachen, 1988.

[52] P. Hagedorn. *Non-Linear Oscillations.* Clarendon Press, Oxford, 1988.

[53] J. Hale. Generic Bifurcation with Applications. In Knops R. I., (ed), *Nonlinear Analysis and Mechanics: Heriot-Watt Symposium*, p. 59–157, Pitman, London, 1977.

[54] J. Hale. Restricted Generic Bifurcation. In *Nonlinear Analysis (Collection of papers in honour of E. H. Rothe)*, p. 83–98. Academic Press, New York, 1978.

[55] M. Hamermesh. *Group Theory and its Applications to Physical Problems*. Addison-Wesley, Reading, 1964.

[56] B. D. Hassard, N. D. Kazarinoff, and Y. H. Wan. *Theory and Applications of Hopf Bifurcation*, vol. 41 of *London Math. Soc. Lecture Note Series*. Cambridge University Press, London, 1981.

[57] W. Herfort and H. Troger. Robust Modelling of Flow Induced Oscillations of Bluff Bodies. *Mathematical Modelling*, 8:251–255, 1987.

[58] G. Herrmann and I. C. Jong. On Nonconservative Stability Problems of Elastic Systems with Slight Damping. *J. Appl. Mech.*, 33:125–133, 1966.

[59] M. W. Hirsch and S. Smale. *Differential Equations, Dynamical Systems and Linear Algebra*. Academic Press, New York – London, 1974.

[60] C. S. Hsu. *Cell-to-Cell Mapping, A Method of Global Analysis for Nonlinear Systems*, vol. 64 of *Applied Mathematical Sciences*. Springer-Verlag, Heidelberg – New York, 1987.

[61] C. S. Hsu, H. C. Yee, and W. H. Cheng. Steady State Response of a Nonlinear System under Impulsive Periodic Parametric Excitation. *J. Sound and Vibration*, 50:95–116, 1976.

[62] K. Huseyin. *Multiple Parameter Stability Theory and Its Applications*, vol. 18 of *Oxford Engineering Science Series*. Clarendon Press, Oxford, 1986.

[63] J. W. Hutchinson. Imperfection Sensitivity of Externally Pressurized Spherical Shells. *J. Appl. Mech.*, 34:49–55, 1967.

[64] G. Iooss and D. D. Joseph. *Elementary Stability and Bifurcation Theory*. UTM. Springer-Verlag, New York – Heidelberg – Berlin, 1980.

[65] K. Jänich. *Analysis für Physiker und Ingenieure*. Springer-Verlag, Berlin – Heidelberg – New York, 1983.

[66] V. Kaçani. *Anwendung der Nichtlinearen Stabilitätstheorie zur Untersuchung der Fahrdynamik eines Lastkraftwagens mit Anhänger*. PhD thesis, TU-Wien, 1987. VDI-Fortschrittberichte, Reihe 12, Verkehrstechnik/Fahrzeugtechnik 100, Düsseldorf, 1988.

[67] V. Kacani, A. Stribersky, H. Troger, and K. Zeman. Dynamics and Bifurcations in the Motion of Tractor-Semitrailer Vehicles. *CMS-Conference Proceedings*, 8:485–499, 1987.

[68] K. Kirchgässner. *Bifurcation in Nonlinear Hydrodynamic Stability*, vol. 17 of *SIAM Review*, p. 652–683. SIAM, 1975.

[69] U. Kirchgraber and E. Stiefel. *Methoden der analytischen Störungsrechnung und ihre Anwendungen*. B. G. Teubner Verlag, Stuttgart, 1979.

[70] E. Klingbeil. Tensorrechnung für Ingenieure. B. I. Hochschultaschenbücher 197/197 a, Mannheim, 1966.

[71] G. H. Knightly and D. Sather. Nonlinear Buckled States of Rectangular Plates. *Arch. Rat. Mech. Anal.*, 54:356–372, 1974.

[72] G. H. Knightly and D. Sather. Buckled States of a Spherical Shell Under Uniform External Pressure. *Arch. Rat. Mech. Anal.*, 72:315–380, 1980.

[73] H. W. Knobloch and F. Kappel. *Gewöhnliche Differentialgleichungen*. Teubner Verlag, Stuttgart, 1974.

[74] R. J. Knops and E. W. Wilkes. Theory of Elastic Stability. In C. Truesdell, (ed), *Handbuch der Physik*, vol. VI a/3, p. 125–302. Springer-Verlag, Berlin – Heidelberg – New York, 1983.

[75] W. T. Koiter. *On the Theory of Elastic Stability*. PhD thesis, Delft University, 1945 (in dutch). English Translation NASA TT-F10, 833 (1967).

[76] W. T. Koiter. Post-Buckling Analysis of a Simple Two Bar Frame. In F. Niordson B. Broberg, J. Hult, (ed), *Recent Progress in Applied Mechanics*, p. 337–354. Almqvist and Wiksell, Stockholm, 1967.

[77] W. T. Koiter. The Nonlinear Buckling Problem of a Complete Spherical Shell Under Uniform External Pressure. *Proc. Kon. Nederl. Acad. Wet. Amsterdam*, B72:40–123, 1969.

[78] W. T. Koiter. *A Basic Open Problem in the Theory of Elastic Stability*, vol. 503 of *Lect. Notes in Math.*, p. 366–373. Springer-Verlag, Berlin – Heidelberg – New York, 1976.

[79] W. T. Koiter. Elastic Stability. *ZFW*, 9:205–210, 1985.

[80] A. N. Kolmogorov and S. V. Fomin. *Introductory Real Analysis*. Dover Publications, New York, 1970.

[81] U. Kosel. Biegelinie eines elastischen Ringes als Beispiel einer Verzweigungslösung. *ZAMM*, 64:316–319, 1984.

[82] E. Kreuzer. *Numerische Untersuchung nichtlinearer dynamischer Systeme*. Springer-Verlag, Berlin – Heidelberg – New York, 1987.

[83] J. La Salle and S. Lefschetz. *Die Stabilitätstheorie von Liapunov*, vol. 194 of *BI-Hochschultaschenbücher*. BI, Mannheim, 1974.

[84] O. E. Lanford. *Strange Attractors and Turbulence*, vol. 45 of *Topics in Applied Physics*, p. 7–25. Springer-Verlag, Berlin – Heidelberg – New York, 1981.

[85] C. G. Lange and G. A. Kriegsmann. The Axisymmetric Branching Behavior of Complete Spherical Shells. *Quarterly Appl. Math.*, 39:145–178, 1981.

[86] H. Leipholz. *Stability Theory*. John Wiley, Chichester, 2nd edition, 1987.

[87] E. Lindtner, A. Steindl, and H. Troger. Generic One-Parameter Bifurcations in the Motion of a Simple Robot. *J. of Computational and Applied Mathematics*, 26:199–218, 1989.

[88] A. E. H. Love. *A Treatise on the Mathematical Theory of Elasticity*. Dover, New York, 4th edition, 1944.

[89] T. S. Lundgren, P. R. Sethna, and A. K. Bajaj. Stability Boundaries for Flow Induced Motions of Tubes with an Inclined Nozzle. *J. Sound and Vibrations*, 64:553–571, 1979.

[90] A. Machinek. *Anwendung der Verzweigungstheorie auf das Beul- und Nachbeulverhalten von Kreisring- und Rechteckplatten*. PhD thesis, TU-Wien, 1986.

[91] A. Machinek and H. Troger. Postbuckling of Elastic Annular Plates. *Dynamics and Stability of Systems*, 3:79–98, 1988.

[92] W. Mack and H. Troger. Zum Reduktionsvorgang in der Nichtlinearen Stabilitätstheorie. In *Proceed. of Symp. Mechanik und Industrie*, p. 249–266. Institut für Mechanik, Univ. Innsbruck, 1985.

[93] K. Magnus. *Schwingungen*. B. G. Teubner Verlag, Stuttgart, 1961.

[94] R. Magnus and T. Poston. *On the Full Unfolding of the von Karman Equations at a Double Eigenvalue*, vol. 109 of *Battelle Mathematics Reports*. Battelle, Geneva, Switzerland, 1977.

[95] S. Majumdar. Buckling of a Thin Annular Plate under Uniform Compression. *AIAA*, 9:1701–1707, 1971.

[96] K. Marguerre. Zur Theorie der gekrümmten Platte großer Formänderung. *Jahrbuch der dt. Luftfahrtforschung*, I:413–418, 1938.

[97] J. E. Marsden and T. J. R. Hughes. *Mathematical Foundations of Elasticity*. Prentice-Hall, Englewood Cliffs, N. Y., 1983.

[98] N. Minorsky. *Nonlinear Oscillations*. van Nostrand, Princeton, 1962.

[99] D. Moelle and R. Gasch. Nonlinear Bogie Hunting. In A. H. Wickens, (ed), *Proc. 7th IAVSD-Symposium in Cambridge (UK) 1981*, p. 455–467, Swets and Zeitlinger B. V., Lisse, 1982.

[100] F. C. Moon. *Chaotic Vibrations, An Introduction for Applied Scientists and Engineers*. J. Wiley and Sons, New York, 1987.

[101] P. M. Naghdi and L. Vongsarnpigoon. A Theory of Shells with Small Strain Accompanied by Moderate Rotation. *Arch. Rat. Mech. Anal.*, 83:245–283, 1983.

[102] P. M. Naghdi and L. Vongsarnpigoon. Some General Results in the Kinematics of Axisymmetric Deformation of Shells of Revolution. *Quarterly Appl. Math.*, 43:23–36, 1985.

[103] A. H. Nayfeh and D. T. Mook. *Nonlinear Oscillations*. J. Wiley and Sons, New York, 1979.

[104] A. I. Neistadt. On Delayed Stability Loss under Dynamical Bifurcation, I. *Differential Equations*, 23:2060–2067, 1987.

[105] A. I. Neistadt. On Delayed Stability Loss under Dynamical Bifurcation, II. *Differential Equations*, 24:226–233, 1988.

[106] H. H. Oberle, W. Grimm, and E. Berger. *BNDSCO, Rechenprogramm zur Lösung beschränkter optimaler Steuerungsprobleme*. Math. Inst., TU-München, Benutzeranleitung, TUM-M8509 edition, 1985.

[107] J. T. Oden and J. N. Reddy. *Variational Methods in Theoretical Mechanics*. Springer-Verlag, Berlin – Heidelberg – New York, 1976.

[108] F. Pfeiffer. Dynamical Systems with Time Varying or Unsteady Structure. *ZAMM*, 71, 1991. Invited paper GAMM-Tagung, Hannover 1990.

[109] J. Pierce. *Singularity Theory, Rod Theory and Symmetry-Breaking Loads*, vol. 1377 of *Lecture Notes in Mathematics*. Springer-Verlag, Berlin – Heidelberg – New York, 1989.

[110] R. H. Plaut. A New Destabilization Phenomenon in Nonconservative Systems. *ZAMM*, 51:319–321, 1971.

[111] T. Poston and I. Stewart. *Catastrophe Theory and Its Applications*. Pitman, London, 1978.

[112] M. Potier-Ferry. *Fundaments Mathematiques de la Theorie de la Stabilite Elastique*. PhD thesis, Universite de Paris VI, 1978.

[113] M. Potier-Ferry. *Foundation of Elastic Postbuckling Theory*, vol. 288 of *Lecture Notes in Physics*, p. 1–82. Springer-Verlag, Berlin – Heidelberg – New York, 1987.

[114] R. H. Rand and D. Armbruster. *Perturbation Methods, Bifurcation Theory and Computer Algebra*, vol. 65 of *Appl. Math. Sciences*. Springer-Verlag, New York – Heidelberg – Berlin, 1987.

[115] E. Reissner. On the Theory of Thin Elastic Shells. In J.W. Edwars, (ed), *Contributions to Applied Mechanics*, (H. Reissner Anniversary Volume), p. 231–247, Ann Arbor, Michigan, 1949.

[116] E. Reissner. On Axisymmetric Deformations of Thin Shells of Revolution. *Proc. Symp. in Appl. Math.*, 3:27–52, 1950.

[117] A. J. Rodrigues-Luis, E. Freire, and E. Ponce. *On a Codimension Three Bifurcation Arising in an Autonomous Electronic Circuit*, vol. 97 of *Int. Series of Numerical Mathematics*, p. 301–306. Birkhäuser-Verlag, Basel, 1991.

[118] J. A. Sanders and F. Verhulst. *Averaging Methods in Nonlinear Dynamical Systems*, vol. 59 of *Applied Math. Sciences*. Springer-Verlag, Berlin – Heidelberg – New York, 1985.

[119] D. H. Sattinger. Bifurcation and Symmetry Breaking in Applied Mathematics. *Bull. AMS*, 3,2:779–819, 1980.

[120] D. H. Sattinger. Branching in the Presence of Symmetry. In *CBMS-NSF Regional Conference Series in Appl. Math.*, vol. 40, SIAM, Philadelphia, 1983.

[121] P. T. Saunders. *An Introduction to Catastrophe Theory*. Cambridge University Press, Cambridge, 1980.

[122] D. Schaeffer. *General Introduction to Steady State Bifurcation,* vol. 898 of *Lecture Notes in Math.,* p. 13–47. Springer-Verlag, Berlin – Heidelberg – New York, 1981.

[123] D. Schaeffer and M. Golubitsky. Boundary Conditions and Mode Jumping in the Buckling of a Rectangular Plate. *Comm. Math. Phys.,* 69:209–236, 1979.

[124] R. Scheidl. *Ein Beitrag zum rotationssymmetrischen Beulen dünnwandiger Kugelschalen.* PhD thesis, TU-Wien, 1984.

[125] R. Scheidl and H. Troger. On the Buckling and Postbuckling of Spherical Shells. In Axelrad E. L. and Emmerling F. A., (ed), *Flexible Shells, Theory and Applications,* p. 146–162, Springer-Verlag, Berlin – Heidelberg – New York, 1984.

[126] R. Scheidl, H. Troger, and K. Zeman. Coupled Flutter and Divergence Bifurcation of a Double Pendulum. *Int. J. Non-linear Mechanics,* 19:163–176, 1983.

[127] H. G. Schuster. *Deterministic Chaos, An Introduction.* Verlag Chemie, Weinheim, 1988.

[128] P. R. Sethna. On Normal Forms, Averaging and Symbolic Manipulators. In *Collection of papers in honour of Yu. Mitropolsky.* Ukrainian Academy of Sciences, Kiev, 1986.

[129] P. R. Sethna and S. W. Shaw. On Codimension-Three Bifurcations in the Motion of Articulated Tubes Conveying a Fluid. *Physica,* 24 D:305–327, 1987.

[130] R. Seydel. *BIFPACK: A program package for calculating bifurcations.* State University of New York, Buffalo, 1985.

[131] R. Seydel. *From Equilibrium to Chaos, Practical Bifurcation and Stability Analysis.* Elsevier, New York, 1988.

[132] S. W. Shaw and P. Holmes. A Periodically Forced Piecewise Linear Oscillator. *J. of Sound and Vibration,* 90:129–155, 1983.

[133] S. W. Shaw and J. Shaw. Nonlinear Dynamics of a Rotating Shaft. In Schneider W., Troger H. and Ziegler F., (ed), *Proceedings STAMM 8,* Longman, London, 1991.

[134] R. T. Shield and A. E. Green. On Certain Methods in the Stability Theory of Continuous Systems. *Arch. Rat. Mech. Anal.,* 12:354–360, 1963.

[135] M. Stein. The Phenomenon of Change in Buckling Patterns in Elastic Structures. Technical Report R-39, NASA, 1959.

[136] A. Steindl. *Numerische Behandlung von Randwertproblemen auf unendlichen Intervallen.* PhD thesis, TU-Wien, 1984.

[137] A. Steindl. Gekoppelte Flatter- und Divergenzverzweigung eines flüssigkeitsdurchströmten Schlauchs. *ZAMM*, 69:T362–T365, 1989.

[138] A. Steindl and H. Troger. Flow Induced Bifurcations to Three-Dimensional Motions of Tubes with an Elastic Support. In Besseling J. F. and Eckhaus W., (ed), *Trends in Appl. Math. to Mechanics*, p. 128–138, Springer-Verlag, Berlin – Heidelberg – New York, 1988.

[139] H. Steinrück, H. Troger, and R. Weiss. *Mode Jumping of Imperfect Buckled Rectangular Plates*, vol. 70 of *Int. Series of Numerical Mathematics*, p. 515–524. Birkhäuser-Verlag, Basel, 1984.

[140] J. Stern. Der Gelenkstab bei großen elastischen Verformungen. *Ing. Archiv*, 48:173–184, 1979.

[141] I. Stewart. Application of Catastrophe Theory to the Physical Sciences. *Physica*, 2D:245–305, 1981.

[142] I. Stewart. *Concepts of Modern Mathematics.* Penguin Books, London, 1981.

[143] E. Stiefel and A. Fässler. *Gruppentheoretische Methoden und ihre Anwendungen.* Teubner, Stuttgart, 1979. Studienbücher Mathematik.

[144] A. Stribersky. *Nichtlineare Stabilitätsuntersuchungen dynamischer Systeme.* PhD thesis, TU-Wien, 1987. VDI-Fortschrittberichte, Reihe 11, Schwingungstechnik 97, Düsseldorf, 1987.

[145] A. Stribersky and H. Troger. Stabilitätsverlust einer Gleichgewichtslage eines gedämpften Schwingungssystems mit zwei Paaren rein imaginärer Eigenwerte. *ZAMM*, 67:T151–T153, 1987.

[146] A. Stribersky and H. Troger. Globales Verzweigungsverhalten am Beispiel eines längselastischen Doppelpendels unter Folgelast. *ZAMM*, 68:T126–T128, 1988.

[147] H. Suchy, H. Troger, and R. Weiss. A Numerical Study of Mode Jumping of Rectangular Plates. *ZAMM*, 65:71–78, 1985.

[148] Y. Sugiyama, Y. Tanaka, T. Kishi, and H. Kawagoe. Effect of a Spring Support on the Stability of Pipes Conveying Fluid. *J. Sound and Vibration*, 100:257–270, 1985.

[149] R. Thom. Topological Models in Biology. *Topology*, 8:313–335, 1969.

[150] J. M. T. Thompson. *Instabilities and Catastrophes in Science and Engineering*. J. Wiley and Sons, Chichester, 1982.

[151] J. M. T. Thompson and G. W. Hunt. *A General Theory of Elastic Stability*. J. Wiley and Sons, London, 1983.

[152] S. Timoshenko and S. Woinowsky-Krieger. *Theory of Plates and Shells*. McGraw-Hill, New York, 1959.

[153] H. Troger. Zur Anwendung der Verfahren von Ritz und Galerkin bei Verzweigungsproblemen. *ZAMM*, 63:T115–T116, 1983.

[154] H. Troger and K. Zeman. Zur korrekten Modellbildung in der Dynamik diskreter Systeme. *Ing. Archiv*, 51:31–43, 1981.

[155] S. A. van Gils and J. Mallet-Paret. Hopf Bifurcation and Symmetry: Travelling and Standing Waves on the Circle. *Proceed. Roy. Soc. Edinburgh*, 104,A:279–307, 1986.

[156] M. M. Wainberg and W. A. Trenogin. *Theorie der Lösungsverzweigung bei nichtlinearen Gleichungen*. Akademie-Verlag, Berlin, 1973.

[157] F. Y. M. Wan and H. J. Weinitschke. On Shells of Revolution with the Love-Kirchhoff Hypothesis. *J. of Eng. Math.*, 22:285–334, 1988.

[158] W. Wedig. Stability and Bifurcation in Stochastic Systems. *ZAMM*, 70:T19–T22, 1990.

[159] R. Weiss. Vorlesungen über Verzweigungstheorie. TU-Wien, 1982.

[160] B. Werner. Eigenvalue Problems with the Symmetry of a Group and Bifurcations. In D. Roose, B. De Dier, and A. Spence, (ed), *Continuation and Bifurcations: Numerical Techniques and Applications*, vol. 313 of *Series C: Mathematical and Physical Sciences*, p. 71–88, Kluwer Academic Publishers, Dordrecht – Boston – London, 1990. NATO ASI.

[161] S. Wiggins. *Introduction to Applied Nonlinear Dynamical Systems and Chaos*. Springer-Verlag, New York – Berlin – Heidelberg, 1990.

[162] G. Xu, A. Steindl, and H. Troger. Global Analysis of the Loss of Stability of a Special Railway Bogy. In W. Schiehlen, (ed), *Nonlinear Dynamics in Engineering Systems*, p. 345–352, Springer-Verlag, Berlin – Heidelberg – New York, 1990.

[163] G. Xu, A. Steindl, and H. Troger. Nonlinear Stability Analysis of a Special Railway Bogy. In *12th IAVSD-Symposium Lyon 1991*, Swets and Zeitlinger, Lisse, 1992.

[164] G. Xu and H. Troger. Zur achsensymmetrischen Schalentheorie von Reissner bei großen Verformungen. *ZAMM*, 69:T332–T334, 1989.

[165] E. C. Zeeman. Catastrophe Theory. *Scientific American*, p. 65–83, April 1976.

[166] K. Zeman. Berechnung der Zentrumsmannigfaltigkeit und der Verzweigungsgleichung am Beispiel eines unsymmetrischen Sattelschleppzuges. *ZAMM*, 63:T133–T135, 1983.

[167] F. Ziegler. *Technische Mechanik der festen und flüssigen Körper*. Springer-Verlag, Wien – New York, 1985.

[168] H. Ziegler. Die Stabilitätskriterien der Elastomechanik. *Ing. Archiv*, 20:49–52, 1952.

[169] H. Ziegler. Trace Effects in Stability. In Leipholz H., (ed), *Instability of Continuous Systems*, p. 238–247, Springer-Verlag, Berlin – Heidelberg – New York, 1971.

[170] R. Zurmühl. *Matrizen, Eine Darstellung für Ingenieure*. Springer-Verlag, Berlin – Heidelberg – New York, 4th edition, 1964.

Index

Christian Menn

Stahlbetonbrücken

Zweite, überarbeitete Auflage

1990. 517 Abb. XVI, 541 Seiten.
Gebunden DM 172,-, öS 1200,-
ISBN 3-211-82115-5

Preisänderungen vorbehalten

Das Buch vermittelt einen Überblick über die Grundlagen und den heutigen Stand der Technik im Stahlbetonbrückenbau. Für die zweite Auflage wurden vor allem die Abschnitte „Beton", „Vorspannung" und „Schiefe Brücke" überarbeitet.
Zur Gewährleistung einer einwandfreien Qualität und zur Vermeidung von Schadenfällen sind eine saubere konstruktive Durchbildung und materialtechnische Maßnahmen im Hinblick auf die erforderliche Dauerhaftigkeit ebenso wichtig wie eine sorgfältige Berechnung. Aus der ausführlichen Darstellung der Entwurfsgrundlagen und -ziele ist ersichtlich, welche Aspekte bei der Projektierung und Ausführung von Stahlbetonbrücken besonders zu beachten sind. Die vorgeschlagenen Berechnungsmodelle und -methoden für den Nachweis der Tragsicherheit und der Gebrauchsfähigkeit sind einfach, übersichtlich und praxisnah dargestellt und liefern ausreichend genaue Ergebnisse. Großes Gewicht wurde auf die Erläuterung und die Lösung der bezüglich Dauerhaftigkeit besonders wichtigen konstruktiven Probleme gelegt. In Anbetracht der vielen grundsätzlichen Überlegungen und Hinweise leistet das Buch auch dem Stahlbetonkonstrukteur, der sich nicht direkt mit Brückenbau befaßt, wertvolle Dienste.

Springer-Verlag Wien New York